# FREEZING
# EFFECTS
# ON
# FOOD
# QUALITY

# FOOD SCIENCE AND TECHNOLOGY

*A Series of Monographs, Textbooks, and Reference Books*

1. Flavor Research: Principles and Techniques, *R. Teranishi, I. Hornstein, P. Issenberg, and E. L. Wick*
2. Principles of Enzymology for the Food Sciences, *John R. Whitaker*
3. Low-Temperature Preservation of Foods and Living Matter, *Owen R. Fennema, William D. Powrie, and Elmer H. Marth*
4. Principles of Food Science
   Part I: Food Chemistry, *edited by Owen R. Fennema*
   Part II: Physical Methods of Food Preservation, *Marcus Karel, Owen R. Fennema, and Daryl B. Lund*
5. Food Emulsions, *edited by Stig E. Friberg*
6. Nutritional and Safety Aspects of Food Processing, *edited by Steven R. Tannenbaum*
7. Flavor Research: Recent Advances, *edited by R. Teranishi, Robert A. Flath, and Hiroshi Sugisawa*
8. Computer-Aided Techniques in Food Technology, *edited by Israel Saguy*
9. Handbook of Tropical Foods, *edited by Harvey T. Chan*
10. Antimicrobials in Foods, *edited by Alfred Larry Branen and P. Michael Davidson*
11. Food Constituents and Food Residues: Their Chromatographic Determination, *edited by James F. Lawrence*
12. Aspartame: Physiology and Biochemistry, *edited by Lewis D. Stegink and L. J. Filer, Jr.*
13. Handbook of Vitamins: Nutritional, Biochemical, and Clinical Aspects, *edited by Lawrence J. Machlin*
14. Starch Conversion Technology, *edited by G. M. A. van Beynum and J. A. Roels*

# FREEZING EFFECTS ON FOOD QUALITY

*edited by*

## LESTER E. JEREMIAH

**Agriculture and Agri-Food Canada Research Centre**
**Lacombe, Alberta, Canada**

**CRC Press**
Taylor & Francis Group
Boca Raton London New York

CRC Press is an imprint of the
Taylor & Francis Group, an **informa** business

CRC Press
Taylor & Francis Group
6000 Broken Sound Parkway NW, Suite 300
Boca Raton, FL 33487-2742

First issued in paperback 2019

© 1996 by Taylor & Francis Group, LLC
CRC Press is an imprint of Taylor & Francis Group, an Informa business

No claim to original U.S. Government works

ISBN-13: 978-0-8247-9350-0 (hbk)
ISBN-13: 978-0-367-40140-5 (pbk)

**Visit the Taylor & Francis Web site at**
**http://www.taylorandfrancis.com**

**and the CRC Press Web site at**
**http://www.crcpress.com**

# Preface

This book presents an overview of the existing knowledge regarding the influence of freezing, frozen storage, and thawing on the quality of different foodstuffs. It provides a valuable reference for food scientists and technologists working in research, academia, and the food industry. Although demand has been shifting from preserved to chilled (fresh) foods at the consumer level, freezing and frozen storage remain primary means for preserving foods. In addition, home and commercial food service requirements have resulted in increased production and distribution of frozen precooked and prepared foods. Consequently, a comprehensive reference summarizing the technologies of freezing, frozen storage, and thawing is presently needed.

The first chapter covers the fundamental aspects of the freezing process. The remaining nine chapters are grouped into four sections, which correspond to the four basic food groups and cover freezing, frozen storage, and thawing of specific food products. These groups are

Muscle food products
Fruits and vegetables
Dairy and egg products
Cereals and baked products

The influence of freezing, frozen storage, and thawing is exerted uniquely in foods in the four basic food groups. Consequently, individual chapters cover specific concerns related to

Fundamentals of the freezing process
Red meats

Poultry and poultry products
Fish and seafood
Cured and processed meats
Fruits
Vegetables
Dairy products
Egg and egg products
Doughs and baked goods

Each chapter addresses such issues as the effects of production, slaughter or harvesting, preparation, and packaging, as well as the effects of freezing, frozen storage, and thawing on

Color
Appearance
Consumer acceptance
Palatability attributes
Nutritional value
Intrinsic chemical reactions
Microbiological quality and safety
Frozen product stability

Each chapter also constitutes an up-to-date, comprehensive summary of the available literature.

The information provided in this book should help stimulate further innovation in the extension of storage life, to provide consumers with the highest quality and most wholesome foodstuffs attainable, given existing physical and commercial constraints.

*Lester E. Jeremiah*

# Contents

### III DAIRY AND EGG PRODUCTS

### IV CEREALS AND BAKED PRODUCTS

# Contributors

**R. Graham Bell**   Meat Industry Research Institute of New Zealand, Hamilton, New Zealand

**Walter Bushuk**   University of Manitoba, Winnipeg, Manitoba, Canada

**M. Pilar Cano**   Instituto del Frío, (C.S.I.C.), Madrid, Spain

**Brian B. Chrystall**   Meat Industry Research Institute of New Zealand, Hamilton, New Zealand

**Paul L. Dawson**   Clemson University, Clemson, South Carolina

**Carrick Erskine Devine**   Meat Industry Research Institute of New Zealand, Hamilton, New Zealand

**H. Douglas Goff**   University of Guelph, Guelph, Ontario, Canada

**Yoshifumi Inoue**   Japan School of Baking, Edogawa-ku, Tokyo, Japan

**Lester E. Jeremiah**   Agriculture and Agri-Food Canada Research Centre, Lacombe, Alberta, Canada

**Simon Lovatt**   Meat Industry Research Institute of New Zealand, Hamilton, New Zealand

**Roger W. Mandigo** University of Nebraska, Lincoln, Nebraska

**Wesley N. Osburn** University of Nebraska, Lincoln, Nebraska

**Michael E. Sahagian** University of Guelph, Guelph, Ontario, Canada

**En. Emilia M. Santos-Yap** College of Fisheries, University of the Philippines in the Visayas, Iloilo, The Philippines

**Joseph G. Sebranek** Iowa State University, Ames, Iowa

**Grete Skrede** MATFORSK—Norwegian Food Research Institute, Ås, Norway

# FREEZING
# EFFECTS
## ON
# FOOD
# QUALITY

# Fundamental Aspects of the Freezing Process

**Michael E. Sahagian and H. Douglas Goff**

*University of Guelph*
*Guelph, Ontario, Canada*

## I.  INTRODUCTION

Over the past two decades, the process of food freezing as a means of preservation has gained widespread attention. Reduction of available water as ice, coupled with subzero temperatures, provides an environment that favors reduced chemical reactions and structural collapse of tissues, hence leading to increased storage stability. However, fundamental factors that accompany ice formation are relatively complex. The freezing process, which includes undercooling, nucleation (homogeneous and heterogeneous), ice crystal propagation, and maturation, is strongly influenced by thermodynamics (heat transfer properties), kinetics (mass transfer properties), and product (e.g., composition and size) variables. Modification of these variables can ultimately lead to large changes in ice distribution and subsequent quality of the product. A thorough understanding of the fundamental aspects of freezing is necessary for a full appreciation of the benefits and disadvantages of freezing. Section II of this chapter will review the basic mechanisms involved during freezing and will define relevant terminology associated with the physics and chemistry of freezing.

In recent years, considerable attention has been given to kinetic rather than thermodynamic processes that occur during freezing of aqueous systems. The formation of a supersaturated state, or glassy domain, resulting from freeze-concentration has potential implications for improvement of subzero temperature stability. This cryostabilization approach to frozen food stability draws nonequilibrium concepts and ideas from glassy polymers; Sec. III will describe the mechanisms involved during formation of the glassy state, kinetic and thermodynamic stability above and below the glass transition temperature ($T_g$), the concept of partial or dilute glass formation, and the process of devitrification. Following this review of nonequilibrium properties, frozen food stability, in terms of the chemical and physical changes as they relate to kinetics and mobility of unfrozen water, will be covered in Sec. IV, with a focus on the collective effects of recrystallization on product quality. Subsequent chapters will review in depth the technology and practical aspects associated with the freezing of various food commodities and products. Therefore, the intent of this chapter is not to discuss freezing and stability of food products per se, but to provide the reader with a thorough but concise reference for the basic concepts upon which the practical technologies for food freezing are founded.

## II. THERMODYNAMIC EVENTS OCCURRING DURING FREEZING

### A. Undercooling

When considering the freezing of pure or aqueous systems, one must be cognizant of the fact both thermodynamic and kinetic factors are present, each dominating the other at a particular point in the freezing process. However, before a crystallization process can occur at the initial freezing point, a significant energy barrier must be surmounted [1]. This energy barrier is demonstrated by the withdrawal of sensible heat below 0°C without a phase change. This process, called undercooling or supercooling, results in a thermodynamic unstable state that initiates the formation of submicroscopic water aggregates leading to a suitable interface (seed) necessary for a liquid-to-solid transformation. The degree of undercooling is dictated by the onset of ice nucleation. In the absence of a stable seed, phase separation is not possible since liquid molecules do not easily align themselves in the configuration of a solid [2]. Therefore, nucleation serves as the initial process of freezing and can be considered the critical step preceding complete solidification. Unfortunately, however, the nucleation phenomenon is complex and still not entirely understood, but in general it involves creating a stable interface via a foreign particle (heterogeneous nucleation) or through a process by which molecules form spontaneously during intrinsic fluctuations (homogeneous nucleation). Independent of the process involved, nucleation theory can be explained as changes in free energy, as discussed in the following section.

### B. Nucleation

### 1. Theory

Classic nucleation theory considers the formation of small stable nuclei as a sequence of bimolecular processes whereby atoms in liquid phase join a growing cluster (embryo). For a cluster of radius $r$, the process is governed by $\Delta G_{l \to s}$, the net free energy of formation accompanying the liquid-solid condensation, and is given by the expression:

$$\Delta G_{l \to s} = \frac{4}{3}\pi r^3 \Delta G_v + 4\pi r^2 \gamma \tag{1}$$

where $r$ is the radius of the particle, $\Delta G_v$ is the difference in free energy between the solid and aqueous phases, and $\gamma$ is the interfacial free energy per unit area between the ice and unfrozen phases [1,3]. In other words, the total $\Delta G_{l \to s}$ is the summation of surface and bulk or volume energy terms.

At temperatures below the initial melting point (undercooling), clusters of molecules have a thermodynamically unfavorable interface with the undercooled liquid. Since the surface-to-volume ratio is large at small cluster size, the total interfacial surface energy constitutes a barrier to growth. However, as cluster size (embryo) increases, surface energy becomes increasingly greater, whereas the volume energy term from Eq. (1) progressively decreases at a lower rate, which favors nucleation. The relationship between the various quantities in Eq. (1) are shown in Fig. 1 demonstrating change in $\Delta G_{l \to s}$ with $r$ of water at $-40°C$. As depicted in Fig. 1, $\Delta G_{l \to s}$ passes through a maximum ($\Delta G^*_{l \to s}$) at which a radius of critical size ($r^*$) exists having an equal probability of growing or disintegrating. Franks [1] has estimated, using the nucleation data of Michelmore and Franks [4], $\Delta G_{l \to s}$ of $0.5 \times 10^{-18}$ J and $r^*$ equal to 1.85 nm, corresponding to $\approx 200$ molecules at $r^*$. The addition of one molecule into the lattice of the cluster at $r^*$ results in a

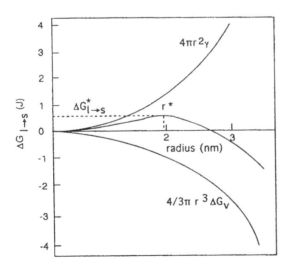

**Figure 1**  Changes in volume free energy of nucleation as a function of cluster radius for pure water at $-40°C$. (From Ref. 1.)

reduction in $\Delta G_{l \rightarrow s}$ (Fig. 1) as the newly formed nuclei gains more bulk energy [first term in Eq. (1)] than that lost in surface energy [2]. Beyond $r^*$, the cluster is considered an active nucleus and the process becomes spontaneous, leading to crystal growth. Factors affecting the size of critical nucleus capable of promoting condensation of molecules from the liquid phase include the interfacial free energy, $\gamma$, the latent heat of fusion, $\Delta H_f$, and the degree of undercooling, $\Delta T = (T_E - T)$, where $T_E$ is the equilibrium solid-liquid freezing temperature and $T$ is the absolute temperature [5]. These variables are related in the following equation:

$$r^* = \frac{2\gamma T_E}{\Delta H_f \Delta T} \tag{2}$$

This relationship relies on several assumptions, including (1) the thermodynamics at the macroscopic level applies to microscopic systems and (2) $\gamma$ is independent of $r$ and exhibits similar properties of an ice/water interface at 0°C. Although these assumptions create uncertainty in the absolute value of $r^*$, it has been shown that $r^*$ is dependent on $\Delta T$ (Fig. 2). As undercooling or the rate of heat removal increases (greater $\Delta T$), smaller nuclei are stable and can exist in equilibrium with the undercooled water. It also follows that with increased undercooling, the number of water molecules at $r^*$ is reduced. Conversely, at minimal $\Delta T$ (close to the equilibrium freezing point), a nucleus approaches the limit of about 1 μm at 0°C and hence displays characteristics of an ice crystal of infinite radius [5].

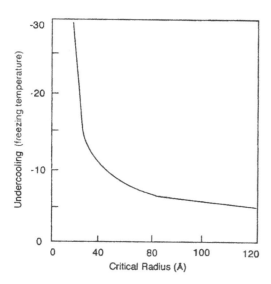

**Figure 2**    Change in critical radius for nucleation of water at various temperatures. (From Ref. 5.)

## 2. Homogeneous Nucleation

In homogeneous nucleation theory, random density fluctuations in the liquid phase result in the formation of nuclei in a pure three-dimensional pattern. Homogeneous nucleation occurs in pure systems absent of impurities of any type of nucleating substance. Probability of homogeneous nucleation of water at 0°C is close to zero; however, as temperature decreases, the probability of a nucleus reaching critical size sharply increases and approaches 1 near –40°C [1,3,6]. This value represents the threshold temperature for the homogeneous nucleation of pure water ($T_{hom}$). However, this limit is influenced by sample size (greater volume, higher $\Delta T_{hom}$), external vibration and physical disturbances (dynamic nucleation), and cooling rate.

An important parameter when studying nucleation of ice is the rate or kinetics at which nuclei appear per volume per unit time: the nucleation rate, $J(T)$. The generalized relationship is:

$$J(T) = A*\exp(B\theta) \tag{3}$$

where $J(T)$ is the steady-state rate of nucleation at temperature ($T$), $A$ and $B$ are constants representing several physical parameters of ice and the aqueous water, and $\theta$ describes the temperature dependence, $((\Delta T)^2 T^3)^{-1}$, where $\Delta T$ is the degree of undercooling and $T$ is the absolute temperature [7]. Since $J$ is dependent with the reciprocal of the square of $\Delta T$ [Eq. (3)], low nucleation rates are seen at small levels of undercooling. However, as $\Delta T$ increases, there is a point characteristic of the system where nucleation rates increase rapidly, followed by a leveling-off point. Measurement of $J(T)$ is limited by the sensitivity of instrumentation for its detection (i.e., where nucleation becomes sufficiently rapid, usually where the probability of a nucleation event occurring within 1 second is high) [8,9]. Since each nucleus promotes the formation of one ice crystal, the degree of undercooling ($\Delta T$), which also is a function of the cooling rate, largely determines the amount and distribution of ice crystals in the product. Based on a thermodynamic framework, the quantity of heat removed during undercooling is proportional to the $J(T)$, the corresponding number of crystals, and initial crystal size. In other words, rapid nucleation rates (favorable during food freezing) are primarily a function of heat removal from the material before phase change [6].

## 3. Nucleation in Aqueous Solutions

When determining $J(T)$ for pure systems, the free energy of activation, $\Delta G^{++}$, related to the viscosity and self-diffusion properties of undercooled water, is incorporated into Eq. (3) [8]. This physical parameter, and the interfacial tension between the water and ice ($\gamma$), are critical parameters and appear to alter the nucleation kinetics significantly, with $\gamma$ having a greater effect than viscosity. However, freezing and nucleation mechanisms in real systems usually involve solutions in which some level of solute has been dissolved. Nucleation studies involving solutions of PEG and other bimolecular materials have shown that the addition of solute, especially high molecular weight materials, reduces $J(T)$ compared to that of water [4,8]. Reduction in $J(T)$ was hypothesized to result primarily from a reduction in the diffusion freedom (coefficient) of the water, directly altering $\Delta G^{++}$. This effect can also be related to cryoprotectant properties of some materials. Rasmussen and MacKenzie [10] demonstrated that $\gamma$ changes with solute type and concentration, showing a reduction with increasing concentration. This result underlines the complexity of nucleation kinetics in aqueous solutions since surface energy is not solute independent and is difficult to measure accurately. Alternative explanations for reduced $J(T)$ include (1) a reduction in the volume fraction of water, (2) the development of a diffusion layer as the chemical potential surrounding the growing cluster increases, (3) a possible increase in critical radius resulting in a change in $\gamma$, and (4) uncertainty with $\gamma$ [8].

Homogeneous nucleation studies utilizing a series of solute concentrations revealed that $T_{hom}$ decreases with increasing solute concentration and freezing point depression $(T_{fr})$ [3]. This phenomenon was initially demonstrated by Rasmussen and MacKenzie [10] and recently by Charoenrein et al. (7) and Ozilgen and Reid (11) for sucrose solutions of varying concentrations. This observation demonstrates the complexity of nucleation in aqueous systems since $T_{hom}$ evolves from several physical parameters, unlike $T_m$ (equilibrium melting temperature), which is solely a function of reduced water activity (see Sec. II.D). Franks (12) has offered an explanation for this relationship between $T_{hom}$ and the colligative properties in the following form:

$$\Delta T_{hom} = K \Delta T_m \qquad (4)$$

where $K$ is approximately 2. For most solutes, the relationship fits fairly well, although deviations are seen with macromolecules where chain entanglements are unstable to diffuse from the cluster surface, therefore changing the interface and increasing $\Delta G^*$ [3,13].

## 4. Heterogeneous Nucleation

Although homogeneous nucleation has been studied extensively, heterogeneous nucleation is more important in the actual freezing process. Nucleation of water by a heterogeneous process is poorly understood, and only recently has attention been directed toward this process. In general, this type of nucleation occurs when water aggregates assemble on a nucleating agent such as the walls of a container or, more commonly, some type of foreign body or insoluble material. In contrast to homogeneous nucleation, the probability of heterogeneous nucleation at smaller $\Delta T$ is high. Water during freezing dominated by a heterogeneous mechanism will nucleate at higher temperatures, since substrate (particles) tend to increase cluster stability, facilitating the process. This translates into a reduction in the activation energy of nucleation at any temperature when compared to homogeneous nucleation and suggests that this process is essentially controlled by some catalyzed mechanism.

It is generally accepted that the geometry and detailed structure of a nucleating material, particularly its surface properties, play an important role in its catalyzing ability. In the case of AgI, a common nucleating agent, a match between its crystal structure and that of ice suggests a molecular alignment. These properties cause nucleating agents to facilitate the organization of molecules into a stable nuclei [14,15]. If the structure of the particle is compatible with the arrangement of water molecules in the cluster, the impurity will facilitate ice crystal growth. Surface properties, such as lattice misfits and dislocations, pit distributions, and overall shape factors, strongly dictate the nucleating capacity and can have noticeable effects on nucleation kinetics [16]. In terms of thermodynamics, the effectiveness of a spherical foreign particle as a nucleating agent is related to its effect on the interfacial free energy at the ice interface and modification of the free energy of formation [3].

Several theories relating to the mechanism of heterogeneous nucleation have been presented in the literature [3,16,17]. The classic theory proposed the presence of a flat nucleator decreased the free energy proportional to the volume of the nucleus. Other researchers extended this examination to spherical, conical, and cylindrical nucleators [17,18]. The reduction in free energy needed to form a nucleus of $r*$ can be related to a heterogeneous nucleation factor, denoted $f(\vartheta)$, where $\vartheta$ represents the contact angle between the nucleator and the nucleus. The contact angle is related to $\gamma$ between the three phases of the system solution, cluster, and substrate by a mathematical expression when $\vartheta$ is <180°. Free energy of formation, $\Delta G_{het}$, is of the form:

$$\Delta G_{het}^* = \Delta G_{hom}^* f(\vartheta) \tag{5}$$

where $\Delta G_{het}^*$ is the change in free energy at the critical nucleus during heterogeneous nucleation and $\Delta G_{hom}^*$ is the change in free energy at the critical nucleus during homogeneous nucleation [19]. Factor $f(\vartheta)$ is actually a mathematical function of the contact angle ($\vartheta$), as shown in Fig. 3. Substrate particles reduce the $\Delta G$ of the cluster according to the factor $f(\gamma)$. For example, for a heterogeneous process in which $\vartheta$ is equal to 90°, the change in free energy necessary to form a cluster of $r*$ on the substrate is approximately half that required for a homogeneous process. As evident from Fig. 3, as $\vartheta$ increases to 180°, the free energy of the process becomes equal to that of homogeneous nucleation. On the other hand, as $\vartheta$ approaches 0°, the $\Delta G$ decreases following the curve in Fig. 3 down to a theoretical point of zero. Nucleators that have inherently small contact angles reduce the number of water molecules necessary to satisfy the requirements for $r*$ and likewise reduce the undercooling required for nucleation. As a result, an increase in nucleation temperature is observed since stable nuclei can exist at higher temperatures. This effect is significant when discussing the mechanism of ice nucleation in cell-containing foods. The nucleation process may depend on the dominant type of intercellular component and its respective $\vartheta$. This implies that at any given level

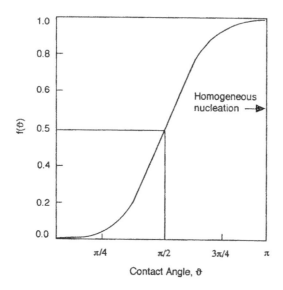

**Figure 3**    Heterogeneous nucleation factor $f(\vartheta)$ as a function of contact angle ($\vartheta$). (From Ref. 19.)

of undercooling, the probability of nucleation will be system specific since the $\vartheta$ may be different. Also, based on the relationship in Fig. 3, relatively small changes in $\vartheta$ may lead to dramatic effects in $\Delta G$ and nucleation behavior.

Studies examining the mechanism of heterogeneous nucleation in aqueous systems have shown that $T_{het}$ decreases with increased freezing point depression, similar to $T_{hom}$, following Eq. (4) [7,11]. Using thermomechanical analysis (TMA), Carrington et al. [20] reported crystallization temperatures of 30% carbohydrate solutions of different molecular weights demonstrating this reduction in $T_{het}$. Ozilgen and Reid [11] recently reported that $\Delta T_{het}$ is dependent on a nucleating agent (template) in solutions of sucrose, D-alanine, and $KNO_3$, the presence of AgI resulting in substantially larger $K$ values when compared to *P. syringae* [Eq. (4)]. Moreover, the effect was solute dependent, since AgI in sucrose solutions led to $K$ values almost twice as large as in the other solutions. These results suggest colligative properties affect heterogeneous nucleation in a linear fashion. However, the type of nucleator and corresponding surface properties, i.e., contact angle, must be considered.

Heterogeneous nucleation kinetics, $J(T)$, in aqueous systems demonstrate the complex behavior often observed in food freezing. Charoenrein et al. [7] reported that the temperature dependence of heterogeneous nucleation can be described using the classical relationship for homogeneous nucleation rates and showed the slope [parameter $B$ of Eq. (3)] of $\ln J$ versus normalized temperature for *P. syringae* nucleation was affected by sucrose concentration. In a related study [11], the effect of solute type and increasing concentration on parameter *B* of Eq. (3), which describes temperature effects on nucleation rates, was investigated and appeared to be dependent on the type of catalytic site involved. For example, in the case of *P. syringae* nucleation, parameter B was only moderately affected by increasing sucrose concentration, whereas with the AgI system, sucrose concentration affected nucleating properties by several orders of magnitude. Furthermore, the effect of concentration on the nucleation mechanism varied with solute type since high concentrations of sucrose alter nucleation parameters of Eq. (3) to a greater extent than similar solutions of $KNO_3$ and D-alanine. This behavior was also dependent on the nucleating agent (template) present. There is evidence to suggest that at high solute concentrations, changes in the catalytic surface properties of the nucleating template may alter the overall nucleation mechanism. Other studies have found polysaccharides to have limited effects on nucleation rate [21], whereas in the presence of hydrophobic amino acids, which exhibit a polar axis [22], nucleation may be readily accelerated. Therefore, in real food systems where it is of prime importance to maximize $J(T)_{het}$, the inherent complex nature makes prediction rather difficult and attests to the relevance of rapid cooling rates during the initial stages of the freezing process.

## 5.  Secondary Nucleation

A third type of nucleation, more relevant with freeze-concentration of fluid foods in batch crystallizers, is called secondary or contact nucleation [9]. Unlike the freezing of foods, where extensive nucleation is favored, large, spherical crystals with a narrow size distribution, which are easily separated, are sought in freeze-concentration processes including the concentration of dairy products [23]. Secondary nuclei may be formed when microscopic attrition between already existing crystals and the walls or impeller of crystallizers causes breakage of the seed crystal. If these fragments, dispersed within the bulk phase, survive, they are considered secondary nuclei and can grow and propagate within the suspension as usual. The net result of secondary nucleation is that nucleation processes and growth can occur at higher temperatures ($\Delta T$ between 0.05 and 0.2°C) than heterogeneous nucleation mechanisms ($\Delta T$ between 5 and 7°C), depending on the solutes present [24].

Hartel and Chung [24] studied the effect of increasing lactose concentration and whey protein on the critical undercooling required to observe secondary nucleation. Increasing lactose concentration had only a slight effect on the critical undercooling temperature. However, the opposite was seen with an increase in whey protein concentration. In other words, as protein concentration increases, the critical undercooling increases. These results suggest that whey proteins act as inhibitors to secondary nucleation, perhaps as postulated by the authors, by altering the surface and/or absorbed layer structure of the crystals, and whey proteins were shown to act as the dominant component in the two fluid dairy products tested. Polymers alone and in solution with other low molecular weight components have also been shown to affect secondary nucleation in a batch crystallizer (25). The nucleation rate constant was dependent on the type and concentration of the polymer, decreasing with increased viscosity of the solution. A complete explanation for this suppression of secondary nucleation is not available, but it may evolve from the polymer interfering with the crystal surface, which would increase the interfacial free energy, possibly to a point at which the seed crystal is unable to stabilize and grow and consequently melts. This may have relevance in the freezing of ice cream due to the presence of polysaccharides in the mix and the large shear forces during freezing. However, to date, this has not been demonstrated.

## C.  Crystal Propagation

During undercooling, water molecules are placed into a thermodynamically unstable state as $\Delta G^*$ between the liquid and solid states increase [1,3]. This difference represents an increase in the dominance of binding energy over disorder energy with decreasing temperature and is the driving force leading to crystallization. Once nucleation and crystal growth are initiated, water molecules move rapidly toward achieving thermodynamic stability as hexagonal ice, $Ih$, the energetically favored structural arrangement [26]. In nature, the tendency is to disorder and form disorganized structures such as liquids and gases. However, this is overcome when substantial energy is lost to entropy, bearing in mind that systems always seek the lowest free energy. Reduction in $\Delta G^*$ is paramount to the ice crystallization process.

Since ice propagation in pure and aqueous systems follows substantially different pathways, they can be dealt with separately. First, consider the freezing curve depicted in Fig. 4. Reduction in temperature from A to B represents the energy of activation (undercooling) prior to nucleation. Once $r^*$ is reached at B, the system will nucleate and release its latent heat at a rate greater than the rate of heat removal, increasing the temperature to 0°C, the equilibrium freezing temperature of pure water. From C to D, with slight undercooling, ice crystal growth occurs, at which time heat transfer is rate-limiting. Under optimal cooling rates, the freezing temperature, $T_f$, between C and D will remain relatively constant as latent heat is being removed. As the amount of ice increases and as the system moves towards point D, the rate of heat transfer will change since the thermal conductivity, diffusivity, and heat capacity ($C_p$) of ice differs greatly from the same properties of water (Table 1). After crystallization is complete, the temperature decreases to E, approaching the external freezing temperature, as sensible heat is removed from the system [27].

Growth of a crystal occurs when the number of water molecules successfully diffusing along the interface and orienting themselves at a growing site is greater than the number departing [5]. A growing unit then becomes tightly bound into the

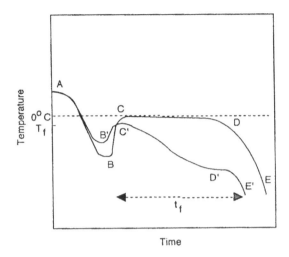

**Figure 4**   A temperature-time curve of water (ABCDE) and an aqueous solution, e.g., 20% sugar (A B′C′D′E′) during freezing. $T_f$ represents the initial freezing point of the solution, and $t_f$ represents the freezing time as determined by such an experiment. (From Ref. 27.)

crystal lattice at a kink site within a step on the crystal surface. The mechanism and rate of crystal growth is dependent on the concentration of kinks and the morphology of the surface. For example, if the surface is rough, many kinks will be present and growth will be continuous, but as the surface becomes smoother, growth rate is reduced and other growth mechanisms become operative [29].

Under conditions of slight undercooling (low $\Delta T$), crystal propagation is thought to be facilitated by crystal defects, namely, point, edge, and screw dislocation

**Table 1**   Various Physical Properties of Water and Ice at 0°C and −20°C

| Property | 0°C | 0°C(ice) | −20°C |
|---|---|---|---|
| Density (kg/L) | 0.9998 | 0.9168 | 0.9193 |
| Viscosity (Pa · s) | $1.787 \times 10^{-3}$ | — | — |
| Vapor pressure (Pa) | $6.104 \times 10^2$ | $6.104 \times 10^2$ | $1.034 \times 10^2$ |
| Heat capacity (J/kg/K) | 4.2177 | 2.1009 | 1.9544 |
| Thermal conductivity (J m · s/k) | $5.644 \times 10^2$ | $22.40 \times 10^2$ | $24.33 \times 10^2$ |
| Thermal diffusivity (m²/s) | $1.3 \times 10^{-5}$ | $\sim 1.1 \times 10^{-4}$ | $\sim 1.1 \times 10^{-4}$ |

defects [5]. In the case of rapid cooling rates, defect mechanisms appear to be less important, since water molecules are more likely to orient correctly in the absence of a step. The mechanisms involved in the development of crystal morphology during freezing are complex and are affected by several factors [5]. The direction and degree of undercooling dictate the type of crystal and plane by which growth is observed. In general, heat may be removed from a material following one of two general schemes: (1) through the undercooled liquid (latent and sensible heat)—if the material is severely undercooled prior to crystallization or the rate of heat removal exceeds the heat released and undercooling is conserved, the temperature gradient is toward the freezing interface—or (2) through the solid phase with a temperature gradient within the ice phase [30]. Growth of ice crystals more commonly follows the latter scheme perpendicular to the $c$-axis, the rate of which is determined by the temperature gradient in the ice and the anisotropy of the various growth velocities in the different directions. At low undercooling, thin planar dendrites form parallel to the seeding crystal, whereas with high undercooling ($\Delta T > 5°$), radial growth follows in the direction of the $a$-axis [30]. This growth will continue with undercooling up to a point where the angle of the growth front reaches a maximum value of 120° and hexagonal dendrite shapes are observed [3,30]. When water is not undercooled, the growth of ice crystals will be predominantly planar [3]. The reader is referred to Hobbs [26] for a more complete examination of the influence of undercooling on the different types of crystal formations and patterns.

## D. Freezing of Aqueous Solutions: The Freeze-Concentration Process

The introduction of a solute greatly increases the complexity of the crystal growth process, since kinetic factors, e.g., mass transfer, play a dominating role at a certain point in the process. Figure 4 shows the general time-temperature relationship for the freezing of an aqueous solution. Point B′, unlike B, occurs at higher temperatures, since heterogeneous nucleation mechanisms will most likely dominate. Due to the release of latent heat during ice formation (nucleation and crystal growth), the temperature rises to C′, which is substantially lower than C as a result of the depressed freezing point of the solution. The freezing point represents the point at which the chemical potential between the solid and liquid phases are at equilibrium at a particular temperature [1–3,13,31,32]. At 0°C, the vapor pressure of pure water and ice are equal, resulting in a freezing point at C. However, the addition of solute reduces the partial vapor pressure of water, so equilibrium between the two phases can only be achieved by a reduction in temperature. In other words, the vapor pressure of the ice is lowered to equilibrate with the solution and represents the freezing point depression (FPD). Based on Raoult's law [1], this relationship is a function of the number of molecules per unit volume and is derived from equilibrium thermodynamics. However, as will be emphasized from this point on, non-ideal behavior occurs at high solute concentrations, and simple expressions predicting FPD no longer hold. A more theoretical discussion can be found in Franks [3] and Hobbs [26]. Using the concept of FPD, a qualitative explanation of the change in temperature between C' and D' can be made.

During the process of ice propagation (C'D'), several variables must be considered. The process can be defined as a combination of equilibrium thermodynamics and mass-transfer effects of freeze-concentration. As ice separates out in pure form, a concurrent reduction in vapor pressure and freezing point of the remaining liquid occurs as the concentration increases. Hence, the freezing temperature must be further lowered, so ice may continue to propagate. This is shown as a continual decrease in freezing temperature to D'. Similar to the freezing of pure systems, heat transfer initially acts as the rate-limiting step. Assuming that moderate cooling rates are employed, the temperature of the freezing interface will be close to the freezing isotherm, resulting in a low level of undercooling at the interface. Like nucleation, undercooling is necessary for growth, although to a lesser extent [2,6]. Under these conditions, the process will be close to, if not at, equilibrium. Ice and the resulting concentrated unfrozen phase (UFP) have similar vapor pressures at the same temperature. As the concentration continues to increase, there is a point at which factors other than thermodynamics are rate-limiting and control further growth.

The influence of kinetics on ice formation can best be demonstrated using the concepts of a modified time-temperature-transformation (TTT) curve, superimposed with a theoretical curve describing the behavior of the liquid phase as temperature is lowered [33,34]. At each temperature, consider two times scales: one representing the time scale for relaxation within the undercooled concentrated liquid ($\tau_{liq}$), and the other time for a fixed volume fraction of ice to form ($\tau_{cry}$). Representative plots showing how these two characteristic times vary with temperature in both homogeneous and heterogeneous systems are shown in Fig. 5. As temperature is decreased, $\tau_{liq}$ increases at a steady rate until a point is reached where $\tau_{liq}$ increases sharply as diffusion-controlled properties become more dominant. The $\tau_{cry}$ plot shows enhanced crystallization kinetics within the higher temperature part of the curve (decreasing $\tau_{cry}$, less time required), followed by a characteristic minimum, denoted $\tau_{nose}$ (maximum ice formation), then an abrupt increase in $\tau$ with a reduction in temperature (Fig. 5). The time scale for ice formation is always greater than the relaxation time scale of the liquid phase at each temperature (Fig. 5) [33]. At a point close to $\tau_{nose}$, relaxation within the liquid phase becomes large, which reflects the inability of the liquids to adjust to the change in temperature. This increase in $\tau_{liq}$ from Fig. 5 has a direct influence on the existence and position of $\tau_{nose}$, since the value of $\tau_{cry}$ (continual increase above $\tau_{nose}$) is controlled by kinetic processes, which influence $\tau_{liq}$. In other words, more time is required for the chosen fraction of ice to transform. Consequently, $\tau_{cry}$ is linked to $\tau_{liq}$, especially

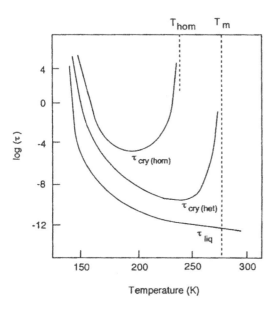

**Figure 5**  Relationship between $\tau_{cry}$ and $\tau_{liq}$ for both homogeneous and heterogeneous nucleating systems as a function of temperature. (From Ref. 33.)

at temperatures below $\tau_{nose}$, and this relationship is dependent on solute type and initial concentration [33,34].

Kadyela and Angell [35] have demonstrated that nucleation rate related to the degree of undercooling (heat removal, freezing temperature) determines the first part of the $\tau_{cry}$ curve (heat transfer limited), whereas the lower temperature $\tau_{cry}$ values are dominated by liquid kinetics. Beyond the critical temperature (also expressed as concentration and $\tau_{nose}$), a situation may arise where the propagation velocity is reduced as water molecules become kinetically hindered, although the freezing point isotherm continues to decrease. The net result is an increase in the level of undercooling at the interface leading up to a point at which propagation velocity decreases [5,7]. Therefore, in real food systems this implies that as concentration increases, crystal growth rates are not a simple linear function of temperature (degree of undercooling) and may involve kinetic factors more than thermodynamic factors.

Another important factor determining propagation rate is the kinetics of diffusion of solute away from the growing interface and into the interior of the liquid phase. This effect becomes more dominating with increasing serum concentration since transport properties (mainly viscosity) of the UFP may exert an influence on the diffusion coefficient ($D$) of the noncrystalline solutes from the interface. The hypothesis that mass transfer and interface kinetics are more rate-limiting than heat transfer during crystal growth in aqueous solutions was tested by Muhr and Blanshard [36] and Blanshard et al. [37]. Using linear crystallization velocities or rates (LCV or LCR) of various sucrose solutions, they reported that calculated diffusion coefficients, representing the quantity of sucrose expelled per volume of ice formed, were much lower than the thermal conductivity when the sucrose exceeded 10%. LCV experiments of sucrose and glucose solutions expressed as dendrite velocity, $V$, (cm/sec) showed relatively linear reductions with increasing concentration $(c_s)$ at constant $\Delta T$. Langer [38] proposed a model that considers two different effects dependent on concentration, namely, a "stability" effect whereby $V$ increases with increasing $c_s$, and a "mass-transfer" effect through which an increase in initial $c_s$ results in a reduction in $V$ at constant $\Delta T$ [37]. Although the Langer model failed to accurately predict $V$, especially at low concentrations, the rapid decline in $V$ was attributed to the increasing mass-transfer resistance resulting from a falling $D$ and rising $c_s$, consistent with this model.

Also important when discussing the freezing of aqueous solutions is the possibility of "constitutional" undercooling resulting from concentration gradients as solutes are permitted to accumulate at the interface [3,5]. During freeze-concentration, the reduction in the liquid-solid equilibrium curve (freezing point) corresponds to this gradient. Constitutional undercooling (CU) is not undercooling in the ordinary sense and occurs when the temperature of the unfrozen solution is above the temperature of the solid-liquid interface but below the equilibrium freezing point. In the presence of an established concentration gradient and temperature gradient through the solid phase, CU may occur, and irregular continuous and/or discontinuous ice formation will be favored over a smooth continuous interface, consequently altering the crystal morphology [5]. This is important in describing the complex variation in crystal structures seen with different solutes, concentration dependence of diffusion coefficients, and, more importantly, LCR.

In contrast to nucleation, the rate of crystal propagation decreases with increased rate of heat removal as the temperature approaches $T_g$. This relationship is shown in Fig. 6 and is highly dependent on solute type and concentration. Hence it is highly likely that during rapid cooling for maximum nucleation frequency, nonequilibrium freezing will occur in the late stages of the process. At high rates of heat removal a point occurs at which the kinetic constraints on growth are sufficiently high (interface highly undercooled) to cause the material to vitrify and crystallization to cease. At this point in the freezing process, a relatively small, freezing rate–dependent amount of unfrozen water will exist. With reference to Fig. 4, the time between D' and E' represents the last stages of crystallization where latent heat release is minimal and the effect of vitrification results in the removal of only sensible heat down to the external freezing temperature. The mechanism of glass formation will be discussed in detail in Sec. III.B.

During slow cooling sample temperature is at or near the decreasing freezing isotherm, hence few nuclei will form, but each will grow extensively. This may

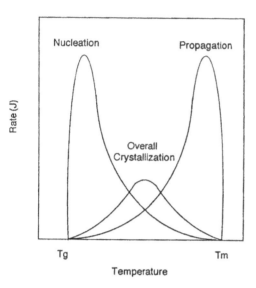

**Figure 6**   Comparison of nucleation and crystallization rate profiles as a function of temperature. (Adapted from Ref. 5.)

facilitate maximum ice growth, however, optimal crystal size is not achieved and substantial structural problems can arise, particularly with tissue foods. These detrimental effects are discussed in detail in Sec. IV.B. Bald [39] has shown, using a mathematical model, that mean crystal size increases with decreasing global cooling rate at the freezing front. This may lead to harmful effects during subsequent storage and distribution. Therefore, a trade-off exists between optimal initial crystal size and the kinetically determined amount of ice that forms in terms of the applied cooling rate.

## E.   Crystal Propagation in Macromolecular Solutions

The effect of high molecular weight stabilizers, alone and in sucrose solutions, has been extensively studied, because of their influence on ice recrystallization in frozen products, such as ice cream [40]. Muhr and Blanshard [36] used spherical flasks to measure the effect of different stabilizers on the LCR of water in sucrose solutions at different bath temperatures. They found that LCR were reduced by a factor of up to two thirds in the presence of xanthan, while guar gum and locust bean gum exhibited a smaller but similar reduction (Table 2). However, at a lower level of undercooling, guar gum had the greatest effect on ice crystal growth. In addition, their study demonstrated that stabilizers do not greatly influence dendrite morphology when compared to sucrose alone (Table 2). In a different but related study, Blond

**Table 2** Effect of Stabilizer Type, Aging Time, and Gel Strength on Ice Growth Velocites (V) in 35% (w/w) Sucrose Solutions Placed in a Bath at –6.5°C

| Stabilizer/gelling system | $V \times 10^3$ (cm/s) | Morphology | Ref. |
|---|---|---|---|
| Sucrose | 6.0 | Straight dendrites | 36 |
| 0.75% Guar gum | 2.7 | Straight dendrites in all directions | 36 |
| 0.75% Locust bean gum | 3.0 | Straight dendrites in all directions | 36 |
| 0.75% Xanthan gum | 1.5 | Tight mass of curved dendrites | 36 |
| 1% Gelatin (aged 3 h) | 1.1 | Angular cloud | 36 |
| 1% Gelatin (aged 24 h) | 0.7 | Angular cloud | 36 |
| 2% Gelatin (aged 24 h) | 0.1 | Cloud with spiky edges | 36 |
| 1% Agar (aged 3 h) | 3.1 | Cloud with well-defined lobes | 36 |
| 1% Agar (aged 24 h) | 2.3 | Cloud with well-defined lobes | 36 |
| 0.75 Alginate (Manucol DM) aged 24 h, 0.01 CaHPO$_4$ | 1.8 | Curved dendrites | 37 |
| 0.75% Alginate (Manucol DM) aged 24 h, 0.05 CaHPO$_4$ | 1.0 | Cloud with large lobes | 37 |
| 0.75% Alginate (Manucol DM) aged 24 h, 0.1 CaHPO$_4$ | 0.4 | Cloud with large lobes | 37 |

[41] studied the effect of low-viscosity carboxymethyl cellulose (CMC) in the absence of sucrose and showed that LCR was substantially reduced as concentration increased. A weak relationship between the viscosity of CMC solutions of varying composition (diffusion of water) and LCR was observed. However, it was shown that PEG, which increased molecular weight, induced a greater effect on LCR. The conclusion that viscosity is not a limiting factor was also stated by Budiaman and Fennema [42]. Therefore, it is possible that above the critical concentration of the macromolecule, entanglement causes a reduction in the ability of the polymer to diffuse from the ice interface, hence exerting a mechanical hindrance, possibly blocking growth sites on the ice front. Moreover, it has been suggested that after a significant amount of ice has separated, the freezing point at the interface may be affected, possibly influencing mass transfer properties [36,37].

The formation of a gel network reduces the LCR to a much greater extent than ungelled solutions [36,37,41]. Reduction in the ice-propagation rate was dependent on type, concentration, and aging time of the gelling system. Gelatin exerted the largest effect, despite having the softest gel (Table 2). Even for a very weak gelatin gel (1% aged 24 h), LCR was substantially reduced to 0.1 cm/s from 6.0 cm/s seen for sucrose. The aging temperature of gelatin gels has also been shown to have a influence on the LCR [41]. It appears that as the number of junction zones increase, either through an increase in concentration or aging (maturing) time, the LCR decreases, suggesting that the level of firmness or rigidity is an important factor. This was confirmed by altering the number of junction zones in alginate [37] and pectin solutions [41] by varying $Ca^{2+}$ content. It is evident from Table 2 that as the number of intermolecular linkages increased, resulting in a gel with a greater storage modulus, $G'$ [41], the LCR dropped greatly, with small changes in morphology. Using HTO as a tracer, it was concluded that whether a gel exists or not, stabilizers, at least alginates, do not affect the diffusion coefficient of water and therefore can migrate freely through the system or gel [37]. Based on this fact, Blanshard et al. [37] have suggested that propagation rate is influenced by transport kinetics resulting in mechanical (physical) hindrance as well as a reduction in the freezing point due to curvature in the ice interface. However, the relevance of this to the stability of frozen stabilized products is still unknown.

## F.  Freezing of Tissue Material

Ice crystallization can cause extensive microstructural changes to tissue foods during freezing [27]. The extent of these changes is largely a function of the location of ice crystals, which depends mainly on the freezing rate and permeability of the tissue [2]. In the case of plant tissue, during slow freezing (low $\Delta T$) ice will form in extracellular spaces, resulting in an ice-rich matrix of low temperature surrounding the cells. It is likely that the extracellular matrix will be close to equilibrium conditions so that the osmolality of the extracellular matrix continues to rise with solute concentration as temperature is reduced. Therefore, this creates an osmotic gradient between the cell components and extracellular matrix, which facilitates a transfer of water through the cell membrane in an attempt to reach equilibrium and minimize undercooling. This translocated water will freeze externally, resulting in large crystals, a shrunken appearance of cells (extensive dehydration), drip loss and tissue shrinkage during thawing, and an overall reduction in quality [2,27,43,44]. Under these conditions, where osmotic transfer of water is high, cell walls may tear and buckle while membranes may rupture and/or fail [45]. On the other hand, if heat removal is rapid compared to the exosmosis of water, small uniformly distributed ice crystals are generally formed in both the intercellular and extracellular matrixes, leading to a superior product. This occurs in tissues with very low permeabilities, since the water within the cell can become undercooled to some critical level where intercellular ice may form. However, if permittivity is high, the cell may dehydrate independently of cooling rate.

Animal cells are different, since the membrane is less effective against ice propagation and hence intercellular ice formation is more prevalent [2]. In general, meat structure is less affected by freezing, due partially to the flexible nature of the fibers as compared to the semirigid nature of plant cells [27]. Intercellular ice formation in biological cells has been actively researched in the field of cryobiology and is known to be catalyzed either by the presence of external ice on the plasma membranes, called surface-catalyzed nucleation, or intercellular particles, called volume-catalyzed nucleation [19,46,47]. Using the models developed in this field of science may provide a better understanding of the mechanism of ice formation in tissue foods, particularly if a material is treated with a cryoprotectant before freezing.

As a result of freeze-concentration, several properties of the UFP, including pH, titratable acidity, viscosity, oxidation-reduction potential, and surface tension, change substantially with increasing concentration. These changes during freezing may lead to several secondary effects, which have the potential to result in a loss of overall quality. In the case of proteinaceous materials, a large increase in ionic strength during freezing (e.g., a 24-fold increase in NaCl at $-21°C$) [48] may lead to conformational changes in structure and hence in stability [1,2,5,49]. As a result, proteins sensitive to ionic strength may become irreversibly denatured and subsequently aggregate and precipitate. Moreover, at high protein concentrations, the probability of cross links forming protein-protein aggregates greatly increases [2]. With meat systems, these effects in combination with the process of Ostwald ripening (discussed in Sec. IV.B) can lead to curdling and drip following thawing [2,27]. As expected, tissues rapidly frozen results in small ice crystals being uniformly distributed within the tissue and less translocation of water compared to slow freezing. Drip loss after storage has been reported to be affected by freezing rate, decreasing with increased heat removal [50]. Unfortunately, the exact nature of denaturation of proteins during frozen storage of muscle is still speculative [51]. In general, it involves major changes to the myosin component, although some evidence suggests that actin also plays a role [51]. The reader is referred to a recent paper by Mackie [51], who gives a thorough review of the factors involved and the strategies used to stabilize muscle proteins during frozen storage.

## G. Freezing Time of Aqueous Systems

The freezing time $(t_{fr})$ of aqueous systems is difficult to assign in comparison to pure systems. In simple terms, it is usually taken as the time for the geometric center of the product to reach some predetermined temperature, i.e., external freezing temperature or $\Delta T$ below the $T_{fp}$ (equilibrium freezing point temperature) (Fig. 4). Additionally, $t_{fr}$ prediction for aqueous systems is difficult since calculation depends on several factors, which include the dimensions (thickness) and shape of the product, $\Delta T$, surface heat transfer coefficient (function of heating medium), thermal conductivity, and the temperature dependence of the specific heat and latent heat of fusion [1,2,52–54]. A large variety of analytically and empirically derived methods for calculation of $t_{fr}$ have been presented in the literature [55]. Within this voluminous array of approaches and methods are models for the determination of $t_{fr}$ for slabs, cylinders, and spherical materials, as well as products with regular and irregular shapes. However, the accuracy of each model is affected by the accuracy of the estimated thermal properties and the assumptions made about the heat-transfer properties at the interface [2]. As a result, many models fail when compared to experimental data, in part because of missing parameters (e.g., initial temperature of the material above $T_{fp}$) and the above-mentioned variables. Therefore, it is important to choose a freezing model that satisfies the geometric and thermal characteristics of the material in question and shows high correlation with experimental data before being considered valid. Moreover, it is worth noting that $t_{fr}$ models are traditionally derived by assuming heat transfer as the rate-limiting step [2,55,56]. From the above and following discussions, it is clear that mass transfer becomes dominating after some point; consequently, this assumption will only increase the inadequacy of the model.

## III.  THE GLASSY STATE IN FROZEN SYSTEMS

In most discussions pertaining to the long-term stability of foods, focus has been directed to water activity ($A_w$) as being the sole parameter for prediction [57]. However, $A_w$ was developed from an equilibrium thermodynamic framework and, in its strictest sense, only applies to solutions of infinite dilution and where solute-solute interactions do not occur [57,58]. Basic concepts in solution chemistry deal with solutions with solute concentrations from infinite dilution to saturation [59]. Such solutions are homogeneous in nature, and relaxation processes tend to be rapid within the time scale of the observation, always reaching a state of equilibrium [59]. Food systems, on the other hand, are very heterogeneous and contain relatively high solute concentrations, which increase through the process of freeze-concentration. Hence, nonideal, nonequilibrium behavior becomes increasingly more dominant as the quantity of ice increases with a decrease in temperature. In recent years, it has come to the attention of many food researchers that, under conditions of low moisture content (i.e., extremely high solute concentration), materials are far from physical or chemical equilibrium but rather are in a kinetically constrained state [60].

Franks [3] presented evidence suggesting the simple model in which added low or high molecular weight solutes compete for and bind water, resulting in low $A_w$, is too simple and does not explain the true dynamics of the aqueous phase. Water molecules in the presence of a polar solute fluctuate rapidly between the bulk phase and solvation sites on the hydrated polymer [3]. Franks has stated that the average resonance time on a solvation site is very short–less than 1 ns [57]–which implies that the hydration molecule is active and is not bound in an energetically covalent sense, but rather behaves as a plasticizer influencing the physicochemical properties of the component in consideration. Variables other than $A_w$ alone are needed to successfully control water mobility, availability, and reaction potential of the highly associated water. The physical state of the dominating component, in terms of its effect on mass transfer and kinetic parameters, may have a larger influence on $A_w$ than equilibrium-bound structures and thermodynamics [27,61].

In response to this new way of thinking, an approach for the study of concentrated nonequilibrium systems known as "food polymer science" has evolved, mainly based upon the pioneering works of Levine and Slade [62] and Slade and Levine [63]. Using relationships developed from the field of synthetic polymer science, functional properties of completely amorphous or partially crystalline food systems during processing and storage can be predicted and explained based on mobility and conceptualized in terms of "water dynamics" or "glass dynamics." Slade and Levine [63] have used these concepts in describing structure-function properties of foods stressing the important role of water, not only as a ubiquitous solvent, but as a plasticizer of natural glass-forming materials. As one would expect, "glass dynamics" deals with the relationships that exist among functional (thermomechanical) behavior, structure, and composition of products. In particular, these time- and temperature-dependent properties are characterized by the glass transition temperature, $T_g$, and the effect of water on the diffusion-limited temperature range (rubbery, liquid states) above $T_g$ during processing and storage [58,62, 63]. The focal point of the approach is that the behavior of food materials is governed by dynamics (mobility, diffusion, kinetics) rather than energetics (thermodynamics). The current interest in this approach to food stability for either foods at subzero temperatures or at low moisture is clearly illustrated by the recent number of publications on the subject [60,64–69]. Of these, Slade and Levine [60] have compiled the most comprehensive and up-to-date survey on glass transitions in foods, and therefore this reference is highly recommended.

The foundation of the glass dynamic concept of food stability evolves around the physical properties of the glassy state and the change in translational mobility at $T_g$. A glass is a nonequilibrium, metastable, amorphous solid of extremely high viscosity (i.e., $10^{10}–10^{14}$ Pa·s). The formation of a glass, known as vitrification, occurs when a liquid of disordered molecular structure is cooled very rapidly below its equilibrium crystallization temperature so that translational mobility decreases to a point where the molecules are unable to achieve their equilibrium conformation and packing. This occurs because of the increasing equilibrium time within the liquid with changing temperature, shown in Fig. 5 as a large increase in $\tau_{liq}$ at temperatures below $\tau_{nose}$ [70–72]. As a result, solidification (vitrification) occurs and the molecules within the liquid undergoing rearrangement are frozen into a disordered state of excess volume and enthalpy, in contrast to the favorable (crystallized) state at the same temperature [71]. Consequently, the glassy state is a nonequilibrium disordered solid (no lattice structure) that is unable to achieve any long-range cooperation relaxation behavior and therefore achieves real "time structural stability" through high viscosity on a practical time scale [59]. It is this intrinsic slowness of molecular reorganization below $T_g$ that the food technologist seeks to create within the concentrated phase surrounding constituents of food materials.

## A.    Principles of Nonequilibrium Materials

The glass transition temperature, $T_g$, is a temperature- and time-dependent change
in physical state from the glassy to rubbery viscous liquid without phase change
[58,71]. It is defined occasionally, although incorrectly, as a second-order thermodynamic transition and is characterized by (1) a change in rate of volume expansion
with temperature (a first derivative of Gibbs free energy, $\Delta G$), (2) a discontinuity
in the thermal expansion coefficient, $\alpha$, and (3) a discontinuity in heat capacity
(first derivative of enthalpy and second derivative of $\Delta G$ with respect to temperature) [71,73,74]. More importantly, $T_g$ can be described in terms of mechanical
relaxation properties, namely, changes in storage or loss moduli ($G'$ or $G''$), loss
tangent, and viscosity with changes in temperature [75].
    A nonlinear decrease in viscosity in the vicinity of $T_g$ is seen during warming,
rather than a discontinuity as in $\alpha$ and $\Delta C_p$. Figure 7 shows the change in viscosity
of a disordered liquid within the temperature range above $T_m$ and below $T_g$. As
temperature is lowered to the $T_m$, common Arreheius kinetics apply, whereas
between $T_m$ and $T_g$, nonlinear kinetics with a large temperature dependence
(particularly near $T_g$) become operative. The change in mechanical properties
within this range is a function of the free volume distribution (space between
molecules) of rubbery viscoelastic liquid and in turn is related to the mean
molecular weight and absolute temperature. At $T_g$, the cooperative mobility of the

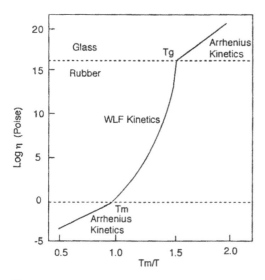

**Figure 7**    Changes in viscosity as a function of reduced temperature showing the regions
where WLF and Arrhenius kinetics apply for a completely amorphous or partially crystalline
material. (From Ref. 77.)

material becomes limited, leading to an increase of several orders of magnitude in viscosity and relaxation times [73]. In general, viscosity, relaxation time, and relaxation frequency at $T_g$ of a glassy material are approximately $10^{13}$ Pa·s, 200 s, and $10^{-3}$ Hz, respectively [72,76]. As temperature falls below $T_g$, the various material properties again follow Arrehenius kinetics, but the rates of diffusion-controlled processes are approximately 14 orders of magnitude greater than at $T < T_m$ [60].

Several variables influence the resultant position of $T_g$ and hence the mechanical properties and stability of amorphous materials. These include the molecular weight (MW) of the glass former, the MW of the plasticizer and, in the case of polymers, the degree of branching and entanglement [70,78]. As the number-average MW, $\overline{Mn}$, increases, $T_g$ increases due to a reduction in free volume up to the region of entanglement, where $T_g$ changes only slightly with further increase in $\overline{Mn}$. Fox and Flory [79] proposed that free volume of a polymer system is a linear function of $1/\overline{Mn}$. Imperfect packing around the ends of low MW materials increases free volume and lowers the $T_g$. In contrast, high MW polymers are packed in a way in which few chain ends exist, resulting in a reduction in free volume and a higher $T_g$. Thus, at constant temperature and pressure, the following equation applies:

$$f_m = f_o + \frac{A}{\overline{Mn}} \tag{6}$$

where $f_o$ refers to the fractional free volume at infinite molecular weight, $f_m$ represents the free volume due to molecular weight, and $A/\overline{Mn}$ represents the additional free volume associated with a pair of molecular ends [70].

Since free volume and $T_g$ are interrelated, Eq. (6) also describes the relationship between $\overline{Mn}$ and $T_g$ knowing the glass transition at infinite molecular weight ($T_{ga}$) in the following form:

$$T_g = T_{ga} + \frac{A}{\overline{Mn}} \tag{7}$$

An increase in $T_g$ represents an increase in the energy component (defined as $\Delta H_{app}$) necessary for a molecular relaxation to occur at a particular temperature. Therefore, a decrease in $\overline{Mn}$, for example, by the addition of a low MW diluent, will cause a reduction in $\Delta H_{app}$ and a concurrent shift of $T_g$ to lower temperature [79]. The fractional free volume resulting from a broad molecular distribution is determined by the adaptivity of the free volume of the different species. Branched polymers greatly increase free volume, and as the degree of branching increases, a decrease in $T_g$ is observed. Although the general correlation between increasing mean saccharide MW and increasing dry $T_g$ was proposed by Levine and Slade [62,80] and confirmed by the works of Orford et al. [81] on binary mixtures of mono-, di-, or oligosaccharides of individual $T_g$, it should be mentioned that deviation from this trend is observed within some compounds of similar MW [60].

Relaxation properties of glassy materials are strongly influenced by added plasticizer, i.e., relaxation times and $T_g$ decrease rapidly with the addition of a small amount of diluent. At constant temperature, the presence of small amounts of water in the form of a miscible plasticizer ultimately creates additional free volume (decreased mean MW of the glass former) within the glass-forming component (i.e., polymer). On the molecular level, an increase in intermolecular space results in decreased local viscosity and increased segmental mobility of side chains in the amorphous regions of the glassy polymer, which, in turn, changes the dynamics/mechanics of the polymer, allowing structural relaxation (primary) at a lower $T_g$ [70]. Increasing temperature at constant moisture parallels this direct effect of increased moisture [80], and therefore either process will facilitate an increase in free volume, enhancing the overall diffusion kinetics and ultimately reducing $T_g$. Prediction of the $T_g$ of mixtures from solution properties and composition is a difficult task since the true nature of the $T_g$ itself is still not completely understood [59,60]. Nevertheless a correlation between mean MW and $T_g$ for varying binary mixtures of individual $T_g$ can most easily be approximated using the Gordon and Taylor [82] equation commonly employed in the synthetic polymer field to predict $T_g$:

$$T_g = \frac{w_1 T_{g_1} + k w_2 T_{g_2}}{w_1 + k w_2} \qquad (8)$$

where $T_g$ is the glass transition temperature of the mixture, $w_1$ and $w_2$ the weight fractions of components 1 and 2, respectively, $T_{g1}$ and $T_{g2}$ their glass transitions in pure form, and $k$ is a constant.

This simple theoretical approach has been used with limited success and is most appropriate to collaborate experimental data. In any case, Slade and Levine [63] have shown, using the glass formed from a molten 1:1 mixture of crystalline sucrose and glucose, that the calculated value is within 1°C of the value measured by differential scanning calorimetry (DSC). Moreover, Roos and Karel [83] used the Gordon and Taylor equation to predict $T_g$ values of a maltodextrin-sucrose binary mixture and observed good correlation between predicted and measured $T_g$. Although in a different context, Roos and Karel [84] used a derivative of the Gordon and Taylor equation to predict the maximum solute concentration of various freeze-concentrated carbohydrate systems. However, in the case of frozen systems with high initial water content, models based on free volume predict $T_g$ values well below experimental values, since the phase separation of the plasticizer (water) into ice and the resulting freeze-concentration are not accounted for [60]. Hence, when one wishes to determine $T_g$ for a system in which the water component will easily crystallize upon lowering temperature, it should be measured rather than predicted.

## B. Factors Affecting the Formation of a Glassy State in Frozen Systems

### 1. Formation Under Equilibrium Conditions

The formation of the glassy supersaturated state in food materials is achieved by the removal of large quantities of water, either by dehyration at high temperature or by the process of freeze-concentration, since water is a readily crystallizable plasticizer ($T_m(K)/T_g(K) \approx 2$) [58,85]. During freezing, the glass-forming constituent of the system (e.g., monosaccharide, disaccharide, or polymeric material) must not precipitate and form a eutectic mixture, otherwise the system will be unable to supersaturate and vitrify. Fortunately, however, eutectic formation rarely occurs largely due to the high viscosities and low thermal energy [1–3,27,59].

Consider the freezing of a 20% sucrose solution (large aqueous phase) having a theoretical $T_g$ close to that of water ($-135°C$) estimated from Fig. 8. The supplementary state diagram showing the solid-liquid coexistence boundaries and glass transition profile for a binary sucrose-water system is shown in Fig. 8. State diagrams of tertiary model systems, which more closely represent real systems, have been reported [86]. State diagrams are commonly employed to describe nonequilibrium systems and illustrate the different physical states that can exist as a function of temperature, composition, and pressure [3,27,58,80]. The initial $T_g$ of this solution

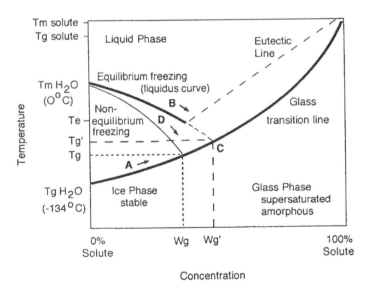

**Figure 8** Supplementary state diagram for an aqueous sucrose solution showing the glass transition line, liquidus curve, theoretical eutectic line, and the various physical states defined within the boundaries. Details of specific points are provided in the text. (From Ref. 27.)

at room temperature before phase separation is indicated by A in Fig. 8. Upon cooling the system much below its equilibrium freezing point (undercooling), nucleation and subsequent crystallization initiate the freeze-concentration process removing water (plasticizer) in its pure form (discussed in detail in Sec. II.D). As ice crystallization proceeds, nonideal behavior quickly sets in and the continual increase in solute concentration (removal of plasticizer) further depresses the equilibrium freezing point of the unfrozen phase in a manner that follows the liquidus curve (B in Fig. 8), while the $T_g$ of the system moves up the glass-transition line with a rapid increase in viscosity in a non-Arrhenius manner, particularly in late stages of the freezing process [84]. At all points on the liquidus line in Fig. 8 a ratio of ice to unfrozen water will exist, increasing with decreasing temperature, to satisfy equilibrium conditions based on the depressed freezing point. Assuming equilibrium freezing conditions are satisfied, which is unlikely during freezing of real food systems, the $T_g$ continues to increase with increased ice formation and the system becomes supersaturated beyond the eutectic curve. However, only up to the eutectic point ($T_e$) does the liquidus curve represent an equilibrium state, while beyond $T_e$ it depicts a state of thermal instability far from equilibrium [48]. As ice is separated from the liquid phase, $T - T_g$ decreases, causing a concurrent increase in viscosity.

When a critical concentration of about 80% [87–89] is reached, the unfrozen liquid exhibits limited mobility and the physical state of the unfrozen phase changes from a viscoelastic "rubber" to a brittle amorphous solid. In Fig. 8, the intersection of the nonequilibrium extension of the liquids curve and kinetically determined nonequilibrium glass-transition curve, denoted as C, represents the solute-specific, concentration-invariant, maximally freeze-concentrated $T_g$, denoted $T_g'$, of the frozen system, where ice formation ceases within the time scale of the measurement [63,75,80]. The corresponding $W_g'$ and $C_g'$ define the maximum (perhaps best defined as practical) amount of plasticizing water unable to crystallize, thereby trapped within the glass, and sucrose concentration at $T_g'$, respectively [58,63,75]. This unfrozen water is not bound in an "energetic" sense, as alluded to earlier, but rather is unable to freeze within practical time frames [1,3,57,58,80,90].

At the $T_g'$, the supersaturated solute (UFP) takes on solid properties because of reduced molecular motion, which is responsible for the tremendous reduction in translational, not rotational, mobility [60,80,84,87,89,91]. The significant reduction in free volume and the associated viscosity increase ($10^{14}$ Pa·s) has been estimated to reduce the diffusion of a water molecule to 1 cm per $3 \times 10^5$ years [92], and therefore, viscous flow of this supercooled liquid takes place at rates of $\mu$m per year. However, warming through the $T_g'$ results in a tremendous increase in diffusion, not only from the effects of the amorphous to rubber transition but also from increased dilution, as melting of small ice crystals occurs almost simultaneously ($T_g' = T_m'$). This is in contrast to polymers where, at $T > T_g$, composition does not change and hence is the source of some difficulty when assessing the effect of temperature above $T_g'$.

## 2. Formation Under Nonequilibrium Conditions

In addition to the thermodynamic driving force to achieve the unfrozen water content corresponding to $W_g'$, one must consider the large kinetic factors that "overtake" the freezing process. At subzero temperatures, the formation of an amorphous state is time-dependent, since the limiting factor of the process (water removal in the form of ice) becomes more difficult as concentration increases. When reexamining Fig. 5, it becomes clear once the freezing process moves beyond $\tau_{nose}$, which is largely dependent upon composition and freezing rate, that the kinetics of ice propagation become progressively more dominated by mass transfer and diffusion of the liquid ($\tau_{liq}$), thereby reducing overall translational mobility. In other words, the exponential effect of viscosity on mass transfer properties acts as the limiting factor for growth. In addition, under conditions where heat removal is rapid, a high level of undercooling at the interface will only add to a further decrease in propagation rate. The net result is that freezing becomes progressively slower as ice crystallization is hindered. Consequently, more time is required for lattice growth at each temperature (increased $\tau_{cry}$). Therefore, the kinetic restriction imposed on the system can result in a situation in which nonequilibrium freezing leading to a partial dilute glass can occur [84,87].

The typical pathway that a system may follow during nonequilibrium freezing is shown in Fig. 8 as the line leading to a $T_g$ (point D) lower than $T_g'$ with a corresponding lower $C_g$ due to excess undercooled water plasticized within the glass. The magnitude of deviation from the equilibrium curve may be regarded as a function of the degree of departure from equilibrium. Systems possessing this undesirable structure may undergo various relaxation-recrystallization mechanisms in order to maximally freeze-concentrate and minimize the unfrozen water content. As a result, during warming, systems formed under these conditions may lead to one or more low-temperature transitions, followed by an exothermic devitrification peak due to crystallization of immobilized water and, finally, the onset of ice melting, $T_m$ [84,87,93,94].

Several researchers have demonstrated that proper annealing protocols within a narrow temperature range between the $T_g$ and onset of melting, $T_m$, are necessary to allow for delayed crystallization [84,87–89,95]. However, it is very difficult to form ice at concentrations greater than 70% solute without extensive annealing (long times) and/or temperature cycling. Therefore it is virtually impossible to maximally freeze concentrate within realistic time frames [3,96–99]. The absolute values of $C_g'$ and $W_g'$ are presently a subject of much debate in the literature. However, the determination of their true values is paramount in the quest to understand glass formation in frozen systems and will be discussed in detail in Sec. III.C [60].

## 3. Methods for Determination of $T_g$

Since glass transition manifests itself as a change in thermal and mechanical properties, thermal analysis techniques are most commonly used to study this phenomena [100]. In particular, measurement of the change in $C_p$ using DSC has become the established method, largely because of its ease of operation and widespread availability. However, a measurable heat change at $T_g$ may be problematic for some glass-forming materials since $\Delta C_p$ can be small depending on the type of glass (i.e., "fragile" or "strong") [101], and the degree of polymerization, as in the case of maltose oligomers, which showed a decrease in $\Delta C_p$ with increasing polymerization [81]. The relationship between $\Delta C_p$ and accompanying $T_g$ is obscure and still poorly understood [59,60]. Other techniques, namely, dynamic mechanical analysis (DMA) [65,69, 84,102] and Thermomechanical Analysis (TMA) [20,27,88,103–107] have been used to characterize phase behavior in frozen systems. Because $T_g$ is a time- and frequency-dependent phenomenon (it occurs over a temperature range), dielectric analysis (DEA) as well as other spectroscopic methods, such as NMR [61,106,108,109] and electron spin resonance (ESR) [110–112], are also useful since they allow for the measurement of relaxation times more applicable to time/frequency-dependent materials. However, as with most investigations, the most effective approach when studying glass transitions and glasses in food systems is the combination of various thermal analysis techniques (DSC, DEA, TMA, DMA).

## C. Interpretation of Subzero Events: The Identification of $T_g{'}$

The practical importance of $T_g{'}$ from a theoretical standpoint is well established and agreed upon by many in the field. However, a wide variation in published $T_g{'}$ values exists mainly as a result of different interpretations of the same DSC-measured transitions of warming curves of many low MW carbohydrates. A typical warming DSC curve for a slowly cooled (5°C/min), 40% sucrose solution is presented in Fig. 9 (dotted line). Based on the original description of $T_g{'}$, consistent with the state diagram, glass relaxation and ice melting should coincide within the time frame of the DSC experiment. However, due in part to the possibility of partial freeze-concentration, biphasic behavior is always evident with a measurable devitrification peak between the transitions once sufficient translational mobility is regained. Ablett et al. [89], using optical microscopy, confirmed that the exotherm in vitrified sucrose solutions was the result of delayed crystallization. Sahagian and Goff [88] have also shown that the magnitude of this exotherm is related to the applied freezing rate (increases with increased rate) and initial concentration, consistent with the effect of imposed kinetic constraints on the process of ice formation. For discussion purposes, these two transitions are labeled as $T_{Tr1}$ and $T_{Tr2}$ in Figure 9 similar to the method adopted by Sahagian and Goff [88].

At present, there are three major interpretations regarding the origin of each transition and the assignment of $T_g{'}$:

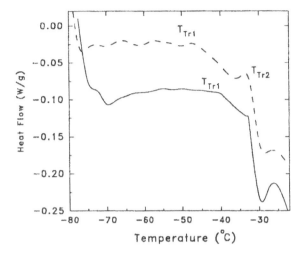

**Figure 9** Thermal behavior of 40% sucrose solutions following slow freezing (- - - -) (2°C min⁻¹) and slow freezing after annealing at –35°C for 60 min (——) showing the onset of transitions $T_{Tr1}$ and $T_{Tr2}$. (From Ref. 119.)

1. The $T_g'$ is represented by the major transition at $T_{Tr2}$ (Fig. 9), while the lower temperature $T_g$ (partial glass) is usually ignored [1,4,8,58,62,63,75,98,99].
2. The $T_g'$ occurs at the lower transition, while the warmer transition is not a glass transition but rather the onset of ice melting (first order transition) initiated soon after the $T_g'$ [65,74,84,87–89,95].
3. The $T_g'$ is close to $T_{Tr2}$ [99] or somewhere between the two transitions [69,97,113], and the complex behavior is partially the result of a relaxation process manifested as an excess enthalpy overshoot close to the $T_g'$.

A completely different explanation than those above with regard to freeze-concentrated ribose solutions was proposed by Wasylyk and Baust [114], who postulated that structural conformation and differences in ordering of hydration molecules may contribute to the coexistence of two $T_g$. In addition, Sahagian and Goff [88] concluded that the lower transition most likely was $T_g'$, even though the occurrence of an overshoot at $T_{Tr2}$ suggested that an amorphous component above –35°C underwent relaxation and therefore must exist. Table 3 represents a more complete summary of the recently published $T_g'$ and $C_g'$ for freeze-concentrated carbohydrate solutions of sucrose, glucose, and fructose. In the case of sucrose, it is evident from the table that there are three different temperature $(T_g')$ ranges: –35 to –32°C, –37 to –42°C, and –46°C and below. Adding to the difficulty is the fact that some reported values are the midpoint of the transition [58,75,99], while others represent the onset [84,87,88]. Moreover, annealing temperatures and times are not consistent and, in

**Table 3** Published $T_g'$ and $C_g'$ Values for Freeze-Concentrated Sucrose, Glucose, and Fructose Solutions

| Glass former | $T_g'$ (°C) | Ref. | $C_g'$ (w/w) | Ref. |
|---|---|---|---|---|
| Sucrose | − 32 | 1,48,62,63,75,86,98, 99,104,110 | ~64 | 1,48,63,75 |
| | − 35 | 95 | ~80 | 74,84,87,95,103 |
| | − 37 | 68 | ~83 | 98,99,116 |
| | − 40, − 42 | 69,88,89,115 | | |
| | − 46 | 74,84,103 | | |
| Fructose | − 42 | 1,48,62,75,108 | ~51 | 1,48,63,75 |
| | − 48 | 20,117 | ~79 | 84,117 |
| | − 53 | 84,116 | ~83 | 116 |
| | | | ~85 | 118 |
| Glucose | − 43 | 1,48,62,75,104,108 | ~71 | 1,48,63,75 |
| | − 36 | 3 | ~80 | 84,116 |
| | − 52 | 84,116 | ~83 | 113,118 |

the case of values reported by Levine and Slade [58,75], $T_g'$ values are from nonannealed systems. In any case, it is evident that clarification of these thermal transitions in frozen systems is a priority, since new technologies and processes that attempt to use this approach must rely on the position of $T_g'$ under specific conditions.

Some of the debate on whether $T_g'$ from interpretation 1 is a glass transition stems from recently reported $\Delta C_p$ values at this transition [88,89,95]. Researchers have reported that $\Delta C_p$ is three times larger than expected, as determined from completely amorphous (no ice) supersaturated systems. Sahagian and Goff [88] have shown that the lower transition has a $\Delta C_p$, which is in close agreement with $\Delta C_p$ for a low-level plasticized sucrose glass of approximately 80%. However, an analysis of $\Delta C_p$ for a galactose-water system by Blond and Simatos [97] demonstrated that changes in heat capacity are associated with the $T_g$ of the freeze-concentrated phase and the second (higher) transition overlaps the melting endotherm. However, as pointed out by Slade and Levine [60] and discussed in detail by Angell et al. [101], "fragile" and "strong" glass formers have substantially different relaxation properties and magnitudes in $\Delta C_p$. In addition, Hatley and Mant [99] have shown that at the $T_g'$ in dilute solutions, a significant contribution from a relaxation process, as well as the onset of ice melting, gives rise to a large distorted $T_g'$. The identification of an enthalpy relaxation at $T_g'$ was previously reported by Blond and Simatos [97] and Simatos and Blond [113] and recently confirmed by isochronal annealing experiments performed by Sahagian and Goff [88]. Furthermore, an increase in plasticizer and mean MW increases free volume, which tends to result in higher $\Delta C_p$ at $T_g$ [87]. Therefore, it appears that there may be no simple relationship between $\Delta C_p$ and $T_g'$, making identification using $\Delta C_p$ rather difficult.

## 1. Effect of Annealing on $T_g$

A significant property of frozen systems is their strong dependency on thermal history (cooling rate, annealing), particularly as seen with the lower transition ($T_{Tr1}$). Changes in $T_{Tr1}$ with annealing temperature ($T_a$) for freeze-concentrated 40% and 60% sucrose solutions are shown in Table 4. The increase in $T_{Tr1}$ has been interpreted as the $T_g$ of diluted glass approaching the $T_g'$ of the system through removal of excess undercooled water during devitrification. Consider an initial $T_g$ of approximately –46°C [84,87,88]. Annealing at –55 and –50°C is within the amorphous state, and therefore little change in the ice fraction is expected or observed. However, above the initial $T_g$ and after annealing for 60 minutes between –45 and –35°C, the $T_{Tr1}$ shows a continual increase but never reaches the specific $T_a$. This result again demonstrates the effect of extreme viscosity on ice formation in highly concentrated solutions. Consequently, more time is required for the UFP to reach a concentration associated with $T_a$. Ablett et al. [89] demonstrated that annealing times of 16 hours were required for the $T_g$ of the unfrozen phase to reach the annealing temperature (i.e., maximal ice formation at that temperature). An

**Table 4**   $T_{Tr1}$ Onset of 40 and 60% (w/w) Sucrose Solutions Slowly Frozen (2°C min$^{-1}$) Following Isothermal Annealing at Various Temperatures for 60 Minutes

| Annealing temperature (°C) | 40% (w/w) | 60% (w/w) |
|---|---|---|
| – 55 | – 54.2 (1.19)[a] | – 54.6 (1.23) |
| – 50 | – 49.4 (0.93) | – 52.1 (1.38) |
| – 45 | – 47.9 (0.85) | – 50.3 (0.95) |
| – 40 | – 46.1 (1.15) | – 48.1 (0.91) |
| – 37 | – 44.5 (0.88) | – 47.4 (0.98) |
| – 35 | – 40.1 (0.78) | – 45.0 (0.99) |
| – 32 | – 48.2 (0.96) | – 46.2 (1.15) |
| – 28 | – 49.1 (0.91) | – 47.4 (0.95) |
| – 20 | – 51.8 (1.19) | – 52.9 (2.23) |

[a]Standard deviation of three determinations.
*Source*: Ref. 119.

important point to note from the annealing study of Sahagian and Goff [88] is that two transitions are still evident, even though the devitrification peak is no longer detectable (Fig. 9, solid line). However, because the $\Delta T$ between observed $T_{Tr1}$ (onset) and $T_a$ is relatively large ($\Delta T \approx 5°C$) and a substantial reduction in $T_{Tr1}$ is seen at $T_a$ greater than $-35°C$ (ice melting increased plasticizer), it is possible that this transition is not the "true" $T_g'$ but rather best defined as very close to experimental maximum freeze-concentration—"practical" $T_g'$. Another possibility is that this small separation is simply the result of a finite warming rate and that infinitely small heating rates may lead to one transition.

These data also suggest that $T_{Tr2}$ has at least a significant first-order component, namely, the onset of ice melting, representing the "achievable" freezing point depression temperature of the UFP. From Fig. 9, the two transitions appear to be strongly related in origin. An increase in $T_a$ with 60% solutions also results in fairly linear increases in $T_{Tr2}$, not the expected decrease as concentration increases [119]. This may suggest that other factors, possibly reduced free volume facilitated by aging, play a role in the position of $T_{Tr2}$ and that the small change in $T_{fp}$ with $C_g$ is not significant enough to be measured at such high concentrations. Another curious result is the occurrence of an overshoot of $T_{Tr2}$ at $T_a$ up to $-35°C$, as shown in Fig. 9, which strongly suggests that an amorphous component or domain above $-35°C$ can undergo relaxation and therefore must exist. The complexity of this phenomenon is exemplified by the fact that a 60% sucrose system does not follow exactly the same trend (Table 4), which may represent an overall change in the kinetics of the system. Therefore, it is clear that our present understanding is far from complete and that time-temperature relationships, crucial to the kinetics of ice formation, are of prime importance when dealing with nonequilibrium materials. Much of the ambiguity that exists may be attributed to a poor recognition of the significance of time/kinetics during freezing.

## D.  Relevance of the Glassy State to Food Stability

Before we begin a discussion of diffusion and mobility of the glassy state and how this translates into kinetic stability, it is necessary to review some of the basics of sub-$T_g$ relaxation. An understanding of the effects of free volume holes below $T_g$ is important since the overall diffusion properties of the glass may have a strong influence on the diffusion of constituents (water) within the glass, which ultimately govern the long-term stability of cryopreserved materials [59,60,69,88,112].

### 1.  Physical Aging and Enthalpy Relaxation at $T < T_g$

At cooling rates greater than the time scale required for molecular relaxation at each temperature, an amorphous state of higher configurational enthalpy and free volume, relative to their equilibrium values, will be induced [120]. Hence, molecules frozen into this unstable, nonequilibrium state will exhibit a strong thermodynamic-relaxation potential to approach the lowest energy state in an effort to achieve equilibrium conditions in volume, enthalpy, and entropy [120]. Structural rearrangement of amorphous materials, facilitated through the process of physical aging (annealing), is governed by complex, nonlinear, and nonexponential conformational changes and is strongly dependent on time and temperature [121–123].

From a structural point of view, the rate at which the system moves toward equilibrium at a particular temperature during annealing is a function of the segmental mobility of the molecules [124]. Figure 10 shows the relationship between free volume, packing, and mobility. A system that is unstable, as a result of excess free volume and disordered molecular packing, will possess a strong rearrangement potential and, therefore, will attempt to rapidly approach equilibrium. As the process proceeds, the change in segmental mobility of the amorphous state becomes reduced and ultimately determines the rate of free volume change and redistribution [125,126]. The following closed-loop scheme represents the relationship:

$$V_f \rightarrow\rightarrow\rightarrow\rightarrow M \rightarrow\rightarrow\rightarrow\rightarrow \frac{dV_f}{dt} \tag{9}$$

Free volume $(V_f)$ determines mobility $(M)$, which determines the rate of change $(dV_f/dt)$, which in turn determines $V_f$. Volume relaxation during annealing proceeds

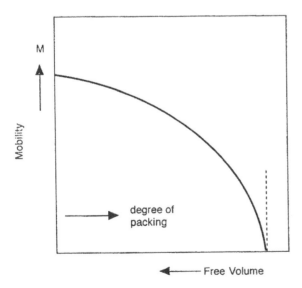

**Figure 10**　Relationship between free volume, molecular packing, and translational mobility of an amorphous material. As free volume decreases, the degree of packing and overall mobility increase and decrease, respectively. (From Ref. 124.)

in a nonlinear fashion produced by a continued decrease in segmental mobility. In addition, the mechanism of structural relaxation is driven by a relaxation potential created by a deviation in the free energy in relation to the equilibrium state corresponding to the same temperature [120,127]. As the temperature decreases from $T_g$, the relaxation potential increases, but the relaxation time becomes longer, resulting in a slower relaxation process (smaller overshoot). At higher temperatures approaching the $T_g$, relaxation times are reduced, and, as such, the relaxation process is greater.

Structural rearrangement of amorphous materials can occur during isothermal annealing, resulting in an enthalpy relaxation process, which, during rewarming, may give rise to an overshoot at the $T_g$. The origin of this overshoot is related to the different relaxation times of the various elements of the material [128]. Sub-$T_g$ relaxations in freeze-concentrated low MW carbohydrate and some organic polymer solutions are thought to occur at much faster rates than with high MW polymers [113]. The characteristic relaxation overshoot overlapping $T_{Tr2}$ ($T_g$) for a 60% sucrose solution rapidly frozen and annealed at –40°C for 60 minutes is shown in Fig. 11. In the case of sucrose, substantial relaxation is evident after only 1 hour of annealing, unlike most polymers, which require much more time. Measurement of the area of the overshoot (excess enthalpy) parallels the extent of physical aging and molecular relaxation that occurs during annealing at a particular temperature

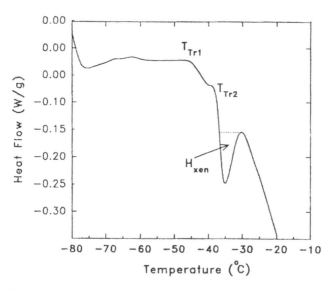

**Figure 11**   Excess enthalpy overshoot of a 60% (w/w) sucrose rapidly frozen and annealed at –40°C for 60 min. $\Delta H_{xen}$ is a measure of excess enthalpy at $T_{Tr2}$ related to the amount of relaxation that occurred during aging. (From Ref. 119.)

[129]. In thermodynamic terms, it represents an enthalpy compensation process for the loss of excess configurational enthalpy and entropy that occurs during aging and reorganization of the highly disordered glass [127].

## 2. Mobility at Temperatures Below $Tg$ or $T_g{}'$

One of the major components of the "glass-dynamics" approach to stability is that within the amorphous phase ($T < T_g$), the extremely high viscosity ($10^{14}$ Pa·s) hinders the diffusion of water molecules to the ice interface, thereby preventing crystallization/recrystallization mechanisms within realistic time frames [1,3,58]. However, evidence exists from NMR and ESR that unfrozen water is relatively mobile, possibly as a result of sample inhomogeneity leading to components or domains within the sample of varying degrees of mobility. Water binding to sites has been found to be highly transient and occurs within the picosecond time scale [130]. Moreover, there is NMR evidence to suggest that rotational and translational properties of water below $T_g$ in amylopectin solutions continue to decline and show no temperature dependence at $T_g$, in contrast to the polymer itself [59,131]. Within the glass, water molecules are substantially hindered and cannot crystallize, but appear to be able to diffuse rapidly in comparison to the glass-forming molecules [59]. Also of particular scientific interest is the migration and removal of unfrozen water from amorphous materials during freeze-drying at $T < T_g$, since it signifies that sufficient translational mobility is present to allow for secondary drying [132]. Therefore, a link may exist between the factors affecting the diffusion of residual water in solid solutions below $T_g$ or $T_g{}'$ and the nature and rates of measurable chemical reactions, in particular those that are not translational diffusion-limited [59,60]. Therefore, storage below $T_g{}'$ may not necessarily assure complete stability.

Slade and Levine [60] have recently suggested an explanation for this behavior. In short, they theorize that below the $T_g$ of the blend but above the $T_g$ of the diluent (water in foods), the free volume of the mixture is such that the critical volume for transport of the diluent is adequately met, and hence it is able to diffuse. Additionally, as the temperature approaches the $T_g$ of the diluent, free volume will continue to decrease toward the hydrodynamic volume of the diluent, at which time diffusion becomes severely hindered. Based on this interpretation, frozen food stability at $T_g{}'$ may depend on the resultant free volume of the unfrozen phase. The kinetics of relaxation processes below $T_g{}'$, which substantially alter the total free volume of the glass, may modify diffusion and subsequent related effects (collapse, recrystallization, and enzymatic activity).

Recent results on the physical aging behavior of freeze-concentrated sucrose solutions at different annealing temperatures are presented in Fig. 12 [88]. Model sucrose systems were used to simulate sub-$T_g{}'$ relaxation behavior of frozen dairy products, particularly ice cream. As previously discussed, excess enthalpy at $T_g$ represents the magnitude of relaxation that occurs during annealing and parallels

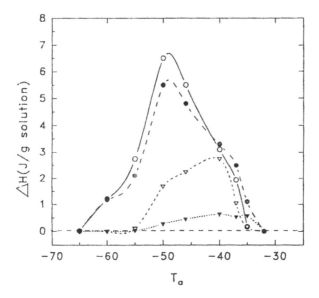

**Figure 12** Excess enthalpy profile defined as $\Delta H_{xen}$ (y-axis) for 40% and 60% (w/w) sucrose solutions rapidly [60% (m); 40% (□)] and slowly (2°C min⁻¹) [60% (∇), 40% (t)] frozen as a function of annealing temperature ($T_a$). Each point represents the mean excess enthalpy of three determinations following isothermal annealing for 60 min. (From Ref. 88.)

the reduction in overall free volume. These results demonstrate that measurable relaxation in sucrose solutions occurs within a relatively short time period (1 h) and that the kinetics and point at which a maximum is reached is largely a function of temperature and freezing rate in a curvilinear fashion, and they suggest that rapid freezing induces an unstable system, which creates a UFP of excess volume. For a system to stabilize, the various domains within the material must physically rearrange to reduce this excess volume [120]. Mobility of low molecular weight constituents may be initially high, until a point is reached during aging in which free volume becomes substantially reduced. However, formation by slow freezing appears to facilitate a more stable glass; consequently the potential for relaxation is lower, but crystal size distribution will suffer as discussed earlier. Sub-$T_g$ relaxation has also been reported for dehydrated sorbitol and aqueous glycerol solutions [127,133,134], glucose solutions [128], freeze-concentrated ribose solutions [114], and freeze-concentrated galactose [96] following annealing.

The underlying complexity of the effects of physical aging to sub-$T_g$ stability has recently been demonstrated by Chang and Baust [135] using frozen sorbitol solutions. After aging below the $T_g$, they found that during subsequent warming the devitrification exotherm moved closer to the $T_g$ with increasing annealing time.

During sub-$T_g$ annealing, the water molecules still had sufficient diffusivity to create enough nuclei to facilitate the devitrification process during warming (lower nucleation temperature). However, as structural relaxation proceeded (longer time), the frequency and rate of ice nucleation was reduced but did not completely cease, due in part to the reduced molecular mobility of the water molecules as free volume decreased. The formation of these nuclei during aging may have been catalyzed by the formation of cracks within the glass [115,136].

With regard to frozen food stability, it must be appreciated that movement of water molecules may occur at appreciable rates and may be a function of initial stability and rate of relaxation, which in turn is dependent on freezing rate and temperature. Blond [69] has shown that 10% dextrin can substantially reduce the degree of physical aging and mobility of a freeze-concentrated sucrose glass. This may be a significant result based on the premise that mobility of the total system can affect the mobility of the constituents. However, the presence of the polymer may simply reduce the free volume enough to limit transport of the glass-forming solute but not the diluent. The question of how macromolecules affect structural rearrangement during aging is one factor that deserves further investigation, since it may lead to better understanding of mobility below $T_g$.

Another study [137] focused on the effect of the glassy state on two diffusion-limited reactions—oxidation of ascorbic acid and an enzyme hydrolysis reaction—as well as the rate of protein insolubilization. Results demonstrated the potential difficulty in assigning a stable region in relation to $T_g{}'$, since stability below $T_g{}'$ was dependent on the glass former. For example, a maltodextrin glass system effectively controlled all reactions below $T_g{}'$. However, the same is not true for a CMC glassy system, where none of the reactions were inhibited at $T < T_g{}'$. This was only partially true for the sucrose system, in which the protein-insolubilizing reaction was effectively controlled close to $T_g{}'$. Results also suggested that other factors may be important (e.g., free volume distribution related to the characteristics of the glass, i.e., degree of polymerization or branching). In an unrelated study, Carrington et al. [138] reported that the presence of CMC resulted in larger numbers of smaller ice crystals in rapidly frozen fructose solutions after storage at $-75°C$ for 2 weeks, in comparison to fructose alone. Although a full explanation is not yet available, one possibility is that CMC stabilized the amorphous phase and reduced recrystallization. Stability below the $T_g$ is not a given and requires experimental analysis to ensure that the chosen storage temperature is adequate.

## IV. FROZEN FOOD STABILITY

### A. Mobility at Temperatures Above $T_g$ or $T_g{}'$

The effects of storage temperature on frozen food stability is of prime importance because of the influence it has on deleterious reaction rates, resulting in losses of nutrients and quality [31,113]. It has been proposed that frozen food stability, particularly the effect of temperature on reaction rates of the UFP, may be modeled using Williams-Landel-Ferry (WLF) kinetics [75]. Unlike the Arrhenius equation, the WLF equation relates the logarithm of time-temperature mechanical properties of amorphous materials to $T - T_g$ instead of 0 Kelvin. The region where the WLF equation may apply is shown in Fig. 7 at temperatures above the $T_g'$. The generalized WLF equation is:

$$\ln a_T = \frac{C_1(T-T_g)}{C_2+(T-T_g)} \tag{10}$$

where $a_T = \tau_T/\tau_{Tg}$ or $\eta_T/\eta_{Tg}$ ($\tau$ and $\eta$ represent relaxation time and viscosity, respectively) and represents the so-called shift factor. $C_1 = -17.44$ and $C_2 = 51.6$, the two universal constants, describe the temperature dependence above $T_g'$ [139]. The WLF constants are mathematically related to the free volume parameters of the amorphous component of the system undergoing relaxation [70,140].

The WLF equation predicts a large decrease in viscosity (increased translational mobility) when the temperature is raised slightly above $T_g$ [70]. However, Blond and Simatos [97] observed that the rate of degradation in various frozen products did not follow WLF kinetics closely, especially the change in crystal growth in beef during storage. Although the WLF equation has successfully been applied to aqueous sucrose-fructose mixtures [85], it appears that a large number of other factors must be considered at temperatures above $T_g'$. Of these, the most important are the effects of ice melting. Specifically, an increase in unfrozen water with increasing temperature adds an additional reduction in viscosity by diluting the systems. As well as increasing the $\Delta T$, the $T_g$ is plasticized to lower temperatures with increasing storage temperatures (increase in unfrozen water to ice ratio). Another important factor is the dilution of reactants, which may largely offset the effect of high viscosity. This dilution would be expected to reverse the large rate enhancement by freeze-concentration discussed earlier. Moreover, a possible change in the order of the reaction kinetics, with concentration [1,49], changing values of $C_1$ and $C_2$ with moisture and composition [58,63], and the fact that diffusivity at or above $T_g'$ may or may not be directly related to increased free volume because of the size of the dilute molecule [113,141] will only add to the complexity of evaluating the adequacy of the WLF equation. Recent TMA results on frozen sucrose solutions clearly demonstrate that translational mobility is greatly enhanced after the onset of melting [88] and show an increase in the rate of collapse above $-10°C$ [103], possibly greater than expected from WLF kinetics [63]. A rough estimate of the contribution of each variable to the overall decrease in viscosity has been presented by Simatos and Blond [113].

## B. Ice Recrystallization at Storage Temperatures

Ice crystals are relatively unstable and undergo changes in number, size, and shape during frozen storage, known collectively as recrystallization. This process is a consequence of the surface energy between the ice and UFP [1]. Although the amount of ice remains relatively constant, over time this phenomenon can be extremely damaging to the texture of frozen products, such as ice cream [142].

At constant temperature, Ostwald ripening involves the growth of large crystals at the expense of smaller ones and results from the system moving toward reduced surface area. During temperature fluctuations, as temperature increases, small ice crystals tend to melt, while on cooling, this proportion of unfrozen water does not renucleate but rather is deposited on the surface of larger crystals [1]. The net result of both processes is that the number of crystals decreases, whereas the average size and distribution will increase [1,2,6,27,39,143]. Recrystallization in frozen beef has been shown to increase with storage temperature leading to greater tissue disruption, protein denaturation, and exudate production [143].

The frozen dairy products industry combats the problem of recrystallization by adding low concentrations of high MW stabilizers to mix formulations [142]. In the presence of these compounds, recrystallizing processes are substantially reduced at normal storage temperatures where the UFP is relatively large. However, the mechanism involved is not well understood, despite numerous studies. Buyong and Fennema [144] studied the effect of gelatin on ice crystal growth in ice cream and concluded that the stabilizer had no effect on amount, size, or shape of ice crystals after freezing or on the rate of recrystallization. The presence of locust beam gum (0.5%) in 35% corn syrup solutions was found to have a limited effect on ice crystal growth during temperature fluctuations [145]. However, stabilizers were found by cryo-scanning electron microscopy to alter both initial crystal size and recrystallization rate after storage in an ice cream study [146].

It has been difficult to demonstrate any real relationship between thermodynamic parameters or LCR and the macroscopic effect of stabilizers [21,36,42,147]. Stabilizers do not significantly ($p > 0.05$) affect either thermodynamic transition after annealing in 20% sucrose solutions during warming in the DSC [102,106] (Table 5). However, under rapid freezing conditions, the presence of xanthan gum appeared to hinder initial nucleation and/or ice crystal formation, as demonstrated by the lower $T_{Tr1}$ in Table 5. This is consistent with the claim that xanthan gum can exhibit a small inhibitory effect on nucleation [21]. In addition, xanthan systems showed significantly ($p < 0.05$) smaller $\Delta C_p$ values at $T_{Tr2}$ [106]. The reason for this is not clear but may stem from a fundamental change in the relaxation mechanism at this temperature.

It has been clearly demonstrated that stabilizers strongly influence mechanical and, thereby, diffusion properties of the highly mobile UFP above $T_{Tr2}$ in sucrose solutions, with xathan having the largest effect [105,106]. Subsequent studies have also shown that stabilizers alter stress relaxation properties of frozen sucrose solutions, thereby suggesting that rearrangement mechanisms for the dissipation of stress are modified [107]. It has been hypothesized that once the critical concentration of the polymer is reached, extensive entanglement occurs, reducing

**Table 5** Thermodynamic Data for 20% (w/w) Sucrose with and Without 0.5% (w/w) Xanthan Gum, Gelatin (275 Bloom), and Guar Gum, Rapidly Cooled (Liquid N₂) or Slowly Cooled (2°C/min) and Annealed at −35°C for 60 Minutes, as Measured by DSC

| Sample | Rapidly frozen | | | | Slowly frozen/Annealed | | | | |
|---|---|---|---|---|---|---|---|---|---|
| | $T_{Tr1}$[a] | $\Delta C_pT_{Tr1}$[b] | $T_{Tr2}$[c] | $\Delta C_pT_{Tr2}$[b] | $T_{Tr1}$[a] | $\Delta C_pT_{Tr1}$[b] | $T_{Tr2}$[c] | $\Delta C_pT_{Tr2}$[b] | $\Delta H_m$[d] |
| Sucrose | −54.6 | 0.75 | −32.2 | 3.4 | −40.5 | 1.23 | −31.4 | 3.81 | 220.2 |
| Sucrose/Xanthan | −58.1 | 0.48 | −34.7 | 2.6 | −42.5 | 1.31 | −32.2 | 3.12 | 219.1 |
| Sucrose/Gelatin | −53.9 | 0.84 | −32.7 | 3.6 | −42.8 | 1.22 | −32.6 | 3.75 | 206.2 |
| Sucrose/Guar | −54.2 | 0.85 | −32.5 | 3.8 | −42.5 | 1.24 | −31.4 | 3.76 | 222.3 |

[a]Onset in °C.
[b]Per gram of sucrose in J/g/K.
[c]Midpoint in °C.
[d]Enthalpy of melt in J/g.
Source: Ref. 106.

free volume and greatly increasing the viscosity relative to a nonstabilized system [27,105–107]. Blond [69] and Simatos et al. [102] have shown that the addition of a polymer, particularly guar gum, to partially frozen sucrose solutions greatly increased the elastic (storage) modulus in DMTA experiments. Entanglement phenomena contributed to these effects on rheological properties. However, these researchers concluded that the ice fraction of the system played a role in the rigidity of frozen systems. Further investigation of the distribution of ice crystals is warranted, since uniformity facilitates ordered packing and, on a theoretical level, greater rigidity. In summary, present understanding of the action of stabilizers is still limited, but, based on experimental evidence, it appears to involve modification of kinetic parameters rather than changes to the thermodynamics of frozen foods.

## C. Influence on Reaction Rates

In the undercooled state, the activity of many enzymes can be stabilized and preserved [148]. However, during ice formation, the kinetics of many enzyme-mediated reactions become strongly influenced by the effects of freeze-concentration, which override the retardant effects of low temperature [48]. The influence of freezing on enzyme activity is complex, as one would expect, particularly when cellular material and/or systems are involved, where the enzyme may be bound to a membrane [1]. Rate enhancement in frozen cellular material is a consequence of cell membranes being disrupted by changes in osmotic pressure, pH, and salt concentration, which allows for substrate/enzyme interaction. In partially frozen systems, the activity of an enzyme may either decrease or increase, depending on enzyme conditions [5]. Various factors are involved, namely, the composition of the medium, viscosity, freezing rate and time, and the complexity of the medium being frozen (e.g., tissue vs. aqueous solution) [49]. No simple relationship exists between the overall enzymatic activity and the effects of freezing, partially because optimal conditions can vary widely between enzyme systems [49].

Whether an enzyme is activated, inhibited, or stabilized depends largely on the nature and concentration of the substances present, pH, and freezing temperature. For example, high salt concentrations may activate and stabilize one enzyme, whereas it may lead to inactivation of another enzyme [1,49]. Deviation from Arrhenius kinetics has been observed for various enzyme systems in the presence of ice during freeze-concentration, but the degree of deviation depends on the type of reaction considered [1]. Several factors, in addition to the above-mentioned effects, may contribute to this deviation. These include the reaction mechanism (i.e., rate-determining step), reaction sequence, stability of the active site, reduction in specificity due to increased hydrogen bonding between water and enzyme active site, increased intraenzymatic bonding, increased viscosity, and rate of thawing. With dilute systems, rapid freezing minimizes enzyme inactivation, and detrimental rate enhancement is much lower at a storage temperatures of –18°C than at 0°C [5]. However, since several enzyme groups may still be active at subzero temperatures (e.g., lipase, lipoxygenase, peroxidase, protease, polyphenoloxidase, and cellulase), blanching before freezing is necessary to maintain and reduce changes in texture, color, flavor, and nutritional quality during frozen storage.

Kinetics of nonenzymatic reactions also show rate enhancement with freeze-concentration. Like enzymatic reactions, changes in reaction rate are primarily a function of the concentration of solutes within the unfrozen phase and freezing temperature [49]. A reduction of temperature usually leads to a reduction in reaction rates, whereas an increase in concentration may either increase or decrease reaction rates depending on conditions and type of reaction [5]. During freeze-concentration, reaction rates increase to a maximum, followed by a decrease with reduced temperature. This trend has been shown with the mutarotation of glucose, where the degree of rate enhancement is related to the initial solute concentration. Enhancement of the rate occurs at low initial concentration [1]. At any given subzero temperature during freezing, the dilute solution contains more ice and exhibits a greater increase in molarity in the unfrozen phase, compared to the initially more concentrated system. However, once equilibrium between the ice/water phase is obtained, both systems have equal molarities but different volumes, and it is likely that reaction rates will converge [49]. Freeze-concentration has been demonstrated to enhance greatly the oxidation of ascorbic acid [48]. An Arrhenius-type reduction in activity has been seen during cooling through the undercooled region, but once ice begins to form, the Arrhenius relationship fails to describe the large rate enhancement as concentration is increased.

## D. Moisture Loss During Frozen Storage

Moisture loss during storage of both tissue and nontissue foods can lead to serious desiccation of the product, rendering the quality unacceptable. This occurs due to vapor pressure gradients (a function of temperature) within the product and between the product and external environment caused by temperature fluctuations during storage [52]. When the surface of the product is colder than the core, moisture moves from the interior of the product toward surface voids. Although the temperature gradient reverses during temperature fluctuations, the moisture does not easily transfer back to the initial location, resulting in an overall tendency for water to leave the denser sections of the material [2]. The use of a tight, water–and vapor–impermeable film will reduce the rate of product deterioration. However, should a space exist between the material and the package, moisture will move into the space and condense inside the package or on the surface of the material when its surface has a lower temperature than the environment [52]. Extensive frost may occur, which accounts for losses of up to 20% of the initial weight [52]. Other degradative effects include a possible increase in the rate of chemical reactions at the surface, major structural deterioration resulting in problems with texture, and off-flavor development. Maintenance of a low, constant temperature is necessary to minimize moisture loss. However, in commercial frozen storage situations this is difficult.

## V. CONCLUSIONS

Freezing involves the separation of water as ice in its pure form. This may appear to be a rather simple task largely controlled by a reduction in temperature. However, as our understanding of freezing has improved, it has become clear that the overall driving force of freezing and stability of frozen products is a complex balance of thermodynamic and kinetic factors. Adding to the complexity, thermodynamic and kinetic factors dominate at different points in the freezing process and greatly influence each other, and this ultimately determines how successful a freezing process is. The formation of a glassy domain is also largely a function of kinetic and thermodynamic parameters of the system, which, in essence, are dependent upon the freezing process itself. From a broader scope, the potential of a glassy state for enhancement of frozen food stability relies on the individual steps in the freezing process that create maximum ice formation; otherwise, the glassy state may be fairly unstable. Therefore, it would be advantageous to develop freezing systems that take into account the relative importance of kinetics and thermodynamics as temperature is lowered so that "equilibrium" conditions can be approached. One such approach may involve a freezing process that employs dual cooling rates: initial rapid freezing followed by slow cooling when the divergence from equilibrium becomes large due to kinetics. Being aware of and respecting the complexities and interplay that exist among variables will ensure that freezing will continue to enjoy widespread success as a tool for the preservation of foods.

## REFERENCES

1. F. Franks, *Biophysics and Biochemistry at Low Temperatures*, Cambridge University Press, Cambridge, 1985, p. 21.
2. D. S. Reid, Basic physical phenomena in the freezing and thawing of plant and animal tissues, *Frozen Food Technology* (C. C. Mallett, ed.), Blackie Academic and Professional, London, 1993, p. 1.
3. F. Franks, The properties of aqueous solutions at sub-zero temperatutes, *Water: A Comprehensive Treatise*, Vol. 7 (F. Franks ed.), Plenum Press, New York, pp. 215–338.
4. R. W. Michelmore and F. Franks, Nucleation rates of ice undercooled water and aqueous solutions of polyethylene glycol, *Cryobiology* 19:163 (1982).
5. O. R. Fennema, Nature of freezing process, *Low-Temperature Preservation of Foods and Living Matter* (O. R. Fennema, W. D. Powrie, and E. H. Marth, eds.), Marcel Dekker, New York, 1973, p. 153.

6. D. S. Reid, Fundamental physicochemical aspects of freezing, *Food Technol.* 37(4): 110 (1983).

7. S. Charoenrein, M. Goddard, and D. S. Reid, Effect of solute on the nucleation and propagation of ice, *Water Relationships in Foods* (H. Levine and L. Slade, eds.), Plenum Press, New York, 1991, p. 191.

8. F. Franks, S. F. Mathias, and K. Trafford, Nucleation of ice in undercooled water and aqueous polymer solutions, *Colloids Surfaces* 11:275 (1984).

9. F. Franks, Nucleation: A maligned and misunderstood concept, *Cryo-Letters* 8:53 (1987).

10. D. H. Rasmussen and A. P. MacKenzie, Effect of solute on ice-solution interfacial free energy; calculation from measured homogeneous nucleation temperatures, *Water Structure at the Water-Polymer Interface* (H. H. G. Jellinek, ed.), Plenum Press, New York, 1972, p. 126.

11. S. Ozilgen and D. S. Reid, The use of DSC to study the effects of solutes on heterogeneous ice nucleation kinetics in model food emulsions, *Lebensm. Wiss. Technol.* 26:116 (1993).

12. F. Franks, The nucleation of ice in undercooled aqueous solutions, *Cryo-Letters* 2:27 (1981).

13. G. Blond and B. Colas, Freezing in polymer-water systems, *Food Freezing: Today and Tomorrow* (W. B. Bald, ed.), Springer-Verlag, New York, 1991, p. 27.

14. L. F. Evans, Requirements of an ice nucleus, *Nature* 206:822 (1965).

15. J. M. McBride, Crystal polarity: A window on ice nucleation, *Nature* 256:814 (1992).

16. P. Wilson, Physical basis of action of biological ice nucleating agents, *Cryo-Letters* 15:119 (1994).

17. N. H. Fletcher, Size effects in heterogeneous nucleation, *J. Chem. Phys.* 29:572 (1959).

18. D. Turnbull, Kinetics of heterogeneous nucleation, *J. Chem Phys.* 18:198 (1950).

19. M. Toner, E. G. Cravalho, and M. Karel, Thermodynamics and kinetics of intracellular ice formation during freezing of biological cells, *J. Appl. Phys.* 67(3):1582 (1990).

20. A. K. Carrington, M. E. Sahagian, H. D. Goff, and D. W. Stanley, Ice-crystallization temperatures of frozen sugar solutions and their relationship to $T_g{}'$, *Cyro-Letters* 15:235 (1994).

21. A. H. Muhr, J. M. V. Blanshard, and S. J. Sheard, Effects of polysaccharide stabilizers on the nucleation of ice, *J. Food Technol.* 21:587 (1986).

22. M. Gavish, J. L. Wang, M. Eisenstein, M. Lahav, and L. Leiserowitz, The role of crystal polarity in α-amino acid crystals for induced nucleation of ice, *Science* 256:515 (1992).

23. R. W. Hartel and L. A. Espinel, Freeze concentration of skim milk, *J. Food Eng.* 20:101 (1993).

24. R. W. Hartel and M. S. Chung, Contact nucleation of ice in fluid dairy products, *J. Food Eng.* 18:181 (1993).

25. Y. Shirai, K. Kazuhiro, R. Matsuno, and T. Kamikubo, Effects of polymers on secondary nucleations of ice crystals, *J. Food Sci.* 50:401 (1985).

26. P. V. Hobbs, *Ice Physics*, Clarendon Press, Oxford, 1974, p. 461.

27. H. D. Goff, Glass formation in food systems and its relationship to frozen food quality, *Food Res. Int.* 25:317 (1992).

28. R. C. Weast, M. J. Astle, and W. H. Beyer, eds., *CRC Handbook of Chemistry and Physics*, 64th ed., CRC Press, Boca Raton, FL, 1983, p. D-192.

29. J. Garside, General principles of crystallization, *Food Structure and Behaviour*, (J. M. V. Blanshard and P. L. Lillford, eds.), Academic Press, Orlando, FL, 1987, p. 35.

30. M. R. de Quervain, Crystallization of water, a review, *Freeze Drying and Advanced Food Technology* (S. A. Goldblith, L. Rey, and W. W. Rothmayr, eds.), Academic Press, New York, 1975, p. 3.

31. N. W. Desrosier and J. N. Desrosier, Principles of food freezing, *The Technology of Food Preservation*, 4th ed. (N. W. Desrosier and J. N. Desrosier, eds.), AVI Publishing Co. Inc., Westport, CT, 1977, p. 110.

32. C. A. Miles, The thermophysical properties of frozen foods, *Frozen Food Technology* (C. C. Mallett, ed.), Blackie Academic and Professional, London, 1993, p. 45.

33. C. A. Angell and Y. Choi, Crystallization and vitrification in aqueous systems, *J. Microsc.* 141:251 (1986).

34. D. R. MacFarlane, R. K. Kadiyala, and C. A. Angell, Homogeneous nucleation and growth of ice from solutions. TTT curves, the nucleation rate, and the stable glass criterion, *J. Chem. Phys.* 79(8):3921 (1983).

35. R. K. Kadyela and C. A. Angell, Separation of nucleation from crystallization kinetics by two step calorimetry experiments, *Colloids Surfaces* 11:341 (1984).

36. A. H. Muhr and J. M. V. Blanshard, Effect of polysaccharide stabilizers on the rate of growth of ice, *J. Food Technol.* 21:683 (1986).

37. J. M. V. Blanshard, A. H. Muhr, and A. Gough, Crystallization from concentrated sucrose solutions, *Water Relationships in Foods* (H. Levine and L. Slade, eds.), Plenum Press, New York, 1991, p. 639.

38. S. Langer, Dendritic solidification of dilute solutions, *Physico-Chem. Hydrodyn.* 1:44 (1980).

39. W. B. Bald, Ice crystal growth in idealized freezing systems, *Food Freezing: Today and Tomorrow* (W. B. Bald, ed.), Springer-Verlag, New York, 1991, p. 67.

40. C. Holt, The effect of polymers on ice crystal growth, *Food Freezing: Today and Tomorrow* (W. B. Bald, ed.), Springer-Verlag, New York, 1991, p. 81.

41. G. Blond, Velocity of linear crystallization of ice in macromolecular systems, *Cryobiology* 25:61 (1988).

42. E. R. Budiaman and O. Fennema, Linear rate of water crystallization as influenced by viscosity of hydrocolloid suspensions, *J. Dairy Sci.* 70:547 (1987).

43. B. Luyet, Basic physical phenomena in the freezing and thawing of animal and plant tissue, *The Freezing Preservation of Foods: Factors Affecting Quality in Frozen Foods*, Vol. 2 (D. K. Tressler, W. B. Van Arsdel, and M. J, Copley, eds.), AVI Publishing Co. Inc., Westport, CT, 1968, p. 1.

44. D. S. Reid, Optimizing the quality of frozen foods, *Food Technol.* 44(7):78 (1990).

45. P. L. Steponkus, Role of the plasma membrane in freezing injury and cold acclimation, *Ann. Rev. Plant Physiol.* 35:543 (1984).

46. J. O. M. Karlsson, E. G. Cravalho, and M. Toner, Intracellular ice formation: causes and consequences, *Cryo-Letters* 14:323 (1993).

47. M. Chandrasekaran and R. E. Pitt, On the use of nucleation theory to model intracellular ice formation, *Cryo-Letters* 13:261 (1992).

48. F. Franks, Freeze drying: From empiricism to predictability, *Cryo-Letters* 11:93 (1990).

49.  O. R. Fennema, Reaction kinetics in partially frozen aqueous systems, *Water Relations in Foods* (R. B. Duckworth, ed.), Academic Press, New York, 1975, p. 539.

50.  M. C. Anon and A. Calvelo, Freezing rate effects on drip loss of frozen beef, *Meat Sci.* 4:1 (1980).

51.  I. C. Mackie, The effect of freezing of flesh proteins, *Food Rev. Int.* 9(4):576 (1993).

52.  International Institute of Refrigeration, General principles for the freezing, storage and thawing of food-stuffs, *Recommendations for the Processing and Handling of Frozen Foods*, 3rd ed., International Institute of Refrigeration, Paris, France, 1986, p. 42.

53.  P. O. Persson and G. Londaiil, Freezing technology, *Frozen Food Technology* (C. C. Mallett, ed.), Blackie Academic and Professional, London, 1993, p. 20.

54.  R. M. George, Freezing processes used in the food industry, *Trends Food Sci. Technol.* 4:138 (1993).

55.  A. C. Cleland, *Food Refrigeration Process; Analysis, Design and Simulation*, Elsevier Applied Science Publ., New York, 1991, p. 45.

56.  S. D. Holdsworth, Physical and engineering aspects of food freezing, *Developments in Food Preservation—4* (S. Thorne, ed.), Elsevier Applied Science Publ., New York, 1987, p. 153.

57.  F. Franks, Bound water: Fact and fiction, *Cryo-Letters* 4:73 (1983).

58.  L. Slade and H. Levine, Beyond water activity: Recent advances based on an alternative approach to the assessment of food quality and safety, *CRC Crit. Rev. Food Sci. Nutr.* 30:115 (1991).

59.  F. Franks, Solid aqueous solutions, *Pure Appl. Chem.* 65(10):2527 (1993).

60.  L. Slade and H. Levine. Glass transitions and water-food structure interactions, *Advances in Food and Nutrition Research* (J. E. Kinsella, ed.), Academic Press, San Diego, (in press).

61.  P. Chinachoti, Water mobility and its relation to functionality of sucrose-containing food systems. *Food Technol.* 1:134 (1993).

62.  H. Levine and L. Slade, A polymer physicochemical approach to the study of commercial starch hydrolysis products (SHPS), *Carbohydr. Polym.* 6:213 (1986).

63.  L. Slade and H. Levine, Non-equilibrium behavior of small carbohydrate-water systems, *Pure Appl. Chem.* 60:1841 (1988).

64.  M. Karel, Y. Roos, and M. P. Buera, Effects of glass transitions on processing and storage of foods, *Glassy State in Food* (J. Blanshard and P. Lillford, eds.), University of Nottingham Press, Nottingham, UK, 1993, p. 13.

65.  W. M. MacInnes, Dynamic mechanical thermal analysis of sucrose solution during freezing and thawing, *Glassy State in Food* (J. Blanshard and P. Lillford, eds.), Nottingham University Press, Loughborough, UK, 1993, p. 223.

66.  D. S. Reid, J. Hsu, and W. Kerr, Calorimetry, *Water in Foods: Fundamental Aspects and their Significance in the Processing of Foods* (D. Simatos, ed.), Elsevier, London, 1993, p. 123.

67.  Y. Roos and M. Karel, Effects of glass transition on dynamic phenomena in sugar containing food systems, *Glassy State in Food* (J. Blanshard, and P. Lillford, eds.), Nottingham University Press, Loughborough, UK, 1993, p. 207.

68.  D. Simatos and G. Blond, Some aspects of the glassy transition in frozen food systems, *Glassy State in Food* (J. Blanshard, and P. Lillford, eds.), Nottingham University Press, Loughborough, UK, 1993, p. 395.

69. G. Blond, Mechanical properties of frozen model solutions, *J. Food Eng.* 22:253 (1994).
70. J. D. Ferry, The WLF equation and the relation of temperature dependence of relaxation times to free volume, *Viscoelastic Properties of Polymers* (J. D. Ferry, ed.), John Wiley and Sons, New York, 1980, p. 297.
71. B. Wunderlich, The basis of thermal analysis, *Thermal Characterization of Polymeric Materials* (E. A. Turi, ed.), Academic Press, Orlando, FL, 1981, pp. 91–234.
72. C. A. Angell, Perspective on the glass transition, *J. Phys. Chem. Solids* 49:863 (1988).
73. A. Eisenberg, The glassy state and the glass transition, *Physical Properties of Polymers* (J. E. Mark, A. Eisenberg, M. W. Graessley, L. Mandelkern, and J. L. Koenig, eds.), American Chemical Society, Washington, DC, 1984, p. 55.
74. Y. Roos, Phase transitions and transformations in food systems, *Handbook of Food Engineering* (D. R. Heldman and D. B. Lund, eds.), Marcel Dekker, New York, 1992, p. 145.
75. H. Levine and L. Slade, Thermomechanical properties of small carbohydrate-water glasses and rubbers: kinetically-metastable systems at subzero temperature, *J. Chem. Soc. Faraday Trans. I* 84:2619 (1988).
76. N. S. Murthy, S. N. Gangasharan, and S. K. Nayak, Novel DSC studies of supercooled organic liquids, *J. Chem. Soc. Faraday Trans.* 89:509 (1993).
77. H. Levine and L. Slade, Water as a plasticizer: physico-chemical aspects of low-moisture polymeric systems, *Water Science Reviews*, Vol. 3 (F. Franks, ed.), Cambridge University Press, Cambridge, 1988, p. 79.
78. L. H. Sperling, *Introduction to Physical Polymer Science*, Wiley-Interscience, New York, 1986, p. 56.
79. T. G. Fox and P. J. Flory, Second-order transition temperatures and related properties of polystyrene. I. Influence of molecular weight, *J. Appl. Phys.* 21:581 (1950).
80. H. Levine and L. Slade, Cryostabilization technology: Thermoanalytical evaluation of food ingredients and systems, *Thermal Analysis of Foods* (V. R. Harwalker and C. Y. Ma, eds.), Elsevier Applied Science, New York, 1990, p. 221.
81. P. D. Orford. R. Parker, and S. R. Ring, Aspects of the glass transition behaviour of mixtures of carbohydrates of low molecular weight, *Carbohydr. Res.* 196:11 (1990).
82. M. Gordon and J. S. Taylor, Ideal copolymers and the second-order transitions of synthetic rubbers. I. Non-crystalline copolymers, *J. Appl. Chem.* 2:493 (1952).
83. Y. Roos and M. Karel, Phase transitions of mixtures of amorphous polysacchaides and sugars, *Biotechnol. Proc.* 7:49 (1991).
84. Y. Roos and M. Karel, Non-equilibrium ice formation in carbohydrate solutions, *Cryo-Letters* 12:367, (1991).
85. T. Soesanto and M. C. Williams, Volumetric interpretation of viscosity for concentrated and dilute sugar solutions, *J. Phys. Chem.* 85:3338 (1981).
86. T. Suzuki and F. Franks, Solid-liquid transition and amorphous states in ternary sucrose-glycine-water systems, *Chem. Soc. Faraday Trans.* 89:2383 (1993).
87. Y. Roos and M. Karel, Amorphous state and delayed ice formation in sucrose solutions, *Int. J. Food Sci. Technol.* 26:553 (1991).
88. M. E. Sahagian and H. D. Goff, Effect of freezing rate on the thermal, mechanical and physical aging properties of the glassy state in frozen sucrose solutions, *Thermochimica Acta.* 246:271 (1994).

89. S. Ablett, M. J. Izzard, and P. J. Lillford, Differential scanning calorimetric study of frozen sucrose and glycerol solutions. *J. Chem. Soc. Faraday Trans.* 88:789 (1992).

90. F. Franks, Metastable water at sub-zero temperature, *J. Microsc.* 141:243 (1986).

91. T. R. Noel, S. G. Ring, and M. A. Whittam, Glass transitions in low-moisture foods, *Trends Food Sci. Technol.* 1:62 (1990).

92. D. Turnbull, Under what conditions can a glass be formed? *Contemp. Phys.* 34:120 (1969).

93. B. Luyet and D. Rasmussen, Study by differential thermal analysis of the temperatures of instability of rapidly cooled solutions of glycerol, ethylene glycol, sucrose and glucose, *Biodynamica* 10:167 (1968).

94. D. Rasmussen and B. Luyet, Complementary study of some non-equilibrium phase transitions in frozen solutions of glycerol, ethylene glycol, glucose and sucrose, *Biodynamica* 10:319 (1969).

95. M. J. Izzard, S. Ablett, and P. J. Lillford, A calorimetric study of the glass transition occurring in sucrose solutions, *Food Polymers, Gels and Colloids* (E. Dickinson, ed.), The Royal Society of Chemistry, Cambridge, UK, 1991, p. 289.

96. G. Blond, Water-galactose system: Supplemented state diagram and unfrozen water, *Cryo-Letters* 10:299 (1989).

97. G. Blond and D. Simatos, Glass transition of the amorphous phase in frozen aqueous systems, *Thermochimica Acta* 175:239 (1991).

98. R. H. M. Hatley, C. van den Berg, and F. Franks, The unfrozen water content of maximally freeze concentrated carbohydrate solutions: Validity of the methods used for its determination. *Cryo-Letters* 12:113 (1991).

99. R. H. M. Hatley and A. Mant, Determination of the unfrozen water content of maximally freeze-concentrated carbohydrate solutions, *Int. J. Biol. Macromol.* 15:227 (1993).

100. V. R. Harwalker, C. Y. Ma, and T. J. Maurice, *Thermal Analysis of Foods* (V. R. Harwalker and C. Y. Ma, eds.), Elsevier Applied Science, New York, 1990, p. 1.

101. C. A. Angell, L. Monnerie, and L. M. Torell, Strong and fragile behavior in liquid polymers, *Structure, Relaxation and Physical Aging of Glassy Polymers* (R. J. Roe and J. M. Reilly, eds.), Material Science Society, Pittsburgh, 1991, p. 3.

102. D. Simatos, G. Blond, and F. Martin, Influence of macromolecules on the glass transition in frozen systems, *Food Macromolecules and Colloids Conference,* March 23–25, Dijon, France, 1994.

103. M. Le Meste and V. Huang, Thermomechanical properties of frozen sucrose solutions, *J. Food Sci.* 57:1230 (1992).

104. T. J. Maurice, Y. J. Asher, and S. Thomas, Thermomechanical analysis of frozen aqueous systems, *Water Relationships in Foods* (H. Levine and L. Slade, eds.), Plenum Press, New York, 1991, p. 215.

105. H. D. Goff, L. B. Caldwell, and D. W. Stanley, The influence of polysaccharides on the glass transition in frozen sucrose solutions and ice cream, *J. Dairy Sci.* 76:1268 (1993).

106. M. E. Sahagian and H. D. Goff, Thermal, mechanical and molecular relaxation properties of stabilized sucrose solutions at sub-zero temperatures, *Food Res. Int.* 28:1 (1995).

107. M. E. Sahagian and H. D. Goff, Influence of stabilizers and freezing rate on the stress relaxation behaviour of freeze-concentrated sucrose solutions at different temperatures. *Food Hydrocoll.* (in press).

108. C. A. Rubin, J. M. Wasylyk, and G. Baust, Investigating of vitrification by nuclear magnetic resonance and differential scanning calorimetry in honey: A model carbohydrate system, *J. Agric. Food Chem.* 38:1824 (1990).

109. S. Ablett and P. Lillford, Water in foods, *Chem. Br.* 11:1025 (1991).

110. M. J. G. W. Roozen and M. A. Hemminga, Molecular motion in sucrose-water mixtures in the liquid and glassy state as studied by spin probe ESR, *J. Phys. Chem.* 94:7326 (1990).

111. M. J. G. W. Roozen, M. A. Hemminga, and P. Walstra, Molecular motion in glassy water-malto-oligosaccharide (maltodextrins) mixtures as studied by convention and saturation-transfer spin-probe ESR spectroscopy, *Carbohydr. Res.* 215:229 (1991).

112. M. Le Meste, A. Voilley, and B. Colas, Influence of water on the mobility of small molecules dispersed in a polymer system, *Water Relationships in Foods* (H. Levine and L. Slade, eds.), Plenum Press, New York, 1991, p. 123.

113. D. Simatos and G. Blond, DSC studies and stability of frozen foods, *Water Relationships in Foods* (H. Levine and L. Slade, eds.), Plenum Press, New York, 1991, p. 139.

114. J. K. Wasylyk and J. G. Baust, Vitreous domains in an aqueous ribose solution, *Water Relationships in Foods* (H. Levine, and L. Slade, eds.), Plenum Press, New York, 1991, p. 225.

115. R. J. Williams and D. L. Carnahan, Fracture faces and other interfaces as ice nucleation sites, *Cryobiology* 27:479 (1990).

116. Y. Roos, Melting and glass transition of low molecular weight carbohydrates, *Carbohydr. Res.* 238:39 (1993).

117. S. Ablett, M. J. Izzard, and P. J. Lillford, I. Arvanitoyannis, and J. M. V. Blanshard, Calorimetric study of the glass transition in fructose solutions, *Carbohydrate Res.* 246:13 (1993).

118. F. Franks, Freeze-drying: From empiricism to predictability. The significance of glass transitions, *Devel. Biol. Standard.* 74:9 (1991).

119. M. E. Sahagian, Sub-zero molecular relaxations in stabilized freeze-concentrated sucrose solutions, M. SC. thesis, University of Guelph, Guelph, Canada, 1993.

120. M. R. Tant and G. L. Wilkes, An overview of the non-equilibrium behavior of polymer glasses, *Polym. Eng. Sci.* 21:874 (1981).

121. I. M. Hodge and A. R. Berens, Calculation of the effect of annealing on sub-Tg endotherms, *Macromolecules* 14:1598 (1981).

122. C. P. Lindsay and G. D. Patterson, Detailed comparison of the Williams-Watts and Cole-Davidson functions, *J. Chem. Phys.* 73:3348 (1980).

123. C. T. Moyniha, S. N. Crichton, and S. M. Opalka, Linear and non-linear structural relaxation, *J. Non-Cryst. Solids* 131:420 (1991).

124. L. C. E. Struik, *Physical Aging in Amorphous Polymers and Other Materials*, New York, 1978. Elsevier, Amsterdam, p. 1.

125. M. H. Cohen and D. J. Turnbull, Molecular transport in liquid and glasses, *J. Chem. Phys.* 31:1164 (1959).

126. D. Turnbull and M. H. Cohen, Free-volume model of the amorphous phase: Glass transition, *J. Chem. Phys.* 34:120 (1961).

127. Z. H. Chang and J. G. Baust, Physical aging of the glassy state: sub-Tg ice nucleation in aqueous sorbitol systems, *J. Non-Cryst. Solids.* 130:198 (1991).

128. S. Matsuoka, G. Williams, G. E. Johnson, E. W. Anderson, and T. Furukawa,

Phenonomenological relationship between dielectric relaxation and thermodynamic recovery process near the glass transition, *Macromolecules* 18:2652 (1985).

129. S. E. B. Petrie, Thermal behavior of annealed organic glasses, *J. Polym. Sci. Part A-2.* 10:1255 (1972).

130. J. W. Brady and S. N. Ha, Molecular dynamics simulation of the aqueous solvation of sugars, *Water Relationships in Foods* (H. Levine and L. Slade, eds.), Plenum Press, New York, 1991, p. 739.

131. M. T. Kaliachewsky, E. M. Jaroszkiewicz, S. Ablett, J. M. V. Blanshard, and P. J. Lillford, The glass transition of amylopectin measured by DSC, DMTA and NMR, *Carbohydr. Polym.* 18:77 (1993).

132. M. J. Pickal, S. Shah, M. L. Roy, and R. Putman, The secondary drying stage of freeze drying: Drying kinetics as a function of temperature and chamber pressure, *Int. J. Pharm.* 60:203 (1990).

133. Z. H. Chang and J. G. Baust, Physical aging of glassy state: DSC study of vitrified glycerol systems, *Cryobiology* 28:87 (1991a).

134. Z. H. Chang and J. G. Baust, Further inquiry into the cryobehavior of aqueous solutions of glycerol, *Cryobiology* 28:268 (1991c).

135. Z. H. Chang and J. G. Baust, Effect of sub-$T_g$ annealing on the critical warming rate to avoid devitrification in 60% w/w aqueous sorbitol solutions, *Cryo-Letters* 14:359 (1993).

136. R. J. Williams and D. L. Carnahan, Association between ice nuclei and fracture interfaces in sucrose: Water glasses, *Thermochim. Acta* 155:103 (1989).

137. M. H. Lim and D. S. Reid, Studies of reaction kinetics in relation to the $T_g'$ of polymers in frozen model systems, *Water Relationships in Foods* (H. Levine and L. Slade, eds.), Plenum Press, New York, 1991, p. 103.

138. A. K. Carrington, H. D. Goff, and D. W. Stanley, Structure and stability of the glassy state in rapidly and slowlycooled carbohydrate solutions, *Food Microstruct.* (in press).

139. M. L. Williams, R. F. Landel, and J. D. Ferry, The temperature dependence of relaxation mechanisms in amorphous polymers and other glass-forming liquids. *J. Am. Chem. Soc.* 77:3701 (1955).

140. P. Lomellini, Williams-Landel-Ferry versus Arrhenius behaviour: Polystyrene melt viscoelasticity revised, *Polymer* 33:4983. (1992).

141. M. Karel and I. Saguy, Effects of water on diffusion in food systems, *Water Relationships in Foods* (H. Levine and L. Slade, eds.), Plenum Press, New York, 1991, p. 157.

142. W. S. Arbuckle, *Ice Cream,* 4th ed., AVI Publ. Co. Ltd., Westport, CT, 1986, p. 1.

143. M. N. Martino and N. E. Zaritzky, Ice crystal size modification during frozen beef storage, *J. Food Sci.* 53(6):1631 (1988).

144. N. Buyong and O. Fennema, Amount and size of ice crystals in frozen samples as influenced by hydrocolloids, *J. Dairy Sci.* 71:2630 (1988).

145. E. K. Harper and C. F. Shoemaker, Effect of locust bean gum and selected sweetening agents on ice recrystallization rates, *J. Food Sci.* 48:1801 (1983).

146. K. B. Caldwell, H. D. Goff, and D. W. Stanley, A low-temperature SEM study of ice cream. II. Influence of ingredients and processes, *Food Struct.* 11:11 (1992).

147. E. R. Budiaman and O. Fennema, Linear rate of water crystallization as influenced by temperature of hydrocolloid suspensions, *J. Dairy Sci.* 70:534 (1987).

148. R. H. M. Hatley and F. Franks, The stabilization of labile biochemicals by undercooling, *Proc. Biochem.* 22:169 (1987).

# 2
# Red Meats

Carrick Erskine Devine, R. Graham Bell, Simon Lovatt,
and Brian B. Chrystall

*Meat Industry Research Institute of New Zealand*
*Hamilton, New Zealand*

Lester E. Jeremiah

*Agriculture and Agri-Food Canada Research Centre*
*Lacombe, Alberta, Canada*

## I.  INTRODUCTION

Meat is a major component of the diets of many people around the world. Animals slaughtered for meat traditionally were (and in some cases still are) immediately distributed, sold, and consumed. Preservation was unnecessary, but as surpluses began to be produced, preservation methods were required so that excess product could be held and used at a later time or at some distant location. Cooling meat with ice was an early method of preservation that did not change the form or state of the product. Freezing, a logical progression, made longer preservation possible. In many countries, locally produced frozen meat is not commercially available. However, a large proportion of the meat consumed in developed countries may have been frozen for transportation.

## A.  Why Meat Is Frozen

Meat is easily frozen and kept in the frozen state until consumed. Therefore, freezing appears to be an ideal way to store meat. Although biochemical and sensory changes are minimized by this process, meat is not ideally presented frozen to the retail consumer. However, the hotel and restaurant trade finds good quality frozen meat convenient.

Other frozen foods are certainly well accepted, so why is frozen meat less acceptable to the consumer? Perhaps it is because meat in large frozen chunks is difficult to manage and, therefore, is not seen by the customer as in a suitable state for cooking until it is further prepared. Frozen meat is not easily cut unless it is warmed to at least −5°C. Moreover, large pieces of meat do not thaw evenly; some parts become soft while the remainder cannot be cut. Perhaps tradition plays a part. Meat is not generally truly "fresh," irrespective of the claims made. It is aged for days and even weeks, albeit at a low temperature, to improve eating qualities before it is cooked. However, towards the end of its storage life meat can become microbiologically compromised. Vegetables, on the other hand, remain alive for a considerable period, and keep for days or weeks, but quality usually deteriorates considerably. Freshness is considered desirable and can be locked in by freezing. Only rarely do microbiological problems occur. Therefore, image is not a problem.

Meat is frequently frozen, not to produce a high-quality product, but for commercial or domestic expediency. Freezing is often applied as a last resort when the chilled product can no longer be traded or used as such. Unfortunately, the prefreezing treatment of such meat may already have resulted in undesirable organoleptic changes, e.g., incipient spoilage and color deterioration, which certainly does not enhance the reputation of frozen meat as a premium product.

Consumers often purchase more meat than needed and put it into normal domestic freezers for later use. Freezing under these conditions is variable and rarely fast enough to preserve quality.

Another "remedial" application of freezing is for the destruction of parasitic cysts in meat and offals. Such treatments are applied to render the product "safe" for human consumption (e.g., trichinosis in pork) or to break an infection cycle if the infested meat is destined for consumption by pets (e.g., hydatids in mutton). Freezing is also used as a treatment to eliminate Salmonella in meat, especially veal.

## B.  Quality of Frozen Meat

When cooked by the end user, the quality of frozen meat is governed by the entire processing chain from the farm through processing, freezing, storage, and thawing to preparation for consumption. With meat, visual characteristics may not be indicative of quality. Tenderness, for example, cannot be determined by visual inspection, and color changes may or may not indicate eating quality. Meat is sometimes purchased at the end of its storage life and then hurried into a freezer at home, where it is subjected to unknown storage temperatures and poor temperature control, and a poor-quality product results in terms of flavor. Under these circumstances, if the consumer has a bad experience, freezing will be blamed, since the meat "looked all right" when purchased. If the meat was near the end of its storage life at the time of purchase and freezing, its quality at the time of thawing, preparation, and consumption cannot please the consumer. In our experience, meat production, processing, storage, and distribution are generally not controlled by strict assurance procedures common in other segments of the food industry, which guarantee product quality.

Can one guarantee that the meat today is better than the antelope killed and eventually eaten in the prehistoric days of our forbears? We have looked at meat production, processing, and preparation from the perspective of trying to assure quality procedures from the farm to the plate. Freezing, storage, and thawing represent three important steps in this sequence. We believe that the totality of the chain needs to be considered.

This chapter is written by scientists and engineers dedicated to understanding the processing, storage, and distribution of meat. New Zealand is economically dependent on a quality frozen meat trade. Situations tolerated elsewhere can in New Zealand eventually produce major problems in the chilling and freezing of meat, which must be shipped long distances to final markets. The New Zealand meat trade has spawned research programs resulting in the eventual development of processes such as "conditioning and aging" of carcasses, electrical stimulation, temperature logging, model-based control of freezing, and vacuum- and carbon dioxide–packaging systems as well as the development of models to predict microbiological growth and tenderness.

## C.  Possible Changes Due to Freezing

Meat is composed of a complex grouping of biochemicals, including soluble and structural proteins, fats, and electrolytes. Its properties, therefore, are more complex than a single-phase water-based system when undergoing freezing. However, many aspects of the physics of freezing of such simple systems are important, and the preceding chapter and standard texts should be consulted to understand the physics involved. The unique combination of these biochemicals imparts to meat certain characteristics that must be considered during freezing storage and thawing. The life of the product before freezing must be strictly monitored if a wholesome, organoleptically desirable and chemically and microbiologically stable end result is to be produced. Consequently, the entire "life" of the product from the farm to the plate must be considered.

## D.  Time at Which Freezing Takes Place

The freezing of meat results in an inhibition of microbiological spoilage and slowing down of chemical reactions. Ideally, from a microbiological perspective, meat should be frozen as soon as possible after slaughter and boning. Such a procedure usually produces disastrous results in terms of product tenderness, but it can work under certain circumstances. This is the concept behind hot boning and immediate freezing. Meat produced under this regime is satisfactory for manufacturing purposes (e.g., emulsion-type products such as sausages or for grinding in the production of hamburgers). Unless the temperatures are precisely conrolled, a hot-boned product will be unsatisfactory for table meats (i.e., whole tissue meats used for frying, roasting, or stewing).

To be most suitable for eating, meat, whether hot or cold boned, must go through processes such as conditioning (preceding rigor mortis) and aging (postrigor). In prehistoric days, time allowed this to happen without thought. Today, with slaughter being conducted on a large scale, the whole process must be carefully controlled for efficient production and to prevent deterioration in product quality. Freezing, when used to produce high-quality products, must be considered an integral part of the overall meat-processing system.

The aging process (described below) takes time and provides an opportunity for microbiological spoilage to compromise product quality. Therefore, the microbiology of the postslaughter, freezing, storage, and thawing processes can be of critical importance to product quality. Many changes, such as aging for tenderness, therefore, need to be essentially complete before the meat is frozen. Since aging potential is modified by preslaughter and postslaughter factors, quality associated with freezing, storage, and thawing is dependent upon the preslaughter history as well as species differences and the processing treatment.

## E.  How Meat Can Lose Quality: One Scenario

The meat is from a process that ensures it is rapidly chilled (possibly because microbiological contamination is suspected), which ensures that meat becomes severely cold shortened (which toughens it, see later). The meat is then left in a chiller operating at high humidity with the lights on, possibly of a type emitting a high ultraviolet component, before the meat is boned out and distributed to a supermarket to experience on excessively filled shelves continuous exposure to the cabinet lighting and subsequent fluctuating display temperature. When a relatively large proportion of the meat does not sell, the price is discounted early the next morning to encourage a rapid sale, and the meat is then purchased by shoppers on their way to work (who cannot leave it in a cool place). The meat, therefore, remains wrapped but unrefrigerated all day and is taken home to be consumed. Unfortunately, another meat dish has been prepared. Consequently, the meat is now placed in a domestic freezer still in its wrapping and the large meat "lump" becomes distorted as it is pushed into an overfull compartment. After slow freezing, the meat is now not in a form to cook and, as a large inconvenient lump, is often overlooked and may even be removed and placed on a bench several times when the freezer is sorted or when defrosting the freezing compartment. Eventually after about 18 months, the meat is required, so it is left to thaw on a bench at room temperature for more than a day and is cooked in the evening as a stirfry. The meal is served, and as the eager family "bites" through the cold-shortened, toughened meat, rancidity/off-flavors are detected as the tough meat is slowly broken down by chewing. The family blames the freezing of the meat for such an unsatisfactory meal. Other possible scenarios (mentioned earlier) include freezing vacuum-packaged meat at the end of its storage life because it has not sold. (Procedures for freezing meat to *ensure* quality will be outlined later in this chapter.)

## II. PRESLAUGHTER FACTORS AFFECTING MEAT QUALITY

### A. Muscle Glycogen

The dominant preslaughter biological factor influencing meat quality is muscle glycogen [1]. Postslaughter, muscle glycogen is converted into lactic acid, which gradually lowers the muscle pH to the final value reached at rigor mortis, termed *ultimate pH*. This process arises from the muscle's attempt to maintain itself in the absence of oxygen, due to removal of the blood supply, through a process termed anaerobic glycolysis, which uses available muscle glycogen reserves. The ultimate pH of a well-fed, rested animal is approximately 5.5. If glycogen levels are low due to preslaughter stress, the decrease in pH is limited. If relatively high stress is encountered preslaughter, the ultimate pH will be >6.0, resulting in dark, firm, and dry (DFD) meat or dark cutting meat [2,3]. Extremes of pH result in tender meat, but intermediate pH values result in tough meat [4–7]. However, if this intermediate-pH meat is appropriately aged, it will usually become tender unless excessively cold shortened [8].

The increasingly dark color of meat that accompanies increasing pH might be better tolerated, despite its inferior flavor and poor cooking characteristics, if not for the fact that high-pH meat does not keep well in vacuum packaging. The glucose content of fresh meat is the critical factor determining the relationship between microflora development and the time to spoilage. With the exception of *Shewanella putrefaciens*, glucose, derived through glycogen breakdown during rigor onset, is the preferred substrate for meat spoilage bacteria. The byproducts of glucose metabolism are relatively inoffensive. As glucose becomes exhausted, microorganisms utilize amino acids and produce ammonia and smaller amounts of malodorous organic sulfides and amines, which are responsible for the appearance of spoilage odors. High ultimate pH, DFD meat has poor keeping quality, because it may be virtually devoid of glucose. Therefore, the adverse consequences of microbial degradation of amino acids become apparent very quickly [9,10].

Also, as meat is displayed it becomes brown in color and eventually becomes microbiologically unacceptable [11,12]. Consequently, brown and dark meat can become merged and confused in consumers' minds, and both are associated with lowered acceptability.

The above factors are applicable to beef, venison, lamb, and pork. Pork has an additional problem associated with a high rate of postmortem glycolysis, resulting in pale, soft, and exudative (PSE) meat. This increased rate of postmortem glcolysis is due to stress immediately preslaughter and occasionally to prolonged electrical stunning procedures. These conditions are exacerbated by genetic predisposition [13]. Great decreases in pH are encountered in other species after electrical stimulation (see below), but conditions analogous to those producing PSE meat are rare. However, under conditions where rigor mortis occurs at high temperatures (30°C), drip is increased.

Consumers generally prefer pork with mild DFD charcteristics and reject pork with PSE characteristics [14]. PSE pork results in substantially higher—and DFD pork lower—drip losses from cuts during chilled storage, and these inherent muscle quality–related differences in drip are exacerbated by the combined effects of freezing, frozen storage, and thawing in water (Table 1) [15]. It has generally been reported that PSE pork also sustains the highest [16–19] and DFD pork the lowest [16,20–23] cooking losses. Early reports indicated that all inherent quality groups (extreme PSE to extreme DFD, Table 1) produced meat well within the acceptable range with respect to palatability. Consequently, with the possible exception of juiciness, inherent muscle quality was of little relevance to eating quality [16,24]. More recent studies, however, clearly document the effects of inherent muscle quality differences on pork palatability [25].

Pork of normal muscle quality has a firm, elastic, and cohesive texture, which is stringy, fibrous, and hard to compress. On chewing, such meat is slow to break down into particles, which tend to be fibrous, grainy, and mealy. The PSE condition, on the other hand, is characterized by a predominance of sour flavor notes, which detract from the overall taste sensation, producing a less balanced and blended flavor. The PSE condition also produces a drier texture as less moisture and fat are released during mastication, resulting in a large amount of moisture being absorbed from the mouth. The DFD condition typically exhibits a predominance of porky, sweet, and fatty flavor notes which enhance the appropriateness and the balance and blend of the flavor. With the DFD condition, greater amounts of moisture are released into the mouth during mastication, resulting in a juicy soft texture, which is less cohesive, fibrous, stringy, and consequently easy to chew. However, as the DFD condition becomes more severe, undesirable character notes contributing to off-flavors become apparent. The texture becomes excessively soft, crumbly, and mushy, contributing to an overall mouthfeel characterized by small, mealy, and/or mushy particles [25].

## III.  PROCESSING FACTORS

### A.  Glycolytic Rates and Effects of Processing Variations

With optimum preslaughter glycogen levels and no significant immediate preslaughter stress, the pH falls in a predictable manner to a value of ~5.5, depending on muscle temperature and on the animal species. Decrease in muscle pH occurs quickly in pigs and more slowly in cattle [26]. The time to reach ultimate pH values varies with chilling conditions, but in a typical chilling environment it takes about 5–10 hours for pigs, 16–24 hours for sheep, and up to 36 hours for cattle. While

**Table 1** Means and Standard Errors for Percent Shrink[a] from Various Cuts with Different Levels of Muscle Quality During Storage

| | Treatment | | | | | | |
| | Fresh | | | Frozen and thawed[b] | | | Effect of storage |
| Muscle quality | n | Mean | SE | n | Mean | SE | |
|---|---|---|---|---|---|---|---|
| **Hams** | | | | | | | |
| Extremely pale, soft exudative | 20 | 1.58a[c] | 0.18 | 21 | 1.59a | 0.17 | NS |
| Pale, soft, exudative | 20 | 1.29a | 0.18 | 20 | 1.78a | 0.18 | * |
| Normal | 20 | 0.82b | 0.18 | 22 | 0.51b | 0.17 | NS |
| Dark, firm, dry | 20 | 0.46b | 0.18 | 20 | 0.20b | 0.18 | NS |
| **Backs** | | | | | | | |
| Extremely pale, soft, exudative | 16 | 4.00ab | 0.78 | 21 | 14.20a | 0.68 | ** |
| Pale, soft, exudate | 23 | 4.55a | 0.65 | 25 | 14.81a | 0.62 | ** |
| Normal color, soft, exudate | 22 | 2.83bc | 0.66 | — | — | — | — |
| Normal | 20 | 1.89cd | 0.70 | 21 | 10.26b | 0.68 | ** |
| Dark, firm, dry | 22 | 1.48d | 0.67 | 33 | 3.74c | 0.54 | ** |
| **Picnics** | | | | | | | |
| Pale, soft, exudative | 20 | 3.39a | 0.55 | 20 | 3.54a | 0.55 | NS |
| Normal | 20 | 2.09b | 0.55 | 20 | 0.31b | 0.55 | * |
| Dark, firm, dry | 20 | 2.69ab | 0.55 | 20 | 0.15b | 0.55 | ** |
| **Bellies** | | | | | | | |
| Pale, soft, exudative | 20 | 1.07a | 0.26 | 20 | 1.63a | 0.26 | NS |
| Normal | 20 | 1.08a | 0.26 | 20 | 0.32b | 0.26 | * |
| Dark, firm dry | 20 | 0.72a | 0.26 | 20 | 0.00b | 0.26 | NS |

[a]Percent shrink (fresh cuts) = (trimmed weight after storage/trimmed weight prior to storage) × 100.
Percent shrink (frozen cuts) = (trimmed weight after storage and thawing/trimmed weight prior to storage) × 100.
[b]Includes thaw-drip loss.
[c]Means in the same column and cut group bearing a common letter do not differ significantly ($p > 0.05$).
NS = Nonsignificant; * = $p < 0.05$; ** = $p < 0.01$.
*Source:* Ref. 15.

the pH is falling but not less than 6.0, the muscle is "alive" and can respond to changes in temperature or other stimuli by shortening: the effect is negligible below 6.0. If the muscle enters rigor in a shortened state, the meat is toughened [27]. When shortening reaches approximately 40%, the meat cannot age, and it remains extremely tough, being virtually inedible [28]. The immediate postmortem period is clearly a time when care is required to maintain temperature control of the processing cycle.

## B. Electrical Stimulation

The process of glycolysis can be accelerated through the use of electrical stimulation [29]. The immediate effect of stimulation is a release of lactic acid and decrease in muscle pH due to intense muscular contraction. This is followed by an increased rate of pH decrease, with an earlier onset of rigor mortis. In sheep, high-voltage electrical stimulation produces rigor mortis within 3 hours postslaughter.

It is difficult to generalize and quantify the effects of stimulation because of the wide variety of systems and wide range of waveforms being used, ranging from 14.3 pulses/s to a 1-second burst of 50 or 60 Hz with 1-second intervals. In a typical electrical stimulation system developed in New Zealand, sheep carcasses, with the pelt removed, pass along a set of electrodes, and a current with approximately a 2.0A peak and a pulse frequency of 14.28 pulses/s at a 1100V peak passes through the carcasses for 90 seconds within 30 minutes of slaughter. The carcasses are then held for up to 8 hours at temperatures greater than 6°C. With cattle and deer, the current is usually applied within 5 minutes of slaughter, but may be delayed for up to 15 minutes if subsequent chilling conditions are appropriate. Lower voltages are used, ranging from 40–100 V with a consequently lower current. Contact is made so that current flows between the legs/anus and the mouth. There are other advantages of electrical stimulation unrelated to pH lowering, since it can reduce postslaughter animal movement and result in a fresher, brighter meat color at 48 hours.

Electrical stimulation is not normally used in pigs to avoid the high temperatures and low pH conditions that cause PSE. However, if carcasses are stimulated at an appropriate time later in the processing chain, electrical stimulation followed by rapid chilling can be used to advantage [30].

## C. Aging

Aging is a proteolytic breakdown of myofibrillar proteins by endogenous enzymes and is most rapid at high temperatures—a situation that arises with electrical stimulation. However, aging can take place until the meat is cooked or frozen. Although opposing views exist, we believe that aging commences only after meat is in rigor mortis [31]. Unpublished data [32] suggest that aging occurs even during the latent heat phase of freezing. While meat is frozen it is not inert, and glycolysis can occur, as evidenced by the disappearance of ATP [33]. In fact, if glycolysis goes to completion, thaw contracture, which would occur if residual ATP remained in the frozen muscle, is avoided. Lipid oxidation also continues in the frozen state (see below). For all practical purposes, therefore, when one considers the attributes of meat prior to frozen storage, even the freezing process itself must be considered a contributor to aging. Once the meat is frozen, many but not all of the characteristics of the meat are held in abeyance.

Predicting tenderness in the case of lamb is possible if one knows the rate of carcass cooling, the rate of pH fall prior to rigor mortis, and whether or not electrical stimulation has been applied [34]. The rate of carcass cooling can be predicted from the ambient temperature and the rate of surface heat transfer using the methods of Lovatt et al. [35] and other techniques. A similar model has been developed for beef omitting the freezing stage [36], and in principle this technology could be extended to other species. Aging also affects the stability of meat color [12].

## D. Hot Boning

The previous statements apply to cold boned meat when entire carcasses are being considered. With hot boning, where the meat is removed prior to rigor mortis, additional factors need to be taken into account. Hot boning procedures generally apply to beef and pork, but even lamb can be hot cut [37] or hot boned. Because of lack of skeletal restraint, the effects of cooling are exacerbated, and cold shortening and consequent muscle distortion become a real problem. In addition, passive shortening during rigor mortis is made worse through various manipulations and removal of the skeletal restraint. Small amounts of shortening do not appear to constitute a problem, since aging still occurs [28]. Therefore, provided cooling is controlled and shortening minimized, hot boning can produce meat similar in tenderness to cold boned meat after appropriate aging [38,39]. Electrical stimulation, which reduces the time required to reach rigor mortis, followed by controlled cooling may be a more useful option [39,40] and is clearly a useful adjunct with hot-boning procedures. Cold shortening is not an issue when meat is used for grinding purposes or when emulsion-type products are produced.

## E. Raw Meat Color

The color of raw meat is determined by proteins, mainly myoglobin and to a lesser extent hemoglobin, which are markedly affected by muscle temperature and pH postslaughter [11,12]. Meat with a high ultimate pH (>6.0) is dark in color, irrespective of the temperature history during rigor mortis (e.g., DFD beef). Since the fibers are able to bind water and swell, no free water is available. Only a small amount of light is reflected, and the meat appears matt and dark. In addition, at high pH the oxygen-utilizing enzymes are relatively active, allowing only a little surface oxygenation, and the reduced purple myoglobin color below the surface dominates. At normal pH values (5.5), the fibers hold less water and scatter light, the oxygen-utilizing enzymes are less active, and the meat is brighter in color and glossier in appearance. If rigor mortis occurs at elevated temperatures, partial denaturation of sarcoplasmic proteins and even myosin occurs and light is scattered to a greater extent resulting in a paler color. This situation is extreme in PSE pork. Once the inherent meat color is achieved, it is changed very little with further processing. However, subsequent storage conditions can affect the color.

Freshly cut meat surfaces exposed to air or oxygen "bloom" due to formation of bright red oxymyoglobin from the purple myoglobin. This red layer spreads downwards into the meat for a few millimeters, intensifying the color. Over time myoglobin oxidizes to brown metmyoglobin, with the rate varying dramatically according to oxygen partial pressure, muscle type, and animal species. Generally, the longer the meat is held in a chilled condition before cutting and display (up to 7 days), the longer the display life, due to the interaction of a decrease in a catalytic oxidative ability with a greater decrease in enzymic reducing ability.

Factors such as electrical stimulation per se have little effect on color deterioration but may modify the rate of color deterioration produced by the temperature and pH conditions during the entire rigor process [41]. Since a wide range of pH and temperature conditions exist in the large muscle groups within animals, it is difficult to characterize the changes accurately when factors such as electrical stimulation and hot boning—with or without rapid chilling—occur [42,43]. The complex interactions are covered by Ledward [12] and Faustman and Cassens, [11,44].

The frozen storage life and subsequent thawing effects on meat color are profoundly affected by postslaughter processing due to these interactions, and in general meat color changes have not been adequately quantified. Meat color does not appear to be significantly affected when only short periods elapse between slaughter and freezing (as in hot boning).

## F.  Oxidation

Grinding of meat prior to freezing would be expected to be ideal for bulk storage, transport, and distribution, but the myriad of cut surfaces, exposing lipid-containing cell membranes to air, ensures the initiation of oxidative rancidity. This is not confined to fresh meat or to minces. It also occurs in cooked meat products and is perceived as "warmed-over flavor." Oxidative rancidity also is found in frozen whole tissue if prefreezing and storage conditions are inappropriate.

To determine the effect of storage temperature on the life of frozen meat based upon oxidative rancidity, two regimes were examined experimentally in lamb when the duration of storage was the same for both but the sequences of storage temperatures were varied [45]. The first sequence in this experiment was at –5°C for 6 weeks before storage at –10°C for 20 weeks, and the second sequence was at

−10°C for 20 weeks before storage at −5°C for 6 weeks. Rancidity initiated at −5°C continued at −10°C, and the meat was noticeably rancid, whereas initiation of rancidity at −10°C was minimal, and when the temperature was raised to −5°C rancidity was initiated, but the duration of storage at this temperature was insufficient to promote rancidity to an unacceptable level [45]. The initiation of rancidity by conditions before freezing evaluated after freezing and thawing, is also very important to meat quality.

## IV. FREEZING OF MEAT

### A. Moisture Loss

Moisture evaporates from any bare meat surface during cooling and storage. This has several consequences. First, the evaporated moisture reduces the total weight of meat, thereby reducing its value if the meat is sold on a weight basis. Second, the meat surface may dry, reducing its appeal to the consumer. Third, a dried meat surface is less hospitable to microbiological growth than a wet surface, making moisture loss microbiologically beneficial.

It is therefore necessary to achieve enough surface drying to reduce microbial growth without reducing meat yields excessively. The degree of weight loss that achieves this balance depends upon the priorities of the meat processor. Weight loss from naked lamb carcasses during chilling and freezing in four typical New Zealand meat-processing plants ranged from 2.4 to 3.1%. Two months in cold storage together with container transport added another 2.0% [46].

The rate of surface drying is enhanced by low humidity and high air flow across the meat, but there is a point where the rate of moisture transfer to the surface cannot keep up with its removal from the surface. Under these circumstances, the rate of moisture loss is controlled by the rate at which moisture will migrate through the desiccated layer. This means that total moisture loss during chilling can be *reduced* by deliberately producing a desiccated layer on the surface.

The rate of moisture transfer to the surface is significantly influenced by the amount of fat cover. Meat with little fat cover lost 43% more moisture during freezing than meat with greater fat cover [46]. This difference was not evident during chilling, since the surface was not sufficiently dehydrated for moisture transfer through the meat to control weight loss.

One solution to the shrinkage (weight loss) problem is hot boning. In this case, meat can be wrapped sooner and there is no weight loss due to evaporation. Unfortunately, this benefit is achieved at the expense of a greater potential for microbial growth unless temperature is sufficiently controlled.

Significant amounts of moisture may be lost during cold storage, giving rise to the phenomenon of freezer burn. A model for moisture loss during cold storage as a function of air velocity, air temperature, relative humidity, and the presence of radiating surfaces has been developed [47]. During long periods of cold storage, moisture loss may exceed that experienced during chilling and freezing by a large margin. In an extreme case, a weight loss of up to 9% was observed after a naked lamb carcass had been in a cold store at −15°C for 12 months with an air velocity of 0.2 m/s [47]. Weight loss from carcasses completely wrapped in close-fitting plastic under similar conditions was virtually nil. With pork, moisture losses ranged from 0.7 to 4.5%, depending upon the cut and the inherent muscle quality (Table 1), and were substantially exacerbated by freezing, frozen storage, and thawing [15].

## B.  Chilled Storage

All frozen products must pass through the chilled state. The duration and conditions imposed on meat during this transition are as varied as the reasons for, and methods of, freezing meat. The changes that occur during the prefreezing period have an important effect on the subsequent quality of the frozen product. Generally, the chilled storage life is determined by the onset of microbial spoilage. However, where microbial growth is limited, the product may become unacceptable because of flavor and textural changes or color deterioration before overt microbial spoilage becomes apparent. Consequently, it is essential that some consideration be given to the microbiology of chilled meat, since it represents an unavoidable but controllable overture to the whole frozen symphony.

Fresh meat is an ideal substrate for the growth of many microorganisms. It is rich in nitrogenous nutrients, minerals, and accessory growth factors, is moist, and has a favorable pH. During slaughter, dressing, and further processing, meat becomes increasingly contaminated with bacteria found on the hide or in the intestinal tract of the animal as well as bacteria found in the processing environment [48,49]. Growth of contaminating microorganisms on meat is rapid providing conditions, particularly if temperature, moisture, and the gaseous environment, remain favorable. Such microbial growth can be prevented if meat is cooled to temperatures below those that permit microbial growth. In practical terms, this is achieved by freezing. However, at the higher temperatures associated with chilled storage, distribution, and retail display (−2.0 to 7.0°C), growth is merely retarded, not prevented.

The microflora developing on meat at chilled storage temperatures and composed of psychrotrophic bacteria. These bacteria are carried on the hides or fleeces of slaughtered animals and are also present on work surfaces and processing implements. They are capable of growth at temperatures close to 0°C but do not have the low optimum growth temperatures characteristic of psychrophiles [50]. The incidence of psychrotrophs in the environment is inversely related to the ambient temperature and declines in summer and with decreasing geographical latitude [51,52]. Colonization of chilled areas, particularly meat-contact surfaces, within meat-processing plants represents a serious hygiene problem that must be overcome if meat is to have an acceptable chilled storage life. Psychrotrophic bacteria most commonly found on meat spoiling at chilled temperatures include species of *Pseudomonas, Acinetobacter, Moraxella, Lactobacillus, Carnobacterium,* and *Brochothrix* and members of the Enterobacteriaceae family.

At chill temperatures, meat spoils most rapidly through the development of aerobic microflora of high spoilage potential, usually dominated by species of *Pseudomonas* [53,54]. Aerobic microorganisms with high spoilage potential produce malodorous or otherwise offensive byproducts during their growth on meat. This type of spoilage, which can occur after only a few days of chilled storage, is usually encountered with meat derived from plants where slaughter hygiene, carcass cooling, and product freezing are less than ideal. Similarly, it may be evident when consumers freeze meat purchased at retail towards the end of its display life (i.e., meat discounted to obtain a quick sale at the supermarket).

Under appropriate commercial conditions, growth of aerobic spoilage bacteria on unwrapped meat, particularly carcasses, is retarded by surface drying as well as by reduction in temperature. When the water content of the surface tissue is reduced from the normal 300% (dry weight basis) to about 85%, no bacterial growth occurs [55,56]. With conditions producing higher rates of desiccation, the number of viable bacteria may actually decrease. However, the control of bacterial growth by surface drying results in significant carcass weight loss and produces appearance changes associated with desiccation [57]. The adverse effects of surface drying are proportionally more serious in smaller carcasses (e.g., lamb compared to beef).

The principal meat spoilage bacteria are strict aerobes. Consequently, their growth can be effectively controlled by manipulating the oxygen concentration in the atmosphere surrounding the meat. These organisms also show a marked sensitivity to elevated concentrations of carbon dioxide in both aerobic and anaerobic storage atmospheres [58,59]. An anaerobic gaseous environment around meat will cause the microflora developing at chill temperatures to undergo a very significant ecological shift and become dominated not by high-spoilage-potential pseudomonads, but by relatively benign lactic acid bacteria [60]. "Lactic acid bacteria" is a general term used to include members of the genera *Lactobacillus, Carnobacterium,* and related cocci, which produce lactic acid as the major byproduct of their growth. Presently the most common method employed to effect this transition to a low-spoilage-potential lactic acid bacteria–dominated microflora is vacuum packaging. The use of vacuum packaging increases the effective chilled storage life of meat to a maximum of about 12 weeks depending on the meat species, storage temperature, and hygienic status of the meat at packaging.

In vacuum packaging, meat is sealed in plastic bags with low gas permeability after a vacuum is created to ensure that no significant amount of oxygen remains within the pack. As the vacuum is created, the flexible packaging collapses around the meat, squeezing most of the air from the pack. Product life of vacuum-packed meat is inversely related to both the storage temperature and the oxygen permeability of the barrier material used [61]. Under ideal conditions—storage at $-1.5°C$ within a packaging film of oxygen permeability $<20$ ml/m$^2$/d/atm—vacuum-packaged meat will eventually spoil due to the accumulation of byproducts produced by lactic acid bacteria. Such spoilage is usually not evident until maximum cell numbers ($10^7$ cfu/cm$^2$) are attained. However, early spoilage may occur as a result of the growth of psychrotrophic Enterobacteriaceae, *Brochothrix thermosphacta*, or *Shewanella putrefaciens*. Preexisting conditions that allow these high spoilage potential bacteria to compete with the faster-growing lactic acid bacteria are initially high numbers in the contaminating microflora, high meat pH ($>5.8$ for the Enterobacteriaceae and *B. thermosphacta* and $>6.0$ for *S. putrefaciens*), oxygen entering the pack, and elevated or fluctuating storage temperatures.

Odors and flavors associated with the growth of various microflora are quite characteristic. Lactic acid bacteria produce a souring of the meat, which later is accompanied by dairy odors and flavors. An off-odour reminiscent of sweaty socks is produced by *B. thermosphacta* under conditions that allow aerobic metabolism (i.e., ineffective oxygen removal or inadequate oxygen barrier). Under anaerobic conditions, *B. thermosphacta* produces innocuous byproducts as do the lactic acid bacteria. Spoilage caused by the growth of *S. putrefaciens* is characterized by the production of hydrogen sulfide and greening of the meat, while the psychrotrophic Enterobacteriaceae produce a most unpleasant sweet putrid odor with just a hint of hydrogen sulfide. Spoilage also imparts to the cooked product an objectionable taste. Unpleasant flavors detected in frozen meat may, therefore, indicate that the product was frozen when it was approaching, or indeed had passed, the incipient spoilage stage during chilling. In some cases vacuum-packed meat, especially beef, may not be microbiologically spoiled before textural deterioration renders it unacceptable to consumers [62].

A further 50% extension of the effective chilled storage life of fresh meat over that attained using vacuum packs (i.e., 18 weeks or more) can be achieved through the use of oxygen-free, saturated carbon dioxide–controlled–atmosphere packaging [63–65]. Under anaerobic conditions the inhibitory effects of carbon dioxide on both the potent spoilage bacteria able to proliferate in vacuum packs and the relatively innocuous lactic acid bacteria component of anaerobic meat microflora increase with increasing carbon dioxide concentration [66]. Carbon dioxide controlled-atmosphere packaging appears to be particularly effective in eliminating enterobacteriaceae, *B. thermosphacta*, and *S. putrefaciens* from the microflora developing on chilled meat. However, its effectiveness in extending product life over that attained in vacuum packs is reduced as storage temperature increases. As with vacuum-packed beef, carbon dioxide packaging may suppress the development of the spoilage microflora to such an extent that textural deterioration or the development of livery flavors resulting from autolytic changes, and not microbial spoilage, renders the meat unacceptable. With pork, chilled storage life under both vacuum and carbon dioxide is limited by the development of predominantly sour and bitter off-flavors, which coincide with lactic acid bacteria approaching maximum numbers [67,68].

Oxygen contamination of the anaerobic atmosphere at pack closure and oxygen ingress into the pack during storage results in deterioration of the visual properties, particularly in pork with PSE properties [69,70]. Freezing such a product will not enhance its consumer acceptability

From this brief review of the microbial spoilage of chilled meat, it is apparent that frozen meat quality can be microbiologically compromised before the product enters the freezer. Consequently, prefreezing treatment must not be left to chance if a high-quality product is to be consistently achieved.

In addition to microbially induced changes, chemical changes also take place during chilled storage. Undesirable chemical changes such as rancidity, if initiated during chilled storage, will continue to take place at an accelerated rate after freezing.

## C.  Frozen Meat Quality

The process of freezing aged meat, if sufficiently rapid, does not necessarily have any demonstrable effect on the cooked color, flavor, odor, or juiciness of that meat. For the first 24 hours of retail display, frozen-stored pork with a variety of protective wraps (196 days, –30°C in still air in the dark) thawed for 24 hours at 2°C was the same in eating characteristics as unfrozen product [71]. Pork chops frozen, stored, and thawed under these conditions had a retail display-life of at least 6 days [72]. Pork chops and roasts stored at –30°C for more than 196 days in a variety of protective storage wraps retained their palatability [73]. An additional advantage was that freezing, frozen storage, and thawing of pork loins produced a slight tenderizing effect, as these cuts became less cohesive and easier to chew. This change was accompanied by a greater moisture release during mastication, which slightly increased perceived juiciness, and a breakdown to more appropriate residual particles. These effects combined to produce an appropriate well-balanced and blended texture [25]. However, the development of off-flavor notes following prolonged storage resulted in a loss of flavor balance and blend [25]. Consequently, within reasonable limits, freezing is generally considered to be an excellent means for preserving meat, since fewer undesirable characteristics develop during prolonged storage than with most other storage methods.

Most nutrients are retained during freezing and subsequent storage. There are problems with soluble proteins being lost with drip during thawing, but the fluid lost with drip approximates the fluid lost when fresh meat is cooked. Once the meat is frozen, however, it is not in an inert state. Potential exists for recrystallization of water, and optimum quality is dependent upon how the meat is protected, freezing rate, length of frozen storage, and freezer storage conditions (temperature stability).

If meat surfaces are not protected, desiccation occurs. As storage temperature decreases, deterioration decreases, and at −70°C (an impractical storage temperature) biochemicals are essentially regarded as being inert. Studies by Moore [74] demonstrated that the display characteristics of chops cut from meat stored in the carcass form were much more acceptable to consumers than those of stored cut chops, due to surface deterioration. Because of changes occurring during storage, it was found that color deterioration of frozen meat on display was actually enhanced after longer storage times (30 weeks) than after short storage times (1 week).

## D.  The Freezing Process

Freezing is simply the crystallization of ice in muscle tissue and includes the consecutive processes of *nucleation* and *growth*. These processes are central to the effects of different freezing rates and subsequent effects on meat quality. Meat does not freeze immediately when its temperature drops below the freezing point and the latent heat (i.e., heat required during the phase change during crystal formation) has been removed. In other words there is a degree of *supercooling*. The greater the supercooling, the greater the number of nuclei formed. The number of nuclei is greatest in the extracellular space, and they are only formed within the cell when the rate of heat removal is higher. As soon as the nuclei form, they begin to grow by the accumulation of molecules at the solid/liquid interface. However, the way they grow depends on the microgeometry and the temperature distribution ahead of the freezing front in a complex way as a consequence of dendrite formation with supercooling in front of the dendrite growth. An important concept is *characteristic freezing time*, which is a measure of the local freezing rate and is defined as the time during which the point under consideration decreases from −1°C (freezing commences) to −7°C (when 80% of the water is frozen). The growth of extracellular ice crystals also takes place at the expense of intracellular water. This leads to partial dehydration of the muscle fibers and subsequent distortion. At high characteristic times (slow freezing), the ice crystals are larger and the tissue distortion is greater.

## E.  Freezing Rate

The rate at which freezing takes place can be considered both at micro and macro levels. At the micro level, freezing rate is described in terms of the speed with which the freezing front moves through the freezing object. At the macro level, the rate at which any given part of the object is cooled determines the temperature profile for that part, and thus has an important bearing on the biochemistry and microbiology of that part.

The undesirable changes in meat during freezing are associated with formation of large ice crystals in extracellular locations, mechanical damage by the ice crystals to cellular structures through distortion and volume changes, and chemical damage arising from changes in concentrations of solutes. The fastest freezing rates are associated with the least damage [75]. Differences in freezing rate modify meat properties. The growth of ice crystals and crystallization in meat tissues is discussed by Calvelo [76].

Freezing commences when the surface temperature of the meat reaches its freezing temperature. A continuous freezing front forms and proceeds from the exterior to the interior. Extracellular water freezes more readily than intracellular water because of its lower ionic and solute concentration. Slower freezing favors formation of pure ice crystals and concentrations of solutes in unfrozen solutions. Intracellular solutions are often deficient in the nucleation sites necessary to form small ice crystals. Such conditions favor the gradual movement of water out of the muscle cells, resulting in a collection of large extracellular ice crystals and a concentration of interacellular solutes. Freezing damage arises from massive distortion and damage to cell membranes. Such effects have implications during thawing as the large extracellular ice crystals produce drip during thawing. The structural changes that occur also obliterate the recognizable muscle structure.

Fast freezing results in small ice crystal formation in both intracellular and extracellular compartments of the muscle and very little translocation of water. Drip losses during thawing are thus considerably reduced, and the surface reflects more light than the surface of slowly frozen meat. Consequently, the cut surface appearance is more acceptable.

## F.  Freezing Methods

The type of freezing method to be used depends upon the freezing rate required and the characteristics of the product. The overall rate of freezing for any object is a function of the driving force and resistances involved in the process.

Internal resistance to cooling results from the finite thermal conductivity of the material and the length of the pathway along which the heat must be conducted. The pathway length is, of course, a function of the object size, and the conductivity is a function of its composition (and the direction of heat flow with respect to the meat fibers for nonhomogeneous meats). The amount of heat to be extracted during freezing is the difference in enthalpy of the meat between the starting and finishing temperatures. This is also a function of composition. The external resistance to cooling is determined by the surface heat transfer coefficient and resistance factors due to layers of packaging materials. The freezing driving force is the difference in temperature between the cooling medium and the object.

Since one cannot readily change the meat composition, the scope for changing the freezing rate of a given quantity of meat is limited to three factors: (1) changing its size or shape by packing several pieces in one package, thereby making the heat-conduction pathway longer, or by cutting the meat into pieces, thereby making the pathway shorter, (2) changing the external resistance by adding or removing packaging or by changing the surface heat transfer coefficient, or (3) changing the driving force (by changing the cooling medium temperature).

Comminution is a successful method for decreasing the time for each piece of meat to freeze. It may be taken to extremes, which results in effectively instantaneous freezing. Its advantages are, however, outweighed by its disadvantages. First, comminution increases the exposed meat surface area, the potential for evaporation rates, and greater total weight loss, even given the shorter time spent in the freezer. Second, repacking densities after freezing are typically low. Third, comminution makes it difficult to identify the original form of the meat. This last factor means that the customer must trust that the meat is what the meat processor claims it to be.

Just as comminution can greatly decrease freezing time, aggregation can greatly increase freezing time. Attempts to freeze meat in very large blocks (e.g., several hundred millimeters thick) inevitably result in high rates of microbiological growth in the center of the package and a large degree of quality variation between the center and the surface. Even in more moderate lump sizes, such as beef frozen in side or quarter form or boned meat frozen in cartons weighing up to 27 kg, the temperature profile of the center can differ significantly from that of the surface, resulting in variation in tenderness and microbiological quality from the surface to the center.

The amount of external resistance to heat transfer in meat freezing ranges from natural convection air freezing (with a typical surface heat transfer coefficient of $2–5$ W/m$^2$K), to forced convection air freezing ($10–30$ W/m$^2$K), liquid immersion freezing ($>100$ W/m$^2$K), cryogenic freezing ($>1000$ W/m$^2$K), or plate freezing (with an infinite rate of surface heat transfer). Thus, one can change the freezing rate very effectively by changing the freezing method.

Freezing technologies affect more than the cooling rate, however. As discussed above, freezing bare meat in air may result in significant weight loss. Meat may therefore have to be packaged prior to chilling or freezing.

Packaging will also be required for meat frozen in a liquid immersion system to prevent the meat from coming into direct contact with the secondary refrigerant. One can avoid using packaging during cryogenic freezing, as the most commonly used cryogens are acceptable for contact with meat. Plate freezing can be done with or without packaging, but the benefits of direct contact between the refrigerating plate and the product are reduced if packaging materials are interposed.

The final parameter that affects the freezing rate is the temperature driving force. The use of lower vs. higher temperatures in freezing and cold storage is very much a matter of economics. All refrigeration systems have lower coefficients of performance at lower temperatures. This makes them progressively more expensive to operate as the refrigerating temperature is reduced.

The ability to supply refrigeration at low temperature is the second factor that makes cryogenic freezing faster than other techniques. However, the increased cost of achieving low temperatures applies to cryogenic systems in the same way as it does to conventional refrigeration systems. The cryogen must be liquified or solidified in some form of refrigerating plant before it is used to freeze the meat, and this process is more expensive for lower-temperature cryogens.

## G. Freezing Effects on Microorganisms

The effect of freezing on microorganisms, with the exception of resistant structures such as bacterial endospores, is nothing short of disastrous. During the freezing process microorganisms are exposed to debilitating and lethal effects of cold shock, intracellular growth of ice crystals, and increasing solute concentration of the surrounding unfrozen meat juices. During the freezing of meat the concentration of solutes is probably the most important cause of microbial death [77]. Almost paradoxically, freezing is, however, a well-recognized method for the preservation of bacteria. The adverse effects of freezing on bacterial viability are markedly influenced by the rate of freezing and the composition of the freezing medium. Consequently, it is difficult to predict the reduction in microbial numbers likely to occur as a result of freezing.

Microorganisms are not equally susceptible to the lethal effects of freezing. Bacterial and fungal spores are extremely resistant to freezing. Generally, the vegetative cells of Gram-positive bacteria, particularly the Gram-positive cocci, are relatively resistant to freezing, Gram-negative bacteria on the other hand are markedly susceptible (78). This differential sensitively will produce a change both in the size and composition of the meat microflora as meat freezes. Typically, the microflora developing on meat held under aerobic conditions before freezing contains 15% Gram-positive and 85% Gram-negative organisms. After freezing to $-30°C$, when the total viable population was decreased to 20% of the prefreezing population, proportions of Gram-positive to Gram-negative organisms reversed to become 70% and 30% respectively (79). Concern has been expressed that freezing may increase the health hazard posed by frozen meat, since it causes a relative enrichment of the microflora in Gram-positive pathogens, particularly *Staphylococcus aureus* and *Listeria monocytogenes.*

The rate of decline in bacterial numbers is usually most rapid immediately after freezing, but then it slows with holding at constant subzero temperatures until viable numbers remain almost constant as the cells surviving freezing enter an induced stationary phase. It is for this reason that freezing is an effective, if not numerically efficient, method of preserving bacteria. Total cell death is directly related to the frozen storage temperature [80]. The decline in viability with temperatures associated with frozen meat storage probably results from continual exposure to concentrated solutes and, thus, represents a continuation of stresses associated with the freezing process itself [81]. Freezing of meat, in practice, can be expected to reduce the numbers of spoilage and pathogenic bacteria. However, the size of that reduction cannot be reliably anticipated because the death rate is determined by conditions present during freezing.

Organisms that are very sensitive to the lethal effects of freezing include parasitic protozoa, cestodes, and nematodes, which may infest meat animals. The cysts of the tapeworms *Taenia saginata* (*Cysticercus bovis*), *Taenia solium* (*Cysticercus cellulosae*), *Taenia ovis* (*Cysticercus ovis*), and *Echinococcus granulosus* (hydatid cyst) are destroyed by storing meat for at least 10 days at $-10°C$ or below [82]. Similarly, nematode larvae (*Trichinella spiralis*) which cause trichinosis in pork, are killed by storage at $-15°C$ for at least 30 days. The storage period can be shortened if lower storage temperatures are used (e.g., 20 days at $-25°C$ or 12 days at $-30°C$ [82]).

## H.   Freezing Effects on Microbial Growth

As meat freezes, ice crystals begin to form and the dissolved solutes are concentrated in the remaining liquid. As a result of the increasing concentration of solutes, any free water has the same water activity as ice at the same temperature [83]. Consequently, frozen meat, as a substrate for microbial growth, poses two major problems: low temperature and reduced water activity. The majority of meat spoilage bacteria are not able to grow on substrates with low water activity (i.e., they are not xerotolerant). Frozen meat at temperatures between $-2$ and $-4°C$ has a water activity approaching that which inhibits growth of meat spoilage bacteria at optimum growth temperatures [84]. The combined effects of low temperature and reduced water activity ensure that in practice bacteria do not grow on meat at temperatures much below $-2°C$. However, growth at lower temperatures has been reported under laboratory conditions in supercooled media. Quality changes caused by bacterial growth are, therefore, associated with prefreezing and postfreezing handling rather than frozen storage of meat.

Although bacterial growth at subzero temperatures is restricted by temperature and water activity to unfrozen substrates, growth of the more xerotolerant molds and yeasts on frozen foods has frequently been observed. Mold spoilage of meat has received considerably more attention than yeast spoilage because it causes dramatic changes in the appearance of the product. Four major forms of mold spoilage are recognized by the meat industry, descriptively referred to as "black spot," "white spot," "blue-green mold," and "whiskers."

The propensity of molds for growth on frozen meat at very low storage temperatures ($-10$ to $-18°C$) may have been exaggerated [81–85]. Fungal spoilage of frozen meat is associated with prolonged storage at marginal temperatures [86]. Most common meat spoilage molds have minimum growth temperatures near $-5°C$. At these temperatures molds grow slowly, forming visible colonies (1 mm in diameter) after several months [87,88].

From a meat-quality perspective, the mold spoilage of greatest concern is black spot, caused by the growth of *Cladosporium herbarum*, *C. cladosporoides*, or

*Penicillium hirsutum.* This condition is commonly associated with frozen meat transported long distances by sea. Black spot is easily recognized as black fungal colonies up to 1 cm in diameter with hyphal penetration into the surface layers of the meat. Although black spot poses no health hazard, considerable financial losses have been incurred when carcasses were downgraded following the necessary trimming to remove the affected areas. White spot (*Chrysosporium pannorum*) and the other forms of mold spoilage are superficial and can simply be wiped off without leaving any visible evidence of growth [86]. Like black spot, they do not present a health hazard, indeed, whisker (*Thamnidium elegans*) is the subject of several patented meat-aging processes claiming not only to tenderize the meat but also to impart a delectable aged flavor [89].

Apart from black spot, fungal growth on frozen meat does not in itself present a quality problem. However, its appearance should be taken as an indication of marginal conditions during frozen storage and transportation. Similarly, the appearance of yeasts on frozen meat is indicative of marginal freezing conditions. At $-5°C$ maximal yeast numbers ($10^6$/cm$^2$) are reached on frozen lamb within 20 weeks [90]. Yeast growth on frozen lamb may not result in adverse sensory changes. However, the quality may be visually compromised by the appearance of colonies. Visual spoilage, particularly of fat surfaces, is exacerbated by the deposition of heme pigments by yeasts, resulting in the appearance of brown spots [91].

## I.  Effects of Freezing Rate on Water-Holding Capacity

The water-holding capacity of meat is a valuable attribute in chilled meat [92] and is modified by the way meat is frozen. Water-holding capacity is dependent upon muscle type, preslaughter factors that affect ultimate pH, area of cut surface exposed, freezing rate, storage conditions, and thawing rate. Calvelo [76] showed that many of the effects of freezing rate on water-holding capacity gave conflicting results because the differences between the thermal histories of the internal and the external portions of meat has not been considered.

Thus, a dependence of ice crystal size on the local freezing times and variation is ensured because a range of freezing times occurs in meat/carcasses in an industrial freezing situation. Even in this situation, ice crystal formation depends on factors such as initial meat temperature, thickness, amount of bone, and the air velocity over the cut/carcass.

The difference in exudate between unfrozen and frozen beef after thawing over a wide range of local freezing times is shown in Fig. 1. The peak exudate is obtained with freezing times of 17–20 minutes, which coincides with the existence of a single ice crystal inside the muscle fibers. This can be interpreted as arising from greater distortion and damage to the myofibrillar and membrane structures. Freezing times slower than 20 minutes result in reduced exudate, because part of the water is outside the muscle fiber. Some water is resorbed during thawing, of course, but the

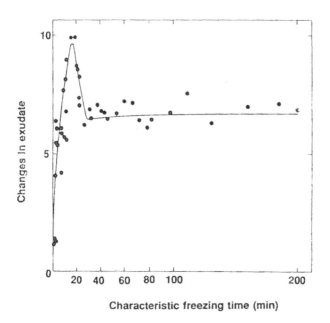

**Figure 1**   Changes in exudate from thawed meat in relation to the characteristic freezing time. The exudate is obtained by centrifugation, and the exudate changes (in frozen (F) and unfrozen (U) states) are defined by the expression:

$$\Delta = \left[ \left( \frac{\text{grams exudate}}{\text{grams meat}} \right)_F - \left( \frac{\text{grams exudate}}{\text{grams meat}} \right)_U \right] \times [10^2]$$

Clearly the freezing time will vary with size of carcass or cut and in large cuts will vary throughout the cut. During thawing, water is reabsorbed over time if it does not drip and become unavailable. (From Ref. 76.)

final exudate remains relatively constant. With such a complex relationship, it is clear that in a practical situation, there will be great variation in the size of samples and hence in freezing and thawing rates. Also, unless identical procedures are followed experimentally, considerable variations will exist.

High ionic strength denatures some muscle proteins [93], and this factor, combined with the translocation of water, largely accounts for the loss in water-holding capacity and failure to resorb exudate.

## V.   FROZEN STORAGE

### A.   Changes During Frozen Storage

During freezing, further denaturation of proteins occurs due to solute concentration effects, a recrystallization of ice occurs, lipids become oxidized, and dehydration of surfaces (freezer burn) occurs. In addition, interactions (e.g., lipid oxidation) will proceed on a larger front after freezer burn occurs. The quality of the meat, while it is affected by protein denaturation and freezer burn, can be partially protected by suitable packaging. The main perceived quality deterioration arises from the exudate that appears during thawing and, as explained above, is to a large extent dependent on the rate of freezing.

The denaturation of proteins is discerned as the phenomenon involving a change in solubility, and, hence, it is related to changes in functional properties. Protein denaturation is generally evaluated in terms of extraction of homogenized tissue by 5% NaCl. The concentration of solutions during ice crystal formation accelerates denaturation of proteins during storage, and maximum denatuation occurs at $-3°C$ in beef [93]. Consequently, at normal storage temperatures below $-10°C$, the rate of insolubility is less.

Water movement in stored frozen meat still takes place due to recrystallization, during which large crystals grow at the expense of small ones [94]. This process is driven by the difference in surface energy between small and large crystals (large and small curvature angles). The lower surface energy of large crystals favors growth. During the freezing process long dendrites are formed along the fiber and the direction of change is for the dendrites to be reduced and large crystals to grow. The rates of change related to temperature and crystal size are detailed by Calvelo [76], who found that a 30-$\mu$m crystal would double its diameter during about 21 days of storage at $-10°C$ or about 50 days at $-20°C$. Such rates are sufficiently rapid to ensure significant changes if above-normal storage durations are experienced.

## B.  Effects of Temperature Variation During Storage

Frozen meat storage generally takes place at temperatures below $-18°C$ [82]. However, it is likely that temperatures will fluctuate above these levels during movement in or out of the cold store. Fluctuations may well occur even in the cold stores. Environmental changes take place over time, and the surfaces of the carcass or box change more rapidly than those of the deeper-lying meat. Temperature variations are probably the main cause of undesirable effects. Consequently, a cold chain should be maintained with as little variation as possible—an ideal that may be achieved in limited cases.

## VI.  THAWING

When meat is thawed, changes initiated before and during freezing are expressed. Thawing should minimize exudate or drip loss, microbial growth, evaporation losses, and further deterioration. The critical temperatures defined for freezing are passed through again as meat is thawed. The higher the thawing temperature, the faster the meat will pass through temperatures between $-2$ and $10°C$ and remain at high temperatures where maximum deterioration of meat proteins occurs. In addition, parts of the meat will also rapidly reach temperatures at which microbial spoilage occurs. Thawing, using liquid immersion, can be achieved at lower temperatures because of a higher heat transfer coefficient, and ideally the meat should be protected by some form of packaging. The size and shape of the meat pieces, therefore, profoundly affect the ultimate quality.

The tenderness of meat thawed during the cooking process is no greater than that of the meat before freezing. However, if meat is thawed and left, it becomes more tender since the meat ages rapidly. Repeated freezing and thawing does not result in an increase in tenderness [95].

## A.   Methods

The physical principles of meat thawing are similar to those for freezing, and many of the same points apply. There are two major differences. First, the thermal conductivity of thawed meat is about one third that of frozen meat. Therefore, heat passing through the thawed layer into the frozen central portion is transferred more slowly when compared with heat passing in the other direction during freezing. Thus, all other factors being equal, thawing will take longer than freezing using the same techniques. Second, it is less practical to use a larger driving force for thawing than for freezing. While one can achieve very large temperature differences in freezing by using a cryogen, the use of similar driving forces during thawing is likely to produce significant changes in the meat structure.

Natural or forced convection air and liquid immersion techniques are used in thawing as they are in freezing. Since the thawing medium temperature is at or above 0°C, however, it is possible to use ordinary water in thawing, and it is not absolutely necessary to use packaging to keep the meat separate from the thawing medium. All of these thawing techniques are commonly used in industrial practice. Direct contact thawing is possible but is rarely used in commercial practice. One exception is the technique of frying a rare steak directly from the frozen state to ensure that it is lightly cooked in the center when served. Thawing technologies that do not have equivalent freezing technologies include vacuum, dielectric, electrical resistance, microwave, and infrared thawing.

Although one cannot apply large temperature differences during thawing, a high heat transfer coefficient can be achieved using vacuum thawing. The surface of meat is heated using condensing steam, but the whole system is maintained under vacuum conditions, so the steam temperature need not be high enough for the meat surface to begin to cook. The surface appearance of vacuum-thawed meat may be poor, however, and the capital cost of the equipment is high. Infrared thawing achieves a high rate of surface heat transfer using the mechanism of radiation to supply heating rather than convection. Care must be exercised, however, to ensure that the surface is not cooked before the center is thawed.

Dielectric, resistive, and microwave thawing are all methods of transferring electromagnetic radiation throughout a piece of meat, where it is transformed into heat energy. These methods differ largely in the frequency of electrical energy used to heat the meat. Resistive heating is applied at low frequencies (e.g., 50 Hz), dielectric heating at moderate frequencies (0.01–0.1 GHz), and microwave heating at high frequencies (0.1–2.5 GHz) [96]. Difficulties with maintaining even heating throughout the meat and the shielding required to protect workers in the presence of strong electromagnetic fields restrict these technologies to thawing relatively small quantities of meat at a time.

In domestic situations meat quantities are much smaller. Therefore, the requirements for maintaining similar heating rates are less stringent. Techniques that may be used include high-temperature air thawing (left overnight in the kitchen), immersion thawing (left in a bowl of water), or microwave thawing. All of these methods must be regarded as being uncontrolled when compared with commercial technologies.

## B. Microbiological Aspects

Thawing, like freezing, can be a very traumatic experience for microorganisms. As meat thaws, the viability of the microorganisms it carries, particularly bacteria, is threatened externally by high solute concentrations and internally by enlargement of intracellular ice crystals resulting from recrystallization [86]. Rapid thawing is more conducive to microbial survival than slow thawing [80]. After thawing, bacterial cells are either in a stationary phase or have suffered sublethal damage as a consequence of the stresses imposed during the freeze-thaw cycle. Consequently, when growth-permissive temperatures are restored, cell growth will not begin immediately. Growth can commence only after cell repair or lag phase resolution, respectively, has been achieved [77,97]. The duration of both of these processes is determined by a complex interaction between the organism, the degree of cell damage, the thawing temperature, and conditions prevailing at the meat surface. Although meat provides a rich medium conducive to the recovery of injured cells, an extended lag phase due to freeze damage can be expected during thawing before microbial growth resumes [97].

With respect to the microbiological quality of meat, the thawing process restores the potential for the proliferation of both pathogenic and spoilage microorganisms. In a hygienically aceptable thawing process, the meat surface time-temperature history prevents microbial proliferation from exceeding acceptable limits. Regulatory authorities generally advocate that commercial thawing be undertaken at low temperatures (<10°C) to ensure that the hygienic status of the meat is not compromised by the growth of mesophilic pathogens. However, significant growth of spoilage bacteria can occur if the surface temperature remains at such a temperature (≤10°C) for the time required to completely thaw large meat masses. The adverse hygienic consequences, particularly those associated with high-temperature thawing, show a direct relationship to the mass of meat being thawed [98,99]. Consequently, the potential health or quality hazard posed by specific thawing regimes requires individual assessment. Predictive microbiology would appear to be a potentially valuable tool for assessing and comparing the hygienic adequacy of thawing procedures [97]. The system developed by Lowry et al. [97] based on the potential for growth of *Escherichia coli* at various temperatures has been adapted to assess the hygienic adequacy of carcass-cooling procedures [100–101].

To effect rapid thawing of large meat masses (e.g., beef quarters or mutton carcasses), water thawing is preferred to air thawing, and the practice is widespread where bulk frozen meat must be thawed to permit further processing. However, the aesthetic and hygienic acceptability of simultaneous thawing of several unwrapped beef quarters or mutton carcasses in the same tank is open to question. Cross-contamination appears to be the matter of most concern to meat hygienists. Although the process itself and the effluent it produces are less than attractive, few if any microbiological problems have been reported with such low-temperature immersion-thawing systems. Commonsense regulations, such as the use of fresh potable water for each thaw cycle and control of thawing times and final water temperatures, appear to be effective to assure the hygienic adequacy of this process.

Thawed meat has long had a reputation for being more susceptible to microbial spoilage than fresh meat [102]. However, the reduction in the size of the surviving microbial population on meat after passing through a freeze-thaw cycle combined with any lag phase effects could be expected to extend the shelf life of thawed meat over that attained with similar unfrozen meat. In theory, this may hold true provided the meat is not further contaminated during the thawing process. Lowry and Gill [86] found no difference in bacterial growth rates between thawed and fresh meat. Notwithstanding the similarity of bacterial growth rate on fresh and thawed meat, if during the prefreezing treatment microbial growth has deplected glucose reserves, then after thawing spoilage onset will occur at lower cell numbers. Since the developing spoilage microflora commences utilization of amino acids earlier, due to glucose exhaustion [9,10]. Detection of microbial spoilage immediately following thawing also occurs. This phenomenon almost certainly indicates that the meat was either already spoiled before it was frozen or spoiled during inefficient freezing and thawing. Such events unfortunately are not an uncommon occurrence. While flaws in the freezing process undoubtedly are a major factor in quality deterioration, failure to conduct any meat-thawing process in a hygienically efficient manner brings with it the risk of further microbial deterioration.

## C. Drip Loss

Drip or exudate contains proteins, peptides, amino acids, lactic acid, purines, B vitamins, and various salts. The potential for exudate has already been set by the freezing parameters. The size of the ice crystals and their formation, whether predominantly intra- or extracellular, will dictate whether denaturation of proteins will take place due to increased solute concentrations during thawing. For example, thawing immediately after freezing consists of the thawing of small thin dendrites corresponding to relatively small ice crystals, where changes in solute concentration and changes in cell shape are minimized. Water-holding capacity in this situation would be expected to be greatest, whereas after prolonged storage resulting in recrystallization, thawing of large intracellular ice crystals dominates and distortion will affect membrane integrity and exacerbate solute concentration effects, which reduce the water-holding capacity. As muscle tissue consists of discrete physical compartments (e.g., myofibrillar or extracellular), some translocation and resorption during thawing can be expected. Rapid thawing, therefore, produces more drip than slow thawing, but this can to some extent be offset by postthawing storage. Clearly if exudate cannot physically be taken up again (i.e., if it forms on a flat tray), then this process is minimal.

## VII.  SPECIES DIFFERENCES

The size of the carcass or meat cuts are the dominant or first-order feature in terms of rate of freezing. Small animals such as rabbits will freeze rapidly, as all portions are a small distance from the surface. Assuming that cold shortening is avoided and aging is appropriate, animals of this size would freeze similarly to cuts of meat with the same dimensions. If rabbits were packaged in cartons, then they would be similar to a block of meat, except that the unavoidable spaces would encourage recrystallization effects even if the carcasses were wrapped separately or in a block surrounded with a plastic film.

With beef, lamb, and pork, the main differences are in size, and this affects postslaughter treatment. While lamb may be frozen as whole carcasses, beef and pork is usually boned out and placed in boxes. This means that lamb can be frozen much earlier than beef and pork and that initiation of rancidity is less likely to take place. For example, in New Zealand, electrically stimulated lamb can be aged above 6°C before being placed in a freezer which is controlled so that the deep leg temperature is less than –4°C in 12 hours. This produces meat with a high degree of tenderness and long storage life without deterioration. The storage duration under such treatment was found to be more than 2 years at –10°C, and this was enhanced with lower storage temperatures [103]. However, even one day of chilled storage prior to freezing reduced the storage life by 25% [103]. There is a trend for lamb to be cut into portions before packaging and freezing, in which case the multiplicity of surfaces may result in some deterioration. Beef and pork carcasses are generally chilled and then boned out 24 hours later, with the boneless meat usually packaged before freezing.

Hot boned beef and pork, however, may be in a freezer within 40 minutes of slaughter, consequently, deterioration from rancidity is probably very low and even lower than that for lamb frozen as whole carcasses. The possibility for evaporative weight losses is also greatly reduced.

Pork processing is variable in that in some instances it may be chilled rapidly and even case-hardened, and in other cases it is slowly chilled. This means that the storage characteristics are related to processing conditions rather than species per se. In pork processing, the skin is usually left intact, protecting the surface from desiccation and oxidative changes during the initial 24-hour chilling period. The skin may then be removed during the boning operation, which creates a different storage profile than when the skin is left intact.

## VIII. SUMMARY AND CONCLUSIONS

To produce high-quality frozen meat products that will receive widespread and continual acceptance, considerable care must be exercised starting with production of the animal on the farm through processing, packaging, freezing, storage, distribution, merchandising, and thawing to final preparation (cooking) in the consumer's home or a food service establishment. Consequently, much more is involved than just the freezing, frozen storage, and thawing operations, even though these operations can exert a substantial amount of influence on ultimate product quality, palatability, and consumer acceptance.

The quality of meat that has been frozen and thawed is not in general limited by its microbiology, although attention to this realm is still required. It is limited by meat processing, prefreezing, and temperature and time factors encountered prior to freezing. Deterioration in terms of oxidative flavor changes commences here. Once these aspects are under control, the freezing rate is set as a slow freeze in practical terms with relatively large ice crystal formation resulting in some tissue damage. Temperature fluctuations during storage cause recrystallization, and this affects the subsequent drip. Oxidation changes, if initiated, continue, but they are slower at low temperatures. Thawing results in drip, especially from cut surfaces, but there is uptake of the drip if postthaw holding takes place, although this generally is impossible on a flat tray. However, much of the drip would be lost during cooking anyway. The whole process, therefore, rather than any single act, affects the final product quality.

## REFERENCES

1.  J. R. Bendall, Post mortem changes in muscle, *The Structure and Function of Muscle*, Vol. 11 (G. H. Bourne, ed.), Academic Press, New York, 1973, p. 244.
2.  T. R. Dutson, J. W. Savell, and G. C. Smith, Electrical stimulation of antemortem stressed beef, *The Problem of Dark Cutting in Beef* (D. E. Hood and P. V. Tarrant, ed.), Martinus Nijhoff Publishers, The Hague, The Netherlands, 1981, p. 253.

3. O. A. Young, D. H. Reid, and G. H. Scales, Effect of breed and ultimate pH on the odour and flavour of sheep meat, *NZ J. Agric. Res.* 36:363 (1993).

4. R. W. Purchas, An assessment of the rate of pH differences in determining the relative tenderness of meat from bulls and steers, *Meat Sci.* 97:129 (1990).

5. R. W. Purchas and R. Aungsupakorn, Further investigations into the relationship between ultimate pH and tenderness for beef samples from bulls and steers, *Meat Sci.* 34:163 (1993).

6. L. E. Jeremiah, A. K. W. Tong, and L. L. Gibson, The usefulness of muscle colour and pH for segregating beef carcasses into tenderness groups, *Meat Sci.* 30:97 (1991).

7. C. E. Devine, A. E. Graafhuis, P. D. Muir, and B. B. Chrystall, The effect of growth rate and ultimate pH on meat quality of lambs, *Meat Sci.* 35:63 (1993).

8. A. Watanabe and C. E. Devine, Tenderness changes during aging of meat at various pH values, *Proceedings of the 40th International Congress of Meat Science & Technology*, The Hague, The Netherlands, S–IV B. 17.

9. K. G. Newton and C. O. Gill, Storage quality of dark, firm, dry meat, *Appl. Environ. Microbiol.* 36:375 (1978).

10. K. G. Newton and C. O. Gill, The microbiology of DFD fresh meats: A review, *Meat Sci.* 5:223 (1980).

11. C. Faustman and R. G. Cassens, The biochemical basis for discoloration in fresh meat: A review, *J. Muscle Foods* 1:217 (1990).

12. D. A. Ledward, Colour of raw and cooked meat, *The Chemistry of Muscle-Based Foods* (D. E. Johnston, M. K. Knight and D. A. Ledward, eds.), Royal Society of Chemistry, Cambridge, UK, 1992, p. 128.

13. P. V. Tarrant, An overview of production, slaughter and processing factors that affect pork quality—general review, *Pork Quality: Genetic and Metabolic Factors* (E. Puolanne, N. J. Demeyer, M. Burrsumen, and S. Ellis, eds.), CAB International, Wallingford Oxon, UK, 1993, pp. 4–6.

14. L. E. Jeremiah, Consumer response to pork loin chops with difference degrees of muscle quality in two western Canadian cities, *Can. J. Anim. Sci.* 75:425 (1994).

15. L. E. Jeremiah and R. Wilson, The effects of PSE/DFD conditions and frozen storage upon the processing yields of pork cuts, *Can Inst. Food Sci. Technol. J.* 20:25 (1987).

16. L. E. Jeremiah, Effects of inherent muscle quality difference upon the palatability and cooking properties of various fresh, cured and processed pork products, *J. Food Qual.* 9:279 (1986).

17. R. A. Merkel, Processing and organoleptic properties of normal and PSE porcine muscle, *Proceedings of the 2nd International Symposium on Condition and Meat Quality of Pigs*, Wageningen, The Netherlands, 1971, p. 261.

18. D. J. Topel, J. A. Miller, P. J. Berger, R. E. Rust, F. C. Parrish, Jr., and K. Ono, Palatability and visual appearance of dark, normal, and pale coloured porcine *M. longissimus*, *J. Food. Sci.* 41:628 (1976).

19. J. D. Kemp, R. E. Montgomery, and J. D. Fox, Chemical palatability and cooking characteristics of normal and low quality pork loins as affected by frozen storage, *J. Food Sci.* 41:1 (1976).

20. M. E. Bennett, V. D. Bramblett, E. D. Aberle, and R. B. Harrington, Muscle Quality,

cooking method and aging versus palatability of pork loin chops, *J. Food Sci.* 38:536 (1973).

21. R. G. Cassens, D. N. Marple, and G. Eikelenboom, Animal physiology and meat quality, *Adv. Food Res.* 21:71 (1975).

22. D. J. Searcy, D. L. Harrison and L. L. Anderson, Palatability and selected characteristics of three types of roasted pork muscle. *J. Food Sci* 34:486 (1969)

23. D. Deethardt and H. G. Tuma, Effect of cooking method on various qualities of pork loin, *J. Food Sci.* 36:626 (1971).

24. L. E. Jeremiah, A note on the infuence of inherent muscle quality on cooking losses and palatability attributes of pork loin chops, *Can. J. Anim. Sci.* 65:773 (1984).

25. L. E. Jeremiah, A. C. Murray, and L. L. Gibson, The effects of differences in inherent muscle quality and frozen storage on the flavour and texture profiles of pork loin roasts, *Meat Sci.* 27:305 (1990).

26. M. Koohmaraie, G. Whipple, D. H. Kretchmar, J. D. Crouse, and H. J. Mersmann, Postmortem proteolysis in longissimus muscle from beef, lamb and pork carcasses, *J. Anim. Sci.* 69:617 (1991).

27. R. H. Locker and C. J. Hagyard, A cold shortening effect in beef muscles. *J. Sci. Food & Agric.* 14:787 (1963).

28. C. L. Davey, H. Kuttel, H. and K. V. Gilbert, Shortening as a factor in meat aging, *J. Food Technol.* 2:53 (1967).

29. B. B. Chrystall and C. E. Devine, Electrical stimulation developments in New Zealand, *Advances in Meat Research*, Vol. 1 (A. M. Pearson and T. R. Dutson, ed.), AVI Publishing Westport, CT, 1986, p. 73.

30. A. A. Taylor and L. Martoccia, The effect of low voltage and high voltage electrical stimulation on pork quality, *Proceedings of the 38th International Congress of Meat Science & Technology*, Clermont-Ferrand, France, 1992, pp. 431–433.

31. C. E. Devine and A. E. Graafhuis, The basal tenderness of unaged lamb, *Meat Sci.* (in press.)

32. A. E. Graafhuis and C. E. Devine, personal communication.

33. C. L. Davey and K. V. Gilbert, Thaw contracture and the disappearance of adenosine triphosphate in frozen lamb, *J. Sci. Food Agric.* 27:1085 (1976).

34. A. E. Graafhuis, S. J. Lovatt, and C. E. Devine, A predictive model for lamb tenderness, *Proceedings of the 27th Meat Industry Research Conference*, Hamilton, NZ, 1992, p. 143.

35. S. J. Lovatt, Q. T. Pham, A. C. Cleland, and M. P. F. Loeffen, Prediction of product heat release as a function of time in food cooling: Part 1—Theoretical considerations, *J. Food Eng.* 18:13 (1992).

36. E. Dransfield, D. K. Wakefield, and I. D. Parkman, Modelling post-mortem tenderisation—I. Texture of electrically stimulated and non-stimulated beef, *Meat Sci.* 31:57 (1992).

37. K. McLeod, K. V. Gilbert, R. Wyborn, L. M. Wenham, C. L. Davey, and R. H. Locker, Hot cutting of lamb and mutton, *J. Food Technol.* 8:71 (1973).

38. G. R. Schmidt and K. V. Gilbert, The effect of muscle excision before the onset of rigor mortis on the palatability of beef, *J. Food Technol.* 5:331 (1979).

39. R. L. J. M. Van Laack, The quality of accelerated processed meats—an integrated approach, Ph.D. thesis, University of Utrecht, Utrecht, The Netherlands, 1989.

40. B. B. Chrystall, Hot processing in New Zealand, *Proceedings International Symposium Meat Science & Technology, Lincoln, Nebraska* (K. R. Franklin and H. R. Cross, eds.), National Live Stock and Meat Board, Chicago, 1986, p. 211.

41. P. S. Sleper, M. C. Hunt, D. H. Kropf, C. L. Kastner, and M. E. Dikeman, Electrical stimulation effects on myoglobin properties of bovine longissimus muscle, *J. Food Sci.* 48:479 (1983).

42. J. R. Claus, D. H. Kropf, M. C. Hunt, C. L. Kastner and M. E. Dikeman, Effects of beef carcass electrical stimulation and hot boning on muscle display colour of polyvinyl chloride packaged steaks, *J. Food Sci.* 49:1021 (1984).

43. D. A. Ledward, R. F. Dickinson, V. H. Powell, and W. R. Shorthose, The colour and colour stability of beef longissimus doris and semimenbranosus muscles after effective electrical stimulation, *Meat Sci.* 16:245 (1986).

44. C. Faustman and R. G. Cassens, Influence of aerobic metmyoglobin reducing capacity on colour stability of beef, *J. Food Sci.* 55:1278 (1990).

45. C. J. Hagyard, A. H. Keiller, T. L. Cummings, and B. B. Chrystall, Frozen storage conditions and rancid flavour development in lamb, *Meat Sci.* 35:305 (1993).

46. G. R. Longdill and Q. T. Pham, Weight losses of New Zealand lamb carcasses from slaughter to market, *Refrigeration of Perishable Products for Distant Markets*, International Institute of Refrigeration Commissions C2, D1, D2, D3, Hamilton, New Zealand, 1982, pp. 125–132.

47. Q. T. Pham and J. A. Willix, Model for food dessication in frozen storage, *J. Food Sci.* 49:1275 (1984).

48. R. B. Haines, Observations on the bacterial flora of some slaughter houses, *J. Hyg.* 33:165 (1933).

49. W. A. Empey and W. J. Scott, Investigations on chilled beef I. Microbial contamination acquired in the meat works, *C.S.I.R.O. Bull.* 126, Cannon Hill, Australia, 1939.

50. B. P. Eddy, The use and meaning of the term "psychrophilic," *J. Appl. Bacteriol.* 23:189 (1960).

51. C. O. Gill and K. G. Newton, The ecology of bacterial spoilage of fresh meat at chill temperatures, *Meat Sci.* 2:207 (1978).

52. A. A. Kraft, Psychrophic organisms, *Advances in Meat Research*, Vol. 2. A. M. Pearson and T. R. Dutson, eds.), AVI Publishing, Westport, CT, 1986, p. 196.

53. J. L. Ayres, The relationship of organisms of the genus *Pseudomonas* to the spoilage of meat poultry and eggs, *J. Appl. Bacteriol.* 23:471 (1960).

54. J. L. Ayres, Temperature relationships and some other characteristics of the microbial flora developing on refrigerated beef, *Food Res.* 25:1 (1960).

55. W. J. Scott and J. R. Vickery, Investigations on beef chilling II. Cooling and storage in the meat works, *Council Scientific and Industrial Research*, Bull. No. 129, Melbourne, Australia, 1939.

56. E. W. Hicks, W. J. Scott and J. F. Vickery, Influence of water in surface tissues on storage of meat, *Bul. IIR. Anexe* 1955-1, International Institute of Refrigeration, Paris, France, 1955, p. 72.

57. R. G. Bell, M. Al-Meshari, and M. Al-Mutairy, Storage life of chilled lamb carcasses either produced in Saudi Arabia or imported by air from New Zealand, *Proc. 34th Int. Congr. Meat Sci. Technol.*, Brisbane, Australia, 1988, pp. 482–484.

58. C. O. Gill and K. H. Tan, Effect of carbon dioxide on growth of *Pseudomonas fluorescens, Appl. Environ. Microbiol.* 38:237 (1979).

59. C. O. Gill and K. H. Tan, Effect of carbon dioxide on meat spoilage bacteria, *Appl. Environ. Microbiol.* 38:317 (1980).

60. K. G. Newton and C. O. Gill, The development of the anaerobic spoilage flora of meat stored at chill temperatures, *J. Appl. Bacteriol.* 44:91 (1978).

61. K. G. Newton and W. J. Rigg, The effect of film permeability on the storage life and microbiology of vacuum-packed meat, *J. Appl. Bacteriol.* 47:433 (1979).

62. R. G. Bell and A. M. Garout, The effective product life of vacuum-packaged beef imported into Saudi Arabia by sea, as assessed by chemical microbiological and organoleptic criteria, *Meat Sci.* 36:351 (1994).

63. W. Davenport, Chilling innovations, *Meat Proc.* 28(13):44 (1989).

64. C. O. Gill, Packaging meat for prolonged chilled storage: The Captech process. *Br. Food J.* 91:11 (1989).

65. C. O. Gill, Controlled atmosphere packaging of chilled meat, *Food Contr* (April):74 (1990).

66. C. O. Gill and N. Penney, The effect of the initial gas volume to meat weight ratio on the storage life of chilled beef packaged under carbon dioxide, *Meat Sci.* 22:53 (1988).

67. L. E. Jeremiah, L. L. Gibson, and G. C. Arganosa, The influence of controlled atmosphere and vacuum packaging upon chilled pork keeping quality, *Meat Sci.* 40:79 (1995).

68. L. E. Jeremiah, L. L. Gibson, and G. C. Arganosa, The influence of inherent muscle quality upon the storage life of chilled pork stored in $CO_2$ at $-1.5°C$, *Food Res. Int.* 28:51 (1995).

69. L. E. Jeremiah, N. Penney, and C. O. Gill, The effects of prolonged storage under vacuum or $CO_2$ on the flavour and texture profiles of chilled pork, *Food Res. Int.* 25:9 (1992).

70. L. E. Jeremiah, C. O. Gill, and N. Penney, The effects on storage life of oxygen contamination in nominally anoxic packagings, *J. Muscle Foods* 3:263 (1992).

71. L. E. Jeremiah, The effects of frozen storage on the retail acceptability of pork loin chops and shoulder roasts, *J. Food Qual.* 5:73 (1981).

72. L. E. Jeremiah, The effects of frozen storage and protective storage wrap on the retail case-life of pork loin chops, *J. Food Qual.* 5:331 (1982).

73. L. E. Jeremiah, Effects of frozen storage and protective wrap upon the cooking losses, palatability, and rancidity of fresh and cured pork cuts, *J. Food Sci.* 45:187 (1980).

74. V. J. Moore, Increase in retail display of frozen lamb chops with increased loin storage time before cutting into chops, *Meat Sci.* 28:251 (1990).

75. R. Gruji, L. Petrovi, B. Pikula, and L. Amilžic, Definition of the optimum freezing rate—1. Investigation of structure and ultrastructure of beef m. longissimus dorsi frozen at different freezing rates, *Meat Sci.* 33:301 (1993).

76. A. Calvello, Recent studies on meat freezing in developments in meat science-2. (R. Lawrie, ed.) Applied Science Publishers, London, UK 1981, p. 125.

77. B. Ray and M. L. Speck, Freezing injury in bacteria, *CRC Crit. Rev. Clin. Lab. Sci.* 4:161 (1973).

78. J. C. Olson and P. M. Nottingham, Temperature, *Microbial Ecology of Foods*, Vol. 1, ICMSF Academic Press, New York, 1980, p. 1.

79. W. Partmann, Effects of freezing and thawing on food quality, *Water Relations of Foods* (R. B. Duckworth, ed.), Academic Press, New York, 1975, p. 505.

80. R. A. MacLeod and P. H. Calcott, Cold shock and freezing damage to microbes, *Survival of Vegetative Bacteria* (T. R. G. Gray and J. R. Postgate, ed.), Cambridge University Press, Cambridge, UK, 1976, p. 81.

81. M. Ingram and B. M. Mackey, Inactivation by cold, *Inhibition and Inactivation of Negative Microbes* (F. A. Skinner and W. B. Hugo, eds.), Academic Press, London, UK, 1976, p. 111.

82. *IIR Recommendations for the Processing and Handling of Frozen Food*, International Institute of Refrigeration, Paris, France, 1986.

83. O. R. Fennema, W. D. Powrie, and E. H. Marth, *Low Temperature Preservation of Foods and Living Matter*, Marcel Dekker, New York, 1973, p. 294.

84. W. Rodel and K. Krispien, Der Einfluss von Kuhl und Gefriertemperaturen auf die Wasseraktivität (a-wert) von Fleisch und Fleischerzeugnissen, *Fleischwirtschaft* 47:1863 (1977).

85. L. Leistner, W. Rodel, and K. Krispien, Microbiology of meat and meat products in high- and intermediate moisture ranges, *Water Activity. Influences on Food Quality* (L. B. Rockland and G. F. Stewart, eds.), Academic Press, New York, 1981, p. 855.

86. P. D. Lowry and C. O. Gill, Microbiology of frozen meat and meat products, *Microbiology of Frozen Foods* (R. K. Robinson, ed.), Elsevier Applied Science Publishers, London, 1985, p. 109.

87. P. D. Lowry and C. O. Gill, Temperature and water activity minimal for growth of spoilage bacteria and of moulds from meat, *J. Appl Bacteriol.* 56:193 (1984).

88. P. D. Lowry and C. O. Gill, Mould growth on meat at frozen temperatures, *Int. J. Refrig.* 7:133 (1984).

89. B. E. Williams, Processes for improving the flavour and tenderizing meat by ante-mortem injection of Thamnidium and Aspergillus, U.S. Patent 3, 128191 (1964).

90. P. D. Lowry and C. O. Gill, The development of a yeast microflora on frozen lamb stored at −5°C, *J. Food Prot.* 47:309 (1984).

91. A. F. Egan and B. F. Shay, Annual Report of CSIRO Meat Research Laboratory, Canon Hill, Australia, 1977, p. 26.

92. K. O. Honikel, The water binding of meat, *Fleischwirt. Int.* (1) 14 (1988).

93. R. M. Love, The freezing of animal tissue, *Cryobiology* (H. T. Meryman, ed.), Academic Press, London, 1966, p. 137.

94. O. R. Fennema, W. D. Powrie, and E. H. Marth, Principles of food science Part II, *Physical Principles of Food Preservation* (M. Karel, O. R. Fennema, and D. B. Lund, eds.), Marcel Dekker, New York, 1975, p. 173.

95. R. H. Locker and G. J. Daines, The effect of repeated freeze-thaw cycles on tenderness and cooking loss in beef, *J. Sci. Food Agric.* 24:1273 (1973).

96. C. Bailey, Thawing methods for meat, *Proceedings of the 6th European Symposium on Food: Engineering and Food Quality*, International Union of Food Science and Technology, Cambridge, UK, 1975, p. 175.

97. P. D. Lowry, C. O. Gill, and Q. T. Pham, A quantitative method of determining the hygienic efficiency of meat thawing processes, *Food Aust.* 41:1080 (1989).

98. S. J. James, P. G. Creed, and T. A. Roberts, Air thawing of beef quarters, *J. Sci. Food Agric.* 28:1109 (1977).

99. P. G. Creed, C. Bailey, S. J. James, and C. D. Harding, Air thawing of lamb carcasses, *J. Food Technol.* 14:181 (1979).

100. C. O. Gill, J. C. L. Harrison, and D. M. Phillips, Use of a temperature function integration technique to assess the hygienic adequacy of a beef carcass cooling process, *Food Microbial.* 8:83 (1991).

101. R. J. Jones, The establishment of provisional quality assurances guidelines for assessing the hygienic adequacy of the lamb carcass cooling process, *NZ Vet. J.* 41:105 (1993).

102. G. Borgstrom, Microbiological problems of frozen food products, *Adv. Food Res.* 6:163 (1955).

103. R. J. Winger, Storage life and eating-related quality of New Zealand frozen lamb: A compendium of irrepressible longevity, *Thermal Processing and Quality of Foods.* (P. Zeuthen, J. C. Cheftel, eds.) Elsevier Applied Science Publishers, London, 1984, p. 541.

# 3
# Poultry and Poultry Products

Joseph G. Sebranek

*Iowa State University*
*Ames, Iowa*

## I.  INTRODUCTION

The poultry industry experienced tremendous growth in the latter part of the twentieth century, and this trend is expected to continue into the next century. Poultry, especially as a meat source, has the distinct advantages of rapid growth rates, short production times, and low unit costs. These advantages make poultry one of the most efficient sources of protein for human consumption of all farm animals [1]. An additional worldwide advantage is that very few stigmas, such as social or religious concerns, exist against poultry consumption. Chicken may well be the most universally accepted and consumed meat in the world. The current movement toward more international trade means that poultry and poultry products may become increasingly more widely distributed.

In the United States, production and consumption of poultry have increased more rapidly than of other meat sources since the 1960s. Consequently, poultry meat has become one of the meats most commonly chosen by consumers and is generally preferred by health-conscious consumers [2]. In 1993, U.S. production of broilers reached 6.65 billion birds, worth almost $10 billion and production of turkeys totaled almost 300 million birds with a value of $2.4 billion. [2]. Both of these represent new records in production volume.

The production, processing, and marketing of such large volumes of a highly perishable food product make temperature control critical for this industry. Freezing and frozen storage constitutes the most effective long-term means of maintaining high-quality, safe products for consumers. Most turkeys are presently sold frozen. However, the majority of broiler chickens are sold unfrozen in the chilled form.

A process of deep-chilling (chilling to about –2°C) to improve shelf life rather than the more traditional ice chills is used for broilers. Freezing equipment is used to achieve this temperature without freezing the tissue. Holding temperatures are also at –2 to –4°C. This process improves shelf life without some of the unique problems associated with frozen broilers. The expected shelf life of deep-chilled broilers is 3–4 weeks, or a possible 6–8 weeks with specialized packaging systems (3). Deep chilling is preferred over freezing for whole broilers because broilers are likely to show a dark discoloration around the bone when frozen. However, this problem is less likely in older birds, such as stewing chickens. With the increased interest in further processed poultry, sold as boneless cuts or slices, more broiler meat is being processed in the frozen form to take advantage of the long-term stability and quality retention achieved by good freezing methods [4]. In addition to boneless, packaged cuts, the poultry industry is providing more and more cooked, microwave-ready, and ready-to-eat products. Most of these products are marketed in frozen form to maintain flavor stability during storage. This is especially true of products used by hotel, restaurant, institutional (HRI) markets.

Other forces at work in the poultry industry include the movement to more international marketing and exports [5]. Consequently, increased transportation and shipping time necessitates extension of product shelf life, and freezing clearly represents the most effective means of achieving long-term shelf stability.

As a result of these current and changing trends in the poultry industry, it appears that frozen products will represent an increasing proportion of the poultry produced. Therefore, it is important to understand how freezing affects poultry products so that the quality of frozen poultry can be maximized.

## II. PREPARATION AND PACKAGING

## A. Definition of Quality

Product quality can be defined using many factors, including appearance, yields, eating characteristics, and microbial characteristics, but ultimately the final use must provide a pleasurable experience for the consumer. The objective of the freezing process is to preserve product characteristics at a desirable level for as long as possible. Therefore, it is important to realize that successful freezing will only retain the inherent quality present initially and will not improve quality characteristics.

Because freezing is capable of maintaining quality but not improving it, the quality level prior to freezing is a major consideration. In fact, it has been suggested the inferior image some consumers have of frozen poultry is due to past experiences with products that were of poor quality or even near spoilage prior to freezing [6]. The use of high-quality initial materials is essential to high-quality frozen products.

## B.  Standards and Grades

Standards and grades constitute an indicator of product quality levels. However, eating quality is not necessarily predicted from them. The U.S. Department of Agriculture (USDA) inspects each bird for wholesomeness and fitness for consumption and also grades ready-to-cook carcasses as A, B, or C quality. Determinants of these grade levels are conformation, fleshing, fat cover, pinfeathers, and defects such as exposed flesh, discoloration, or broken bones. Freezing effects are also considered, since a "freezing defects" category is used, with darkening, pock marks, or freezer burn resulting in lower grades. Poultry is also categorized by market class depending on species, age, and sex. Broilers, for example, are young chickens (7–12 weeks) of either sex, while stewing chickens are hens aged over 10 months. Broilers are defined by USDA as 7–12 weeks of age, but many are currently marketed at about 6 weeks. Turkeys of either sex are classified as fryer-roasters if younger than 16 weeks but are categorized as hens or toms if they exceed 4 months of age. Gender is not a factor in eating quality at young ages, but becomes important at older ages [1]. Likewise, age is well recognized as a factor for eating quality and cooking requirements. Therefore, age categorization is quite important. While tenderness may change somewhat with age, it has been suggested that flavor intensifies in mature birds [6].

Other USDA grading standards apply to frozen poultry, since products intended for government and institutional markets (military, etc.) are required to be frozen to –18°C within 72 hours of chilling and packaging [3]. Raw stuffed poultry products, if frozen, must reach –18°C within 24 hours [7].

## C.  Production and Processing Factors

Other considerations prior to freezing may include genetics, dietary factors, slaughter and processing effects, and initial chilling operations. Poultry genetics, particularly for broiler chickens, do not contribute very much variability in quality. Not only have different genetic strains been evaluated and shown to be similar in eating quality [6], but the genetic base of the broiler industry is not very broad. Only a few genetic types have been selected by the primary breeders who supply the broiler industry [8]. One area where genetic strains can have an impact is on total meat yield from carcasses and achieving a high yield of breast meat. The latter is particularly true for U.S. markets, where breast meat is strongly preferred. Another area where genetics may prove to be important is for cooking losses, where recent reports indicate differences between different broiler strains [9].

Diets of live birds generally do not influence meat quality, with the exception of the well-recognized effect of highly unsaturated oils, such as fish oils. This effect is similar to that observed in other nonruminant species, where large amounts of highly unsaturated fatty acids in the diet result in more unsaturated carcass fat and frequent off-flavors and softness problems. A possible positive contribution of dietary factors may come from vitamin E (tocopherol), which has been shown to be effective in frozen and frozen/thawed beef for improving color stability and reducing oxidative changes [10]. Similar effects of tocopherol have been observed in poultry [11] for suppressing lipid oxidation. However, the use of high levels of tocopherol has not been investigated as extensively for poultry as for other species.

One interesting aspect of the potential feeding of tocopherols to poultry is the observation that tocopherol deposition is significantly greater for chickens than turkeys [11]. Since fatty acid profiles are similar in chickens and turkeys, the greater tocopherol content associated with membrane phospholipids would explain why chicken fat is less likely to oxidize during frozen storage than is turkey fat. However, feeding tocopherols to turkeys is still effective for reducing lipid oxidation in at least some cases [12,13], even if less effective than with chickens.

Slaughter, processing, and handling can affect carcass quality by inducing bruises, tears, or even broken bones. Most of these defects will be obvious and result in downgrading. Incomplete bleeding may result in reddening of some areas of the carcass and has some potential for producing undesirable flavors [1]. Struggling during the bleeding phase can also result in broken wings. Struggling is reduced by electrical stunning, but if stunning is not carefully controlled, rib cage fractures may occur.

Following slaughter and evisceration, rapid and well-controlled chilling of poultry carcasses is essential to maintaining microbiological quality and tenderness. It is undesirable to freeze poultry meat immediately following slaughter, since meat tenderness is likely to be decreased. Electrical stimulation of carcasses (similar to that done for beef) can be used to speed up postmortem tenderization [3], if deemed necessary. Poultry must be chilled to a temperature of 4°C (40°F) or less (required by the USDA) within a time limit depending on carcass weight. A large majority of poultry carcasses are chilled in ice water slushes because of relatively fast chilling and because some moisture is absorbed by the carcasses, thereby increasing yields. Maximum tolerances based on carcass weight for absorbed moisture have been established by the USDA and range from 4.3% for heavy turkeys to 8% for lightweight chickens. The absorbed water is significant to the freezing process because it can influence freezing rate and the extent of ice crystal formation.

A number of mechanical chilling systems employing ice slushes combined with agitation, tumbling, or oscillating movement are used to accelerate the chilling rate [6]. Chill water including 20 ppm chlorine may help to further reduce surface bacteria [14]. However, two effects of accelerated chill systems can greatly influence frozen poultry. First, considerably more moisture absorption is likely, mostly in or under the skin. This excess moisture must be drained away before packaging and freezing to prevent excessive drip or separated water from occurring during thawing. The second and perhaps more critical consideration for freezing is that adequate time before freezing must be allowed for the meat to achieve optimum tenderness—6–8 hours for chickens and 12–24 hours for turkeys (Table 1) [3,6]. The muscle most susceptible to toughness problems is the breast, which is in greatest demand by U.S. consumers and, therefore, is the most valuable.

Product preparation prior to freezing may include cutting, deboning, slicing, and other operations to provide greater convenience. Preparation of products by cooking prior to freezing is becoming increasingly more popular as a greater variety of poultry products are offered to consumers. These include breaded and fried portions, cured and smoked products, and items in marinades or broths. Poultry products according to the USDA Food Safety Inspection Service must be either raw or fully cooked prior to freezing. Fully cooked means reaching an internal temperature of 71°C (160°F) for uncured products or 68°C (155°F) for cured products. Breaded products such as patties may be "fried" to set the breading without further cooking of the product [3].

## D. Packaging for Freezing

Following product preparation, packaging requirements become a consideration. The package is one of the most important factors in maintaining quality during frozen storage and is especially important for cooked products. The package chosen performs two basic functions. First, it is the marketing vehicle by which the product is delivered to customers, which means that the package must effectively communicate its contents in an attractive way. At the retail level, packaging must enable the product to compete successfully against other consumer choices. Second, the package must provide protection for its contents, including protection from external contamination during shipping and handling, but even more importantly protection from chemical and physical changes in the product during storage. Most important are protection from exposure to oxygen and from loss of moisture. As a result, packages must provide a good barrier to oxygen to prevent off-flavor development and to moisture to avoid dehydration or freezer burn.

**Table 1** Time Required for Tenderization of Poultry Prior to Freezing

| Kind of poultry | Time for maximum tenderization (h) |
| --- | --- |
| Chicken broilers | 6 |
| Chicken, all others | 8 |
| Turkey, young and mature hens or toms | 12 |
| Turkey, fryers | 24 |

It has been suggested that the packaging system is of paramount importance to the quality of frozen products, and of even greater importance than the freezing treatment itself [15], because the stability of the product during storage is dependent on the protection provided by the package. Use of PPP (product, process, and packaging) concepts has been suggested to be the most important overall consideration for frozen product quality [15]. A good package may be equivalent to as much as 10–12°C (20°F) lower frozen storage temperatures [6,15].

To provide the greatest protection, a package must be well evacuated of air (oxygen) using a vacuum or gas flush system and provide an adequate barrier to both oxygen and moisture. For whole birds such as turkeys, a heat-shrink film bag that fits very tightly to the carcass surface is the traditional packaging approach. This package provides attractive, high-barrier protection. With the increased development of further processed poultry, specialized packaging films and applications are being used, and many are presently available. A recent publication [16] lists more than 20 different films with a several hundred–fold range for permeability of oxygen and water vapor for chilled and frozen foods. Various packaging materials can also be put together in laminates of different combinations to provide heat sealing, clarity, strength, rigidity, and other properties in addition to specific permeability requirements. An almost unlimited number of combinations are possible for specific product uses. Thus, it becomes important to work closely with packaging suppliers to determine the most suitable application. Table 2 lists some examples of packaging materials and permeability differences. Aluminum foil provides a superior barrier in most cases as long as no pinholes exist, and it is used in cases where an ultimate barrier is needed. Aluminum foil is observed frequently as part of the package for cooked products where protection from oxygen is critical to preventing flavor loss. However, it is important to realize that the permeability of most flexible films will change if temperature or relative humidity changes. Consequently, permeability may vary somewhat during different applications if environmental conditions vary.

Vacuum packaging of both uncooked and cooked poultry products is the most common approach to packaging needs, although modified atmosphere packages (MAP) using carbon dioxide or nitrogen are available. Most MAP applications have been used for unfrozen products, where the atmosphere (especially $CO_2$) plays a role in microbiological inhibition. For frozen products, particularly cooked products, a complete vacuum coupled with an adequate barrier package that adheres well to the product surface is the most common choice. Preventing headspace between the product surface and the package aids in preventing ice recrystallization and frost formation inside the package during frozen storage. It has been suggested that packaging materials with an oxygen permeability of less than 15 $cm^3$ $m^2$ be used for vacuum packages [16]. It is imperative that cooked products be packaged with a high vacuum and an adequate barrier film to prevent oxidative flavor changes.

**Table 2** Oxygen and Water Vapor Transmission Rates of Selected
Packaging Materials

| Packaging film (25 μm) | Oxygen transmission rate $(cm^3\ m^{-2}\ day^{-1}\ atm^{-1})$ | Water vapor transmission rate (38°C:90% RH) |
|---|---|---|
| Aluminum foil | neg.[a] | neg.[a] |
| Ethylene vinyl alcohol | 0.2 – 1.6[b] | 24 – 120 |
| Polyvinylidene chloride | 0.8 – 9.2 | 0.3 – 3.2 |
| Modified nylon | 2.4[b] | 25 |
| Polyethylene terephlalate | 50 – 100 | 20 – 30 |
| Modified polyethylene terephlalate | 100 | 60 |
| Unplasticized polyvinyl chloride | 120 – 160 | 22 – 35 |
| Plasticized polyvinyl chloride | 2,000 – 10,000[c] | 200 |
| Oriented polypropylene | 2,000 – 25,000 | 7 |
| High-density polyethylene | 2,100 | 6 – 8 |
| Polystyrene | 2,500 – 5,000 | 110 – 160 |
| Oriented polystyrene | 2,500 – 5,000 | 170 |
| Polypropylene | 3,000 – 3,700 | 10 – 12 |
| Polycarbonate | 4,300 | 180 |
| Low-density polyethylene | 7,100 | 16 – 24 |
| Ethylene vinyl acetate | 12,000 | 110 – 160 |

[a]Dependent on pinholes.
[b]Dependent on moisture.
[c]Dependent on moisture and level of plasticizer.
*Source*: Ref. 16.

## III. COLOR, APPEARANCE, AND CONSUMER ACCEPTANCE

### A. Surface Appearance

Poultry, like red meats, requires a freezing process that is rapid enough to minimize surface dehydration during freezing, avoid the chemical changes that can occur at temperatures just below the freezing point, and decrease drip losses during thawing. However, poultry is unique in that a light surface color for carcasses is considered important and is best achieved with a rapid surface-freezing treatment. This is particularly true for young birds containing little subcutaneous fat. Rapid surface freezing generates a smooth chalky white surface, which is considered ideal for poultry, by supercooling the product and forcing nucleation of a high number of small ice crystals. These crystals stay small because there is little water migration to already formed crystals during a fast process. Numerous small ice crystals cause the surface to reflect light and appear white in color. Rapid processes resulting in smaller ice crystals may also have the advantage of less thaw drip without affecting consumer eating quality significantly. Cryogenic freezing (–196°C) with liquid nitrogen has been shown to improve color and reduce drip 3% in frozen chicken thighs [17]. Immersion freezing at –29°C has been found to produce surface color similar to that achieved by air blast freezing at –73°C [18]. In addition, the desirable surface appearance in that study was well maintained during 20 weeks of storage at –29°C, but was maintained only 2 weeks at –7°C. When surface temperature exceeds –7°C, the light surface appearance is lost [3]. Even if cryogens or immersion systems are not available, increasing air blast freezing rates by decreasing temperatures and increasing air velocity will improve the surface appearance of frozen turkey [3,19]. Despite differences in appearance, many reports indicate that the effects of typical freezing processes on the eating quality characteristics of poultry and poultry products are minor [3,7,15].

## B.  Bone Darkening

A second aspect unique to poultry freezing is the bone-darkening effect observed in young birds (dark discoloration around the bone, which becomes obvious when cooked). Bone darkening is caused by the leaching of hemoglobin from bone marrow to adjacent muscle as a result of the freeze/thaw treatment. Leaching only occurs in carcasses from relatively young birds because the bones are not completely calcified and are more porous than in mature birds. While eating qualities do not change, the appearance constitutes a negative factor in consumer acceptance. Many attempts have been made to eliminate bone darkening in young frozen birds. While some reduction can be achieved, only removing the bone marrow or cooking prior to freezing will eliminate this defect [3,20]. As a result, young birds such as broilers destined to be sold raw are not commonly frozen in carcass form but are deep chilled (–2°C) instead.

## C.  Freezing Rates

Freezing processes used to achieve the desired carcass surface appearance can take several forms. The actual rate is dependent on the temperature, the medium, and the velocity of the medium (if appropriate). Air blast is most common, and temperatures of –29°C or less combined with air velocities of at least 600 fpm have been recommended to achieve the most desirable carcass appearance [3,6,19]. An alternative approach is to crust freeze the outer part of the carcass rapidly using liquid brine immersion, spray systems, or cryogenics like liquid nitrogen or carbon dioxide, and then to move the partially frozen bird to air blast or cold storage for the remainder of the process. Sprays and immersion systems require 20–40 minutes, depending on bird size, at –20 to –29°C to adequately crust freeze [18]. The time required for the outer edge of the carcass to reach –7°C (20°F) determines the color; the remainder of the freezing process can be done more slowly without altering surface color. For acceptable surface lightness, the time to reach –7°C should not exceed about 3.5 hours [3]. Cryogens can also be used for crust freezing to realize greatest efficiency.

The freezing rate required, regardless of the medium used, determines the effectiveness of the freezing process. Many different attempts have been made to express freezing rates, including time to reach a prescribed temperature at the product center and time required for the product center to pass through a freezing temperature zone, such as from +5°C to –5°C. These apply to specific products and do not consider product size or shape, packaging, or other variables. One of the best ways to express freezing rates is by the rate of movement of the ice front. As water crystallizes, the point of crystallization migrates from the cold exterior to the warmer interior of a product. Faster freezing methods achieve a more rapid migration of the ice front. In muscle foods, a freezing front migration rate of 2–5 cm/h is recommended to achieve "fast" freezing effects. Such effects are achieved with freezer temperatures of –40 to –60°C with air movement and relatively small product size [21,22]. A minimum of 1.0 cm/h on the surface for poultry has been suggested to be necessary to achieve a light surface color [15]. As previously noted, temperatures of –29°C in air blast (600 fpm or more), which decreases the external surface below –7°C in less than 3.5 hours, has been reported to be suitable to produce acceptable surface color.

Freezing rates for other poultry products such as cooked and microwave-ready products are not as critical to color and appearance. Such rates need only be sufficient to maintain quality levels, because surface color is not a major concern as it is with carcasses. Freezing characteristics, including the freezing temperature required for cooked and processed products, are often variable, because these products often vary considerably in water content, salt content, size, etc. This may also be true for some raw products that have been injected with flavorings or other adjuncts.

## D. Frozen Storage

Frozen storage of all products is generally recommended to be at about –18°C (0°F) or lower with a minimum of temperature fluctuations. This temperature is adequate to minimize appearance changes and has been reported to give a practical storage life (time until loss of consumer acceptability) of 23 months for broilers [23]. Some typical expectations for storage life of various poultry products at –18°C, assuming adequate packaging is used, are shown in Table 3. It has been suggested shelf life is likely to change by a factor of 3.5 for each 10°C change (increase or decrease) in storage temperature [3].

Temperature fluctuations (such as defrost cycles during storage) encourage ice recrystallization and may change product appearance over time. A recurring change of 10°C or more is likely to result in significant recrystallization [24] and should be avoided. Fluctuating storage temperatures have been observed to almost double the frost formation associated with frozen turkeys, though organoleptic

**Table 3**   Expected Storage Life of Frozen Poultry Products at –18°C

| Product | Storage time (months) |
| --- | :---: |
| Consumer pack turkeys | 12 – 15 |
| Canner pack turkeys (if individually packaged) | 8 – 12 |
| Canner pack fowl (if individually packaged) | 12 |
| Consumer pack chicken roasters | 12 – 15 |
| Consumer pack chicken broilers, whole | 16 |
| Chicken parts (poly wrapped) | 12 |
| Mechanically deboned chicken neck and back meat | ≥3 |
| Mechanically deboned turkey frame meat | ≥2 |
| Turkey parts, consumer pack | 12 to 15 |
| Ground turkey | 4 – 6 |
| Turkey roasts, uncooked | ≥12 |
| Breaded and fried poultry | ≤9 |
| Ducks and geese, consumer packed | 12 – 15 |

*Source*: Ref. 3.

differences were not great [25,26]. Fluctuating temperature during storage has been found to eliminate differences in drip losses associated with different freezing rates [27]. Packaging that eliminates space between the product surface and the package helps to minimize frost from ice recrystallization.

## E.   Thawing Effects

Thawing of frozen poultry generally does not influence appearance or color providing temperature abuse is avoided. Drip loss is an obvious concern during thawing but is affected more by processing (e.g., cut-up parts compared to intact carcasses) and freezing rate than thawing treatments. Drip loss has been reported to be less when rapid freezing processes are used when compared to slow freezing rates, but a difference of 1–2% drip is not likely to be of significance to consumers. Comparison of thawing treatments for turkeys, including cooking from the frozen state, has shown no difference in cooked yields [28]. Frozen and thawed cut-up, deboned, or sliced poultry is more susceptible to higher thaw drip losses because of the greater surface area. Broken or cut muscle cell walls (such as cross-sectional slices or ground meat) permit higher increased drip losses during freezing and thawing.

## IV.   PALATABILITY ATTRIBUTES

## A.   Effects of Freezing on Palatability

Of all of the considerations regarding freezing, frozen storage, and thawing of poultry, palatability changes are the most critical because of the need to ultimately provide consumers with a pleasurable eating experience. The impact of freezing and freezing rates on palatability attributes appears to be an area of disagreement among researchers. Jul [15], for example, suggests that product properties, processing treatments, and packaging (PPP concept) are at least as important as freezing time/temperature, if not more so. Palatability differences in poultry resulting from freezing treatments are often reported as small or nonexistent. For example, Marion and Stadelman [29] compared several freezing rates and found that, while surface color differed, cook yield or palatability differences were not found. Similarly, no difference in tenderness was found for chicken breasts frozen at temperatures from −18 to −68°C [30]. The flavor of chicken fried before freezing was found to be similar whether freezing was done in a household freezer or cryogenically using liquid nitrogen [31]. However, thiobarbituric acid (TBA) values were lowest with the most rapid (liquid nitrogen) method. Because TBA values are indicative of rancidity, differences imply that palatability may have changed. Other reports, however, have demonstrated no effect of freezing rate on TBA [32]. TBA values are very likely to change in response to storage conditions. The most consistently reported change in poultry frozen at various rates is in thaw drip losses, with fast freezing rates resulting in less drip [33,34].

Tenderness of fried broiler thighs and breasts improved when they were frozen with liquid nitrogen compared to when slower methods were used [35], and protein solubility was somewhat higher in slow-frozen chicken muscle [36]. Other reports indicate no difference between freezing rates in tenderness or shear values of turkeys [37]. Thus, differences in tenderness characteristics produced by freezing rates are likely to be small and dependent on the product size and other variables. Many of these changes (texture, tenderness), when observed, are undoubtedly related to the ultimate physical and structural effects of freezing on muscle tissue.

## B.  Protein Changes

Because greater amounts of poultry meat are being used in further processed products, understanding the functional properties of the proteins is becoming an important consideration. Protein functionality determines product texture and moisture retention. Maximizing the functional properties of frozen meat is important in producing processed products, because frozen meat is generally considered to be somewhat less functional. Some protein denaturation and solubility changes are known to occur as a result of freezing, but the practical significance of these changes is not clear. Comparison of thermal gelation properties of poultry dark meat frozen at different rates showed little effect of freezing or freezing rates on rheology or gel strength [38]. Freezing, regardless of rate, resulted in somewhat greater water-holding capacity, which might be caused by increasing charged sites on the meat proteins. However, freezing does not appear to be a major detriment to processing functionality of poultry meat.

## C.  Mechanically Deboned Meat

A poultry meat ingredient that has become widely used in various processed products is mechanically deboned meat (MDM). MDM is the recovered meat that adheres to bones following preparation of boneless products; mechanical deboning produces a very finely comminuted product that is a valuable ingredient which can deteriorate very quickly [39]. Deterioration may be due to microbial growth or rapid oxidative change resulting in loss of palatability [40,41]. Rapid chilling and freezing are critical to the quality of this poultry meat ingredient. Even then, storage stability is limited (Table 3) because of rapid oxidative changes. Attempts to achieve rapid chilling and/or freezing with $CO_2$, however, have been reported to result in increases in TBA numbers and potential palatability losses [42,43]. Freezing with $CO_2$ has been reported not to be as detrimental as chilling, but differences in TBA values still occurred [44]. However, these changes were not observed with other chill/freeze methods.

## D.  Changes During Frozen Storage

The effects of frozen storage conditions on palatability attributes of poultry have been thoroughly studied under a variety of conditions. One of the major detrimental changes occurring during normal frozen storage is lipid oxidation, which is closely related to storage time, temperature, and packaging. Research on frozen poultry generally indicates that a storage temperature of about $-18°C$ or lower will minimize deterioration if products are well packaged and temperature fluctuation is minimal [45–47]. Over a range of $-10$ to $-30°C$, a reduction of $10°$ in storage temperature will apparently double or triple the number of days that broilers retain stability and/or acceptability [3,48]. Chicken dark meat is more susceptible to lipid oxidation than is white meat during frozen storage [49].

## E.  Cooked Products

Cooked products are likely to exhibit greater increases in lipid oxidation than raw products during frozen storage. Vacuum packaging combined with temperatures of $-18°C$ or less are critical. Antioxidants are very effective in cooked chicken during frozen storage [50,51]. Cooked products probably show greater lipid oxidation, because cooking alone produces oxidative change and higher TBA values [52], making the product more susceptible to further oxidative change during frozen storage. Consequently, the storage life of cooked frozen products will be shorter than · that of uncooked products, with all other factors being equal (Table 3).

## F.  Thawing and Drip Losses

Drip losses may increase during frozen storage [53]. In most cases, changes in drip are slight but may become significant if storage temperatures are significantly above $-18°C$. Increased drip loss and greater cooking loss may produce products that seem less juicy.

However, the effects of thawing on palatability have been found to be relatively small. While thawing rates can be important for some food products, poultry and most meat products are not greatly influenced [21] by different thawing treatments. Relatively slow thawing results in somewhat less drip and improved poultry meat texture [15,54]. Comparison of sensory characteristics of broiler breast meat thawed by a wide range of thawing rates, however, showed very little significant difference in taste, texture, or juiciness [15]. A study of repeated freezing and thawing treatments for broilers did not drastically affect palatability characteristics [55].

## V.  NUTRITIONAL VALUE

### A.  Protein

Poultry meat products are considered to be valuable nutritive sources of high-quality protein and of some B vitamins. Retention of nutritional components in foods is a concern when any type of preservation method is used, but freezing is probably the least destructive. Freezing may induce some protein denaturation, as evidenced by research on freezing rates, structural changes, and drip losses [15]. However, changes in digestibility and nutritive value of proteins, even in denatured form, are very slight and appear to be of no practical significance. The drip losses that occur during thawing and cooking can include water-soluble protein, vitamins, and minerals, but the amount lost relative to that remaining is small.

### B.  B Vitamins

B vitamins are among the most labile nutritive components, and evidence exists that thawing and cooking may lead to significant losses of vitamin $B_6$ [56]. However, an earlier study with frozen turkey, which demonstrated rancidity development and increased drip loss as a result of frozen storage, failed to show changes in B vitamin levels [57]. Research with meat from other species on B vitamin retention during and after freezing is more extensive than for poultry. Results are variable, but the vitamins are retained in most cases. Apparently B vitamin losses may be significant from frozen poultry products, but the losses are produced largely by the subsequent thawing and cooking treatments rather than by the freezing per se.

## VI.  INTRINSIC CHEMICAL REACTIONS

Chemical reactions in frozen foods are usually suppressed due to temperature effects. However, the concentration of solutes that occurs as water is converted to ice places reactants in close proximity and increases reaction rates. Consequently, chemical reactions in frozen products are greatest at temperatures just below the freezing point.

## A. Lipid Oxidation

The chemical reactions of greatest concern with frozen poultry, those involving lipid oxidation, constitute a major determinant of frozen product shelf life [6]. Because poultry lipids are relatively unsaturated, susceptibility to oxidation is high. Freezing results in concentration of solutes, which catalyze the initiation of oxidative reactions. The greater the concentration of these catalysts, the greater the acceleration of changes. Freezing also disrupts and dehydrates cell membranes, exposing membrane phospholipids to oxidation. Membrane phospholipids are highly unsaturated and have been demonstrated to be the initiation point of oxidation in muscle tissue [49,58]. The initiation step involves abstraction of a hydrogen ion from the susceptible (unsaturated) lipid, thereby producing a free radical. Initiation is the rate-limiting step and is the point where a catalyst, such as heme or nonheme iron, becomes involved. Free radicals formed react quickly with oxygen, forming hydroperoxy radicals, which in turn abstract hydrogen ions from lipid molecules forming a hydroperoxide and another free radical. The free radical reacts once more with oxygen, while the hydroperoxide decomposes to form flavor and aroma compounds characteristic of rancid foods. As a result of this sequence, the chemical oxidation, which probably begins with membrane phospholipids, can quickly spread to triglycerides and other lipids.

The role of membrane lipids and catalysts becomes clearly obvious when the use of antioxidants is considered to decrease oxidation. Phenolic antioxidants such as BHA and BHT are of little value in intact muscle cuts because distribution is difficult. Consequently, they are more effective if mixed or chopped into ground products. Antioxidants, such as the tocopherols, are of limited effectiveness even when blended with meat but are much more effective if fed as nutritional supplements [59]. Because tocopherols are deposited in membrane locations as a result of dietary intake, they are much more effective in preventing the initiation step with phospholipids. It has been demonstrated that chickens deposit tocopherols more efficiently and completely than turkeys, and this appears to be the reason that chicken is less likely than turkey to develop lipid changes and flavor problems during frozen storage [60].

The effectiveness of compounds that provide antioxidant contributions by reacting with catalysts demonstrates another means of controlling oxidation reactions. Phosphates, as pyrophosphates, tripolyphosphates, or hexametaphosphates, are effective metal chelators and prevent iron and other catalysts from acting as initiators. Phosphates work well for all meat products, including battered and breaded poultry [61]. The use of nitrite in cured products provides strong antioxidant properties by complexing heme iron and by providing nitric oxide as a free radical scavenger. Therefore, cured poultry products are relatively stable to lipid oxidation.

However, frozen poultry products should be well protected from oxygen to prevent hydroperoxy radical formation. Cooked products are often immersed in sauce or gravy to decrease oxygen contact as well as being packaged using vacuum or gas flush–high-barrier packaging. This approach helps to minimize off-flavor compounds produced by peroxide decomposition, even if some initiation reactions occur due to catalysts.

Lipid oxidation may be of some significance to nutrition and product safety, as well as to flavor changes. Lipid oxidation can contribute to destruction of fat soluble vitamins A and E but specific information regarding the influence of freezing is not available for poultry. In addition, several of the end products of lipid oxidation, including peroxides, oxidized cholesterol and malonaldehyde have been shown to be potential human health risks, but the practical significance of the levels found in foods is not clear (58).

## B.  Protein Denaturation

Chemical reactions during freezing and frozen storage may also contribute to some protein denaturation and decreased solubility. Decreases in sulfhydryl groups and ATP-ase activity occur during frozen storage and are indicative of protein changes [6,62]. Peptides and amino acids are also increased in the drip fluid, as are nucleic acids, indicating protein changes and structural cellular damage, respectively [63]. Moreover, cryoprotectants, such as sorbitol and starch or sucrose, improve the functionality of frozen chicken myofibrillar proteins [64], probably by reducing protein denaturation. However, most of these changes are relatively small, and the chemical changes induced in proteins by freezing treatments are not a major practical concern unless thaw drip becomes excessive.

## C.  Enzyme Activity

Biochemical reactions involving muscle enzymes may occur at very slow rates in frozen poultry, depending upon storage temperature. For example, glycolysis may continue if not completed before freezing and may reach completion if enough time is available. Phosphatase activity is not completely inhibited by frozen storage. Therefore, it has been suggested that a progressive decrease in acid-soluble nucleotides, which are flavor enhancers, may be produced [65]. Muscle ATPase activity may continue during frozen storage [66], and some proteolytic enzyme activity has been suggested [63]. Myofibrillar fragmentation has been demonstrated to increase slowly during the frozen storage of hen carcasses, implying proteolytic activity by the specific enzyme systems that tenderize postmortem muscle [67]. Biochemical activity can continue at temperatures as low as about

−80°C (−112°F) [6]. However, rates are so slow below −18°C (0°F) that biochemical and enzymatic reactions are usually of little practical significance.

## VII.  MICROBIOLOGICAL QUALITY AND SAFETY

Poultry and poultry products are similar to other muscle foods in terms of the general microbial flora present. However, the relative proportions of some organisms may be different. While viruses, fungi, and parasites can cause microbiological problems in poultry, bacteria are far and away of the greatest practical concern. Bacteria of concern on poultry products include the nonpathogenic organisms that cause spoilage and pathogenic organisms that can cause human illnesses.

Spoilage organisms on poultry are similar to those found on red meats, including *Pseudomonas, Achromobacter, Micrococcus, Alcaligenes, Flavobacterium*, and others. For poultry spoilage involving odor and slime formation, *Pseudomonas* and *Alcaligenes* have been reported to be most common [1,6]. Pathogenic organisms include *Staphylococci, Clostridium, Campylobacter, Escherichia*, and others. The most common pathogen associated with poultry, however, is *Salmonella*. Poultry is considered to be the largest single source of this organism, although other meat sources have also been found to be causative agents of *Salmonella* outbreaks [68].

## A.  Effects of Freezing on Microorganisms

The effects of freezing and frozen storage on microbiological characteristics are variable, depending upon the organisms considered. Minimum temperatures for growth of various microorganisms vary from 10 to −10°C. Thus, growth of microorganisms at typical frozen temperatures is not a concern, because below −10°C, microbial growth will not occur [69]. The main concern are organisms that are likely to survive the freezing treatment and grow when the product is thawed. Freezing and frozen storage have the potential to induce sublethal or lethal injury in microbial cells, but microorganisms differ considerably in their sensitivity to freezing. For example, spore-forming organisms such as *Clostridium* and *Bacillus* are among the most resistant to freeze damage [70]. Gram-positive organisms can suffer some damage, but *Micrococci* and some *Streptococci* survive relatively well. Other gram-positive organisms such as *Lactobacillus* and *Pediococcus* used in starter cultures are considered more sensitive. Yeast and fungi also survive freezing treatments very well.

As a group, gram-negative organisms are the most sensitive to damage or death by freezing. This group includes many spoilage organisms, such as *Pseudomonas*, and many pathogens, such as *Salmonella*. Freezing and frozen storage has been demonstrated to reduce counts of most organisms on turkeys, including *Salmonella, Staphylococcus*, and coliforms [71]. Reductions may be in the order of

97–99% [72]. However, reduction of counts by freezing is usually variable regardless of freezing method used and, at best, achieves only partial reduction of the most sensitive organisms [73]. Even with low initial numbers of *Salmonella*, this organism has been observed to be present in chicken patties after 30 days at −10°C [74]. A survey of commercial frozen ground meat demonstrated that frozen poultry contained a significant number of viable *Salmonella* (isolated in 33% of the samples) [75]. Frozen ground poultry also exhibited higher contamination (11%) of *Staphylococcus aureus* than did ground beef (2%). A wide range of pathogenic organisms has been isolated from frozen ground turkey, including *Escherichia coli*, *Clostridium perfringens*, *Staphylococcus aureus*, and *Salmonella* [76]. More than one third of the samples (38%) included *Salmonella*. Other pathogens found on chilled and frozen poultry include *Yersinia enterocolitica* and *Listeria monocytogenes* [77]. Although these organisms are noted for their ability to grow at temperatures below 5°C and have minimum growth temperatures of slightly less than 0°C, they also seem to be relatively sensitive to damage and inactivation by freezing processes [78].

Despite the fact that freezing reduces microbial counts to some extent and slow freezing appears to be more detrimental than rapid freezing to microbial survival [79], some organisms usually survive any freezing process. Also, freeze/thaw treatments may occasionally result in higher bacterial counts, depending on thawing conditions [80]. Brown [69] pointed out that freezing rates and local environments will differ within food products during freezing, which means that the effects of freezing on microbial populations are likely to be different at various locations within the same product.

## B. Storage Effects on Microorganisms

Storage temperature can play a role in how well microorganisms survive freezing. Storage temperatures above −10°C are more effective in reducing microbial counts [69]. Temperature fluctuations including freeze/thaw cycles have been shown to reduce *Salmonella* on chicken by as much as 99% but to have much less effect on *S. aureus* [81,82]. A study of *Campylobacter jejuni* in fresh and frozen poultry and red meat showed similar results, since frozen products were significantly less contaminated than fresh products. However, the organism was still present in some frozen samples (2–3%) [83].

Freezing and frozen storage damage to microbial cells is also increased by addition of salt and by a lowered pH [69]. These effects are probably due to increased cellular dehydration contributing to freeze damage. On the other hand, some food components such as sugars and peptides can serve as cryoprotectants and increase microbial survival. Consequently, further processed products that include additional ingredients may have altered microbial survival rates during freezing treatments.

## C. Thawing and Microorganism Growth

The effect of thawing on microbial survival in poultry is not clear. Very little research has been done on thawing effects alone, and virtually no research has been conducted on poultry and poultry products. Brown [69] concluded that the little work published did not agree on whether thawing contributed to any of the freezing-process damage observed in microbial cells. Thawing can, however, contribute to significant microbiological proliferation if warm temperatures are used and the product surface is allowed to stay above 5°C for an extended period of time. Thawing of poultry often produces drip, which can accumulate on or around the product and provide an ideal environment for microbial growth. Concern regarding microbial growth constitutes the reason for the common recommendation for thawing in a refrigerated environment of 5°C or less [54]. Rapid thawing using microwave equipment or conventional heat can be used as long as product temperatures are well controlled. Unique systems involving heating under vacuum have been developed with claims of decreased drip and improved quality, but these concepts have not been widely adapted.

## D. Control of Microorganism Numbers

Consideration of the effects of freezing, frozen storage, and thawing on microbiological quality and safety clearly shows that the process is injurious to many microorganisms. However, even though microbial counts of specific organisms may be reduced, they are seldom completely eliminated. From the quality standpoint, reduction of spoilage microorganisms during freezing has not been shown to have a major impact on spoilage of poultry after thawing. Comparison of spoilage rates for fresh (never frozen) and frozen-thawed broilers has shown no difference [84]. Microbiological quality prior to freezing remains a major determinant of postthaw quality. From the safety standpoint, freezing processes can achieve a significant reduction of some pathogens, such as *Salmonella*. However, because there is also usually significant survival, other methods must be used to ensure elimination of pathogenic organisms from frozen poultry.

## VIII. STABILITY OF FROZEN PRODUCTS

### A. Stability Time and Acceptability Time

Frozen product stability can be defined in more than one way but is determined primarily by the organoleptic changes that occur during storage. There is a slow progressive change in organoleptic quality during storage, which does not become objectionable for some time. Consequently, some definitions of stability depend on the length of time before change becomes detectable, while others depend on the time until change becomes objectionable. Jul [15] defined these periods as "stability time" (time to first noticeable change) and "acceptability time" (time to loss of consumer acceptability). Figure 1 shows a comparison of stability and acceptability of chicken at different storage temperatures and with different packaging systems.

## B. Factors for Stability and Acceptability

The factors important in determining stability and acceptability are those that influence oxidative and organoleptic change. As discussed previously, the amount of unsaturated lipids and the oxidative/antioxidant environment surrounding membrane phospholipids are major factors. Also important are process treatments such as cooking, which initiates oxidative change and decreases practical frozen storage life by about 50% [49]. Addition of salt as well as chopping or grinding also greatly shortens frozen storage life [15]. The effects of packaging are critical, as shown in Fig. 1. High barrier vacuum nearly doubles storage life in most cases when compared to a low-barrier polyethylene film. The final major determinant is storage temperature, the effect of which can be clearly seen in Fig. 1.

In general, intact poultry carcasses and cuts will remain acceptable for at least

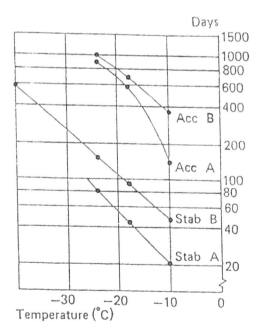

**Figure 1**  Duration of stability (Stab) and acceptability (Acc) for chicken portions, normal stability. (A) Packed in polyethylene film; (B) vacuum-packed. (From Ref. 43.)

12 months if they are well packaged and held at −18°C (Table 3). Lowering storage temperature increases stability. Thus, 24 months of acceptibility may be expected with temperatures of −25°C or less. Cooked poultry products such as fried chicken may be expected to have a practical storage life of 6–9 months at −18°C, more than 9 months at −25°C, and 12 months or more at −30°C. Stability of frozen products can be summarized as dependent upon the product characteristics, the processes used for the product, and the packaging system chosen [15].

## IX. SUMMARY

Poultry and poultry products are frequently frozen to achieve long-term storage with minimal loss of quality. Poultry accommodates freezing treatments relatively well, and the quality level at the time of freezing can be well maintained as long as the freezing process and storage conditions are adequate. Freezing requirements include a relatively rapid rate of migration of the ice front (1 cm/h), which can be achieved with air blast freezing at −29°C (600 fpm air velocity) or other methods that achieve a similar freezing rate. Storage needs to be at −18°C or less using an appropriate high-barrier package with minimal temperature change during the storage period. Thawing may be done in a variety of ways as long as product temperature and potential microbial growth are well controlled.

## ACKNOWLEDGMENTS

The author gratefully acknowledges helpful suggestions for this chapter by Dr. Glenn Froning, Dr. Art Maurer, and Dr. William Stadelman.

## REFERENCES

1.  G. J. Mountney, *Poultry Products Technology*, 2nd ed., AVI Publishing Co., Westport, CT, 1976, p. 27.
2.  *Meat Proc.* 32:20 (1993).
3.  Poultry, raw, chilled and frozen, *Commodity Storage Manual*, The Refrigeration Research Foundation, Bethesda, MD, 1993.
4.  S. Bjerklie, Showell shows off, *Meat Poultry* 39(8):14 (1993).
5.  J. Kelly, Prepared, preserved and profitable, *Meat Poultry* 39(4):18 (1993).
6.  D. deFremery, A. A. Klose, and R. N. Sayre, Freezing poultry, *Fundamentals of Food Freezing* (N. W. Desrosier and D. K. Tressler, eds.), AVI Publishing Co., Westport, CT, 1977, p. 240.
7.  Poultry products, cooked, chilled and frozen, *Commodity Storage Manual*, The Refrigeration Research Foundation, Bethesda, MD, 1993.
8.  S. Bjerklie, A fine scattering of genes, *Meat Poultry* 39(12):19 (1993).
9.  Y. L. Xiong, A. J. Pescatore, A. H. Cantor, S. P. Blanchard, and M. L. Straw, Influence

of genetic strain, pH and salts on cooking properties of processed light and dark broiler meat, *Int. J. Food Sci. Technol.* 28:429 (1993).

10. M. C. Lamari, R. G. Cassens, D. M. Schaefer, and K. K. Scheller, Dietary vitamin E enhances color and display life of frozen beef from Holstein steers, *J. Food Sci.* 58:701 (1993).

11. E. P. Mecchi, M. F. Pool, G. A. Bekman, A. Hamachi, and A. A. Klose, The role of tocopherol content in the comparative stability of chicken and turkey fat, *Poultry Sci.* 35:1238 (1956).

12. M. A. Uebersax, L. E. Dawson, and K. L. Uebersax, Storage stability (TBA) of meat obtained from turkeys receiving tocopherol supplementation, *Poultry Sci.* 57:937 (1978).

13. I. Bartov, D. Basker, and S. Angel, Effect of dietary vitamin E on the stability and sensory quality of turkey meat, *Poultry Sci.* 62:1224 (1983).

14. K. N. May, Chilling poultry meat. 3. Changes in microbial numbers during final washing and chilling of commercially slaughtered broilers, *Poultry Sci.* 53:1282 (1974).

15. M. Jul, *The Quality of Frozen Foods,* Academic Press, Orlando, FL, 1984. pp. 33–153.

16. B. P. F. Day, Chilled food packaging, *Chilled Foods: A Comprehensive Guide* (C. Dennis and M. Stringer, eds.), Ellis Horwood Limited, Chichester, UK, 1992, p. 147.

17. K. C. Li, E. K. Heaton, and J. E. Marion, Freezing chicken thighs by liquid nitrogen and sharp freezing process, *Food Technol.* 23:107 (1969).

18. C. P. Lentz and L. van den Berg, Liquid immersion freezing of poultry, *Food Technol.* 11:247 (1957).

19. A. A. Klose and M. F. Pool, Effect of freezing conditions on frozen turkeys, *Food Technol.* 10:34 (1956).

20. A. W. Kotula, J. E. Thomson, J. F. Novotny, and E. H. McNally, Bone darkening in fryer chicken as related to calcium and phosphorus levels in the feed, *Poultry Sci.* 42:1009 (1963).

21. E. Harder, *Blast Freezing System for Quantity Foods*, CBI Publishing Co., Boston, MA, 1979, p. 12.

22. L. Petrovic, R. Grujic, and M. Petrovic, Definition of the optimal freezing rate—2. Investigation of the physicochemical properties of beef M. *longissimus dorsi* frozen at different freezing rates, *Meat Sci.* 33:319 (1993).

23. L. Boegh-Sorensen and J. H. Jensen, Factors affecting the storage life of frozen meat products, *Int. J. Refrig.* 4:139 (1981).

24. R. Rust and D. Olson, Freezing tips, do's and dont's, *Meat Poultry* 34(5):10 (1988).

25. A. A. Klose, M. F. Pool, and H. Cineweaver, Effects of fluctuating temperatures on frozen turkeys, *Food Technol.* 9:372 (1955).

26. A. A. Klose, M. F. Pool, A. A. Campbell, and H. L. Hanson, Time-temperature tolerance of frozen foods. IX. Ready-to-cut-up chicken, *Food Technol.* 13:477 (1959).

27. M. J. Lachanski, The effect of freezing rate and storage conditions on some quality factors of frozen broilers, M. S. thesis, Purdue University, Lafayette, IN, 1977.

28. L. H. Fulton, G. L. Gilpin, and E. H. Dawson, Turkeys roasted from frozen and thawed states, *J. Home Econ.* 59:728 (1967).

29. W. W. Marion and W. J. Stadelman, Effect of various freezing methods on quality of poultry meat, *Food Technol.* 12:367 (1958).

30. W. O. Miller and K. N. May, Tenderness of chicken as affected by rate of freezing, storage time and temperature and freeze drying, *Food Technol.* 19:1171 (1965).

31. J. G. Berry and E. E. Cunningham, Factors affecting the flavor of frozen fried chicken, *Poultry Sci.* 49:1236 (1970).

32. M. C. Tomas and M. C. Anon, Study on the influence of freezing rate on lipid oxidation in fish (salmon) and chicken breast muscles, *Int. J. Food Sci. Technol.* 25:718 (1990).

33. E. M. Streeter and J. V. Spencer, Cryogenic and conventional freezing of chicken, *Poultry Sci.* 52:317 (1973).

34. J. C. Cregler and L. E. Dawson, Cell disruption in broiler breast muscle related to freezing time, *J. Food Sci.* 53:248 (1968).

35. J. N. Butts and F. E. Cunningham, The effect of freezing and reheating on shear press values of precooked chicken, *Poultry Sci.* 50:281 (1971).

36. C. S. Huber and W. J. Stadelman, Effects of freezing rate and freeze drying on the soluble proteins of muscle. 1. Chicken muscle, *J. Food Sci.* 35:229 (1970).

37. L. P. Pickett and B. F. Miller, The effects of liquid nitrogen freezing on the taste, tenderness and keeping quality of dressed turkey, *Poultry Sci.* 46:1148 (1967).

38. S. Barbut and G. S. Mittal, Influence of the freezing rate on the rheological and gelation properties of dark poultry meat, *Poultry Sci.* 69:827 (1990).

39. G. W. Froning, Mechanical deboning of poultry and fish, *Adv. Food Res.* 27:110 (1981).

40. L. E. Dawson and R. Gartner, Lipid oxidation in mechanically deboned poultry, *Food Technol.* 37(7):112 (1983).

41. S. H. Barbut, H. Draper, and P. D. Cole, Effect of mechanical deboner head pressure on lipid oxidation in poultry meat, *J. Food Prot.* 52:21 (1989).

42. K. L. Uebersax, L.E. Dawson, and M. A. Uebersax, Storage stability (TBA) and color of MDCM and MDTM processed with $CO_2$ cooling, *Poultry Sci.* 57:670 (1978).

43. M. G. Mast, P. Jurdi, and J. H. MacNeil, Effects of $CO_2$ snow on the quality and acceptance of mechanically deboned poultry meat, *J. Food Sci.* 44:346 (1979).

44. S. Barbut, Y. Kakuda, and D. Chan, Effects of carbon dioxide freezing and vacuum packaging on the oxidative stability of mechanically deboned poultry meat, *Poultry Sci.* 69:1813 (1990).

45. M. Ristic, Shelf life of poultry parts independent of time of preparation, *Thermal Processing and Quality of Foods* (P. Zeuthen, J. C. Cheftel, C. Eriksson, M. Jul, H. Leniger, P. Linko, G. Varela, and G. Vos, eds.), Elsevier Applied Science Publishers, London, 1984, p. 647.

46. H. L. Hanson and L. R. Fletcher, Time-temperature tolerance of frozen foods. XII. Turkey dinners and turkey pies, *Food Technol.* 12:40 (1958).

47. H. L. Hanson, L. R. Fletcher, and H. Lineweaver, Time temperature tolerance of frozen foods. XVII. Frozen fried chicken, *Food Technol.* 13:221 (1959).

48. L. Boegh-Soerensen, TTT and PPP tests for broiler chicken and chicken parts, *Fleischwirtschaft* 55:1587 (1975).

49. J. O. Igene, A. M. Pearson, L. R. Dusan, Jr., and J. F. Price, Role of triglycerides and phospholipids on development of rancidity in model meat systems during frozen storage, *Food Chem.* 5:263 (1980).

50. J. E. Webb, C. C. Brunson, and J. D. Yates, Effects of feeding antioxidants on rancidity development in pre-cooked, frozen broiler parts, *Poultry Sci.* 51:1601 (1972).

51.  P. P. Jantawat and L. E. Dawson, Stability of broiler pieces during frozen storage, *Poultry Sci.* 56:2026 (1977).

52.  J. Pikul, D. E. Lesczynski, P. J. Bechtel, and F. A. Kummerow, Effects of frozen storage and cooking on lipid oxidation in chicken meat, *J. Food Sci.* 49:838 (1984).

53.  S. O. Awonorin and J. A. Ayoade, Texture and eating quality of raw- and thawed-roasted turkey and chicken breasts as influenced by age of birds and period of frozen storage, *J. Food Service Syst.* 6:241 (1992).

54.  G. Cano-Munoz, *Manual on Meat Cold Store Operation and Management*, Food and Agriculture Organization (FAO), of the United Nations, Rome, Italy, 1991, p. 18.

55.  R. C. Baker, J. M. Darfler, E. J. Mulnix, and K. R. Nath, Palatability and other characteristics of repeatedly refrozen chicken broilers, *J. Food Sci.* 41:443 (1976).

56.  P. P. Engler and J. A. Bowers, Vitamin $B_6$ content of turkey cooked from frozen, partially frozen and thawed steaks, *J. Food Sci.* 40:615 (1975).

57.  B. B. Cook, A. F. Morgan, and M. B. Smith, Thiamin, riboflavin and niacin content of turkey tissues as affected by storage and cooking, *Food Res.* 14:449 (1949).

58.  H. M. Brown, Non-microbiological factors affecting quality and safety, *Chilled Foods: A Comprehensive Guide* (C. Dennis, M. Stringer, eds.), Ellis Horwood Limited, Chichester, UK, 1992, p. 261.

59.  M. Mitsumoto, R. N. Arnold, D. M. Schaefer, and R. G. Cassens, Dietary versus postmortem supplementation of vitamin E on pigment and lipid stability in ground beef, *J. Animal Sci.* 71:1812 (1993).

60.  W. L. Marusich, E. DeRitter, E. F. Ogring, J. Keating, M. Mitrovic, and R. H. Bunnell, Effect of supplemental vitamin E in control of rancidity in poultry meat, *Poultry Sci.* 54:831 (1975).

61.  E. Brotsky, Automatic injection of chicken parts with polyphosphate, *Poultry Sci.* 55:653 (1976).

62.  A. W. Khan and L. van den Berg, Changes in chicken muscle proteins during cooking and subsequent frozen storage and their significance in quality, *J. Food Sci.* 30:151 (1964).

63.  A. W. Khan, Changes in non-protein nitrogenous constituents of chicken breast muscle stored at below freezing temperatures, *Agric. Food Chem.* 12:378 (1964).

64.  T. G. Uijttenboogaart, T. L. Trziszka, and F. J. G. Schreurs, Cryoprotectant effects during short time frozen storage of chicken myofibrillar protein isolates, *J. Food Sci.* 58:274 (1993).

65.  J. D. Daubin, Freezing, *Technology of Meat and Meat Products* (J. P. Girard, ed.), Ellis Horwood Limited, Chichester, UK, 1992, p. 261.

66.  A. W. Khan and L. van Den Berg, Biochemical and quality changes occurring during freezing of poultry meat, *J. Food Sci.* 32: 148 (1967).

67.  K. Yamamoto, K. Samejima, and T. Yasui, A comparative study of the changes in hen pectoral muscle during storage at 4°C and –20°C, *J. Food Sci.* 42:1642 (1977).

68.  N. H. Bean and P. M. Griffin, Foodborne disease outbreaks in the US, 1973–1987: Pathogens, vehicles and trends, *J. Food Prot.* 53:804 (1990).

69.  M. H. Brown, Microbiological aspects of frozen foods, *Chilled Foods: A Comprehensive Guide* (C. Dennis, M. Stringer, eds.), Ellis Horwood Limited, Chichester, UK, 1992, p. 15.

70.  C. A. White and L. P. Hall, The effect of temperature abuse on *Clostridium perfringens*

and *Bacillus cereus* in raw beef and chicken substrates during frozen storage, *Food Microbiol.* 1:97 (1984).

71.   K. V. Reddy, A. A. Kraft, R. J. Hasiak, W. W. Marion, and D. K. Hotchkiss, Effect of spin chilling and freezing on bacteria on commercially processed turkeys. *J. Food Sci.* 43:334 (1978).

72.   A. A. Kraft, J. C. Ayres, K. F. Weiss, W. W. Marion, S. L. Balloun, and R. H. Forsythe, Effect of method of freezing on survival of microorganisms on turkey, *Poultry Sci.* 63:128 (1963).

73.   K. W. B. Gunaratne and J. V. Spencer, Effect of certain freezing methods upon microbes associated with chicken meat, *Poultry Sci.* 53:215 (1974).

74.   N. A. Cox, J. S. Bailey, C. E. Lyon, J. E. Thomson, and J. P. Hudspeth, Microbiological profile of chicken patty products containing broiler giblets, *Poult. Sci.* 62:960 (1983).

75.   A. Mates, Microbiological survey of frozen ground meat and a proposed standard, *J. Food Protect.* 46:87 (1983).

76.   L. S. Guthertz, J. T. Fruin, R. L. Okoluk, and J. L. Fowler, Microbial quality of frozen comminuted turkey meat, *J. Food Sci.* 42:1344 (1977).

77.   S. J. Walker, Chilled foods microbiology, *Chilled Foods: A Comprehensive Guide* (C. Dennis and N. Stringer, eds.), Ellis Horwood Limited, Chichester, UK, 1992, p. 165.

78.   D. A. Golden, L. R. Beuchat, and R. E. Brachett, Inactivation and injury of Listeria monocytogenes as affected by heating and freezing, *Food Microbiol.* 5:17 (1988).

79.   B. Ray and M. L. Speck, Freeze injury in bacteria, *CRC Crit. Rev. Lab. Sci.* 3:161 (1973).

80.   C. R. Rey and A. A. Kraft, Effect of freezing and packaging methods on survival and biochemical activity of spoilage organisms on chicken, *J. Food Sci.* 36:454 (1971).

81.   V. M. Olson, B. Swaminathan, and W. J. Stadelman, Reduction of numbers of *Salmonella typhimurium* on poultry parts by repeated freeze-thaw treatments, *J. Food Sci.* 46:1323 (1981).

82.   C. A. White and L. P. Hall, The effect of temperature abuse on *Staphylococcus aureus* and salmonellae in raw beef and chicken substrates during frozen storage, *Food Microbiol.* 1:29 (1984).

83.   N. J. Stern, S. S. Green, N. Thaker, D. J. Krout, and J. Chiv, Recovery of *Campylobacter jejuni* from fresh and frozen meat and poultry collected at slaughter, *J. Food Prot.* 47:372 (1984).

84.   J. V. Spencer, E. A. Sauter, and W. J. Stadelman, Effect of freezing, thawing and storing broilers on spoilage, flavor and bone darkening, *Poultry Sci.* 40:918 (1961).

# 4
# Fish and Seafood

En. Emilia M. Santos-Yap

*College of Fisheries*
*University of the Philippines in the Visayas*
*Iloilo, The Philippines*

## I. INTRODUCTION

Spoilage starts as soon as a fish dies as a result of a complex series of chemical, physical, bacteriological, and histological changes that occur in the muscle tissue. These interrelated processes are usually accompanied by the gradual loss or development of different compounds that affect fish quality.

These quality changes are influenced by many factors, the most important of which is temperature. If not properly controlled, exposure of fresh fish to temperature abuse can cause serious deterioration in fish quality.

Commercially, icing or chilling continues to play a major role in slowing down bacterial and enzymatic activities in fish. However, this process is not designed to totally eliminate changes in quality, since it only offers protection for 2–3 weeks, depending on the species.

Freezing is considered an excellent process for preserving the quality of fish for longer periods of time (commercially, up to 18 months or more). Freezing and subsequent cold storage are particularly useful in making seasonal species of fish, like herring and mackerel, available all year round. In addition, freezing finds its application in a number of different products made from different fish species. For example, tuna is frozen on board large commercial fishing vessels, brought to land, and then thawed for canning. In the production of various value-added fishery products, freezing is applied to breaded and battered fish sticks, fillets, steaks, or nuggets. Likewise, high-quality fish are usually filleted, frozen, and eventually sold to consumers.

Ideally, there should be no distinguishable differences between fresh fish and frozen fish after thawing. If kept under appropriate conditions, fish in the frozen state can be stored for several months or more without appreciable changes in quality. However, it is now well recognized that deteriorative changes take place in fish and seafood during freezing, frozen storage and thawing, which influence the quality of the finished product. Interestingly, the major determinants of these changes are the conditions that prevail during the several phases of processing. Considerably more knowledge of the basic structure of fish muscle and its chemical composition is essential to understanding these changes that occur during processing.

## II. NATURE OF FISH MUSCLE

Fish muscle has a unique arrangement of muscle fibers. It is divided into a number of segments called myotomes, which are separated from one another by a thin sheath of connective tissue called the myocomma or myoseptum (Fig. 1). The

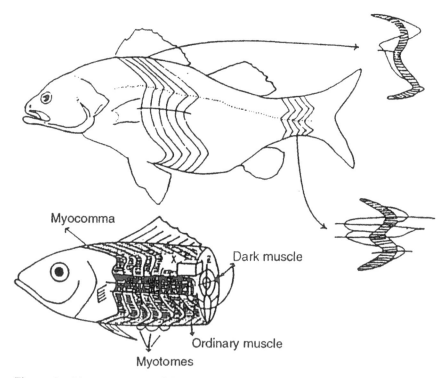

**Figure 1** Myotome pattern of the musculature of bony fish, with detailed lateral views of a single myotome. (From Ref. 87.)

number of myotomes in fish is dependent on the size of the fish, while their diameters vary from head to tail [1].

There are two major types of fish skeletal muscles: white and red. The red or dark muscle lies along the side of the body next to the skin, particularly along the lateral lines, and may comprise up to 30% of the fish muscle, depending on species [2]. Since the cells in this muscle are packed with more lipids than those in the white muscle [3,4], red muscle is basically employed by fish for sustained swimming activities, functioning aerobically using lipids for fuel. Also, red muscle has more mitochondria [5] but less sarcoplasmic reticulum than white muscle [6]. In addition, red muscle has a large supply of oxygen and a high content of myoglobin, the colored compound that gives its red color. These characteristics, coupled with the presence of the large amount of lipid in red muscles, particularly among the fatty species, present a serious problem because of increased susceptibility of this muscle to lipid oxidation.

The red muscle of some species has also been reported to contain enzymes that are responsible for chemical reactions such as lipid oxidation and the conversion of trimethylamine oxide (TMAO) to dimethylamine (DMA) and formaldehyde (FA) [7,8]. The shape of the red muscle area in different species varies considerably. Lean species of fish such as flounder, hake, sole, cod, pollock, and whiting have a very small amount of red muscle, which lies along the fish skin, whereas the fatty and semi-fatty fish species have larger areas of red muscle.

White muscle, on the other hand, constitutes the majority of fish muscle. Unlike red muscle, it has minimal myoglobin and restricted blood supply [5]. White muscle, often referred to as the "fast" tissue [9], is used for anaerobic activities such as short bursts of swimming activities. This muscle exhibits rapid but powerful contractions, the energy for which is produced by reducing glycogen to lactic acid anaerobically [5].

Intermediate between these two types of muscle are intermixed red and white muscle, commonly referred to as "mosaic" muscle [10]. In some species of fish this is a thin layer of muscle that separates the red from white muscle. However, in other fish, such as salmon, carp, and trout, this muscle is scattered throughout the body of the fish.

The chemical composition of fish varies depending on several factors, such as age, species, maturity, method of catch, fishing grounds, and other seasonal and biological factors. Even within a single species, chemical composition may vary significantly. Generally, fish contain a considerable amount of protein, lipid, and water and small amounts of vitamins and minerals. Other components such as nonprotein nitrogenous compounds are also present in the muscle. These include urea, taurine, peptides, free amino acids, and nucleotides like inodine and hypoxanthine [11]. These compounds, together with the macronutrients found in fish muscle, may be particularly important to fish processors, since they are frequently used as spoilage indices.

## III.  EFFECTS OF PREPARATION AND PACKAGING ON FROZEN FISH

Product preparation and packaging significantly affect the quality and shelf life of frozen fish. If not properly controlled, these processes result in some deleterious effects after prolonged storage.

### A.  Product Preparation

Product preparation, in particular, produces a considerable effect on shelf stability of frozen fish. Whole and eviscerated fish have longer shelf stability than fillets, while minces can usually only be stored for a much shorter period of time. Crawford et al. [12–14] observed this difference during several studies using hake. Minced blocks exhibited reduced quality and accelerated deterioration during storage when compared to intact fillets. This characteristic of minces, which is more apparent among the gadoids, is probably due to the mincing action applied to the fish flesh, which results in tissue damage and subsequently more rapid deterioration. In addition, mixing of red and white muscle during mincing may also result in the dispersion of lipids and some of the enzymes present in the red muscle, leading to greater susceptibility of the minced tissue to deteriorative changes.

### B.  Packaging Materials and Methods

An efficient packaging system is essential to offset the detrimental quality changes that occur during frozen storage. Packaging materials and methods are obviously designed not only to protect the product from microbial and chemical contamination, dehydration, and physical damage, but also to protect the environment from the packaged product. Fish and seafood can leak gases or unsightly fluids, which may have unpleasant odors. Therefore, the choice of appropriate packaging materials and methods for frozen fish is a critical factor in terms of shelf-life extension.

Studies have shown that packaging systems affect the quality and shelf stability of frozen fish. For instance, vacuum packaging is well established as a method to provide an oxygen-free environment to minimize the problems associated with lipid oxidation and dehydration during frozen storage.

Several studies have shown the effectiveness of this method for frozen storage of some species of fish. For example, it has been reported that frozen blocks of fillets vacuum packed in moisture-proof films showed high degrees of acceptance and desirable frozen characteristics [12]. Likewise, Santos and Regenstein [15] reported the effectiveness of vacuum packaging for inhibiting lipid oxidation in frozen mackerel fillets.

Ahvenainen and Malkki [16] examined the influence of packaging on frozen herring fillets stored at different temperatures. They found that vacuum-packed product covered with metallized cardboard had a longer shelf life than a product vacuum packed and stored without cardboard.

Vacuum packaging, on the other hand, need not be used if lipid oxidation is not the limiting factor affecting the shelf life of a product. Although the effect of absence of oxygen in packages on some fish species must be considered, other packaging methods such as glazing and the use of heat-sealable packaging films should also be considered.

Pacific hake minced blocks stored in moisture-proof, vapor-proof packaging films exhibited superior quality over glazed samples [12]. Likewise, Colakoglu and Kundacki [17] observed frozen mullet packed in plastic films with low permeability to oxygen and moisture to have a longer shelf life than when unpacked in the glazed form.

However, it should be noted that glazing is still considered to be the cheapest means of protecting frozen fish during storage and transport. Glazing provides a continuous film or coating that adheres to the frozen product, which retards moisture loss and the rate of oxidation.

Many different glazes are available, including (1) those with inorganic salt solutions of disodium acid phosphate, sodium carbonate, and calcium lactate, (2) alginate solution, otherwise known as the "Protan" glaze, (3) antioxidants such as ascorbic and citric acids, glutamic acid, and monosodium glutamate, and (4) other edible coatings such as corn syrup solids [18]. Ice glaze is particularly important in handling frozen fish in developing countries.

For products intended for short-term storage, glazing can be practically utilized as a viable alternative to storage without a protective covering. For instance, Jadhav and Magar [19] concluded that glazing was a cheaper alternative to expensive packaging systems for glazed Indian mackerel (*Rastrelliger kanagurta*) stored at –20°C. Glazed samples had a shelf life of 6 months, while samples without a protective covering lasted only 3–4 months.

## IV. EFFECTS OF FREEZING, FROZEN STORAGE, AND THAWING ON COLOR, APPEARANCE, AND CONSUMER ACCEPTANCE

One problem encountered during handling, freezing, and storage of fish is the difficulty in retaining the color and appearance of the meat. Changes in color and appearance of fish occur even immediately after catch. Blood pigments become noticeably discolored to various degrees after some period of time. The natural oils in fish play an important role in these color changes. The color of these oils is produced by the colored pigments dissolved in them which vary from one species to another. These pigments are subjected to considerable oxidation when the fish is frozen and stored. This then results in meat color darkening to either dark brown or, in some cases, black. This discoloration occurs especiallly when the fish is stored for an extended period of time.

Some fish, like tuna, develop discoloration during frozen storage, reportedly due to oxidation of myoglobin to metamyoglobin in fish blood [20]. Other species, such as salmon, swordfish, and shark, also exhibit color changes during storage. Salmon has a pink meat, but when subjected to oxidation its color slowly fades and, in extreme cases, may completely disappear after prolonged storage. Swordfish, on the other hand, develops green discoloration beneath its skin during frozen storage, which according to Tauchiya and Tatsukawa [21] is due to the development of sulfhemoglobin, a product of oxidation. Shark flesh also discolors and occasionally develops off-odors during storage, most probably as a result of the presence of high amounts of trimethylamine oxide. Interestingly, these marked differences in the color and appearance of frozen fish are quite noticeable in fish sold either as steaks or as fillets, especially when cross-sectional cuts of the fish are made, which permits a comparison of the color of the exposed fish surface with that of the inner portion.

In shrimp, the rapid formation of black pigments, widely known as "melanosis," occurs within a few hours after death and is enhanced by exposing the shrimp to air (oxygen). It can occur within just 2–12 hours of exposure. The oxidation reaction leading to the formation of these black pigments can occur at 0°C, however, at –18°C, no visible spots were detected at up to 3 months of storage [22]. Below this temperature, it is believed that melanosis can still positively occur. It should be noted, however, that although black spots do not necessarily make shrimp unfit for human consumption, such discoloration is usually associated with spoilage, resulting in a decrease in market value.

In other shellfish, such as crab and lobster, the development of blue or black discoloration, otherwise known as "blueing," is one of the most troublesome problems. Blueing may occur after freezing or during frozen storage, or it may appear after thawing and subsequent air exposure or even shortly after cooking.

Needless to say, these changes in color and appearance of fish and shellfish significantly affect consumer acceptance. When consumers select frozen fish, if these products can be seen through the packaging material used, the color and appearance of the frozen product provide an indication as to its degree of quality. As shown in Table 1, undersirable appearance and discoloration of samples have been observed in different frozen whole fish and fillets obtained from U.S. supermarkets [23]. Preventing such quality changes is of great commercial importance, since they detract not only from the consumer acceptability of the products but their shelf stability as well.

Thawing also influences the color and appearance of frozen fish and, inevitably, its consumer acceptability. Depending on the thawing technique used, discoloration may occur in fish and other seafood. For instance, when shrimp are thawed at temperatures higher than 0°C, black discoloration or melanosis may occur. This is due to the unnecessary exposure of the shrimp to air, leading to oxidation.

**Table 1** Characteristics of Some Frozen Fish Purchased in U.S. Supermarkets

| Fish | Thawed-state characteristics |
| --- | --- |
| Herring, whole | Skin and meat show rusting |
| Mackerel, whole | Surface dehydrated, skin and meat show rusting, spongelike meat |
| Mackerel, whole | Rancid smell in skin and meat |
| Chinese pompret, whole | Dehydration at lower part of belly and fin |
| Chinese pompret, whole | Head and belly parts yellowish discolored, spongelike meat |
| White pompret, whole | Skin and meat show rusting, spongelike meat |
| Jew fish, whole | Slight rancid smell in skin |
| Lemon sole, whole | Surface dehydrated, spoiled and rancid smell, spongelike meat |
| Haddock, fillet | No smell, spongelike meat, cracks |
| Cod, fillet | No smell, spongelike meat |
| Flounder, fillet | No smell |

*Source*: Ref. 23.

A phenomenon known as "shimi" occurs in frozen-thawed fish meat. Shimi are the undesirable blood spots observed in the belly portion of carp on thawing and are also the distinguishable spots tainting frozen-thawed tuna meat [24]. The latter condition is probably due to the blood vessels that remain in unbled tuna meat prior to freezing. When thawed, these blood vessels produce unsightly spots in the meat.

Interestingly, it is possible to determine if the product has been properly thawed and then refrozen. This is particularly noticeable in packaged frozen fish, where spaces on the sides of the package may be filled with a frozen cloudy liquid, known as thaw drip. Such muscle drip was originally attributed to the rupturing of cell walls caused by ice crystal formation during freezing, resulting in excess drip during thawing. However, it has been postulated that drip or exudate formation is directly related to the capacity of the fish protein to hold moisture [25]. This unsightly exudate from fish muscle indicates, among other things, inappropriate handling, prolonged ice storage prior to freezing, frozen storage at inappropriate cold-storage temperatures, or improper thawing,

If not properly controlled, freezing, frozen storage, and thawing generally result in quality changes in fish and seafood that in most cases render the product unacceptable to consumers.

## V. EFFECTS OF FREEZING, FROZEN STORAGE, AND THAWING ON PALATABILITY ATTRIBUTES

Changes in the texture, odor, and flavor of fish and seafood affect their palatability. Fresh fish have a distinct succulence and a delicate odor and flavor, which are characteristic of the species. These attributes change noticeably when fish is frozen and stored for prolonged periods of time. Interestingly, the changes that influence the palatability of frozen fish and seafood can all be measured organoleptically and, to some extent, chemically.

## A. Changes in Texture

Frozen fish gradually loses its juiciness and succulence after freezing and subsequent frozen storage. Such textural changes, reportedly caused by protein denaturation [26–29], are more pronounced in some species of fish, specifically the gadoids. In these species, the chemical breakdown of TMAO to DMA and FA and the subsequent cross-linking of FA to muscle proteins [30] produce the textural breakdown in the gadoids and result in a "cottony" or "spongy" texture. Fish muscle that has undergone such changes tends to hold its free water loosely like a sponge. When eaten, the fish muscle loses all its moisture during the first bite, and subsequent chewing results in a very dry and cottony texture.

In some species devoid of TMAO-degradation products, muscle fibers also tend to toughen and to become dry during freezing and storage. This is particularly true for most of the nongadoid species and for crab, shrimp, and lobster when stored for prolonged periods.

In contrast, the effect of the thawing method on the texture of fish muscle basically depends on the product form. For instance, whole fish, when thawed, exhibits less textural change than filleted fish, basically as a result of the presence of the backbone, which serves as structural support for the flesh. In terms of the effect of the thawing method, it has been reported that microwave thawing results in higher gel strength of minced samples, when compared to samples thawed under running water (20°C) and samples thawed at room temperature [31]. Consequently, the extent of textural changes depends upon the species of fish and upon the condition of handling, freezing, duration of frozen storage, and the thawing method used.

Several methods have been developed to objectively measure such textural changes, in addition to the gathering of comparative data from sensory evaluations. From texture analysis of minced fish. Borderias et al. [32] concluded that hardness as measured by the Kramer Shear cell and puncture (penetration) tests were highly correlated with the sensorily perceived firmness of raw samples, while a compression test was found to be a valuable technique for characterizing the cohesiveness and elasticity of both raw and cooked fish minces.

Alterations in the texture of frozen fish fillets, on the other hand, are difficult to measure objectively, mainly due to the textural variability that exists within the fish fillets, which is associated with the flakiness and the orientation of the muscle fibers [33]. Several attempts have been made to determine the extent of textural changes in fish fillets, including those tested using fish minces [32]. However, significant correlations were not obtained.

An instrumental method that may work on fish fillets is the deformation test using the Instron Universal Testing Machine equipped with a flat compression plate. As a nondestructive test [34], it can potentially be modified to conform to the irregular shape and the segmented structural orientation of fish fillets.

## B. Changes in Odor and Flavor

Other important changes that affect the palatability of frozen fish include changes in the flavor and odor of fish and seafood. Fish are often described as having a "fishy" odor and flavor. Although the term sounds unpleasant, it can also be used to describe the pleasing taste and odor characteristics of freshly caught fish. Such pleasant, palatable characteristics may be retained as long as the fish are promptly and properly frozen, stored, and thawed. However, the transformation of these attributes to unpleasant and unacceptable traits occurs very rapidly in some fish species, particularly the fatty fish species.

Changes in the delicate flavor of fish and seafood generally occur in three distinct phases during frozen storage: (1) the gradual loss of flavor due to loss or decrease in concentration of some flavor compounds [35,36], (2) the detection of neutral, bland, or flat flavor, and (3) the development of off-flavors due to the presence of compounds such as the acids and carbonyl compounds that are products of lipid oxidation. These phases, however, only apply to those species with originally delicate, sweet, and meaty flavors. Other species, such as hakes, have an originally bland flavor [37], but develop off-flavors during prolonged frozen storage.

Changes in odor occur in two phases: the loss of characteristic odor and the development of off-odors, which render the frozen product unacceptable. Generally, fish and seafood initially have a fresh, seaweedy odor, which can be retained even after freezing and frozen storage. However, gradually such odor is lost, and eventually an unpleasant odor is given off, particularly when abused with inappropriate storage temperature. The development of unpleasant odors is due either to lipid oxidation, a reaction more apparent among the fatty fish species that results in the production of a strong oily, blown oily, or rancid odor, or to the degradation of TMAO, which leads to the production of an unpleasant ammoniacal odor. Other species such as white hake (*Urophysus tenuis*) initially give off weak odors of sweet, boiled milk, but when frozen storage is extended, hake assumes weak off-odors (often described as milk jug odor) followed by a sour milk odor.

## VI. EFFECTS OF FREEZING, FROZEN STORAGE, AND THAWING ON NUTRITIONAL VALUE

Considerable emphasis has been given to the influence of freezing, frozen storage, and thawing on quality indices such as appearance/color, texture, flavor, odor, and the chemical reactions that accompany such organoleptic changes. Less attention has been given to yet another useful area, i.e., the influence of such treatments on the nutritional value of frozen fish and seafoods.

Put simply, considerable attention is given to sensorily perceived attributes because if consumers reject a frozen product on display, it is not purchased or eaten regardless of its nutritional value. Conversely, if consumers are attracted to a frozen product, they tend to buy it whether it has the needed nutrients or not. However, as the market shifts to the development and merchandising of products to meet the demands of health-conscious consumers, the nutritional value of frozen fish becomes of great importance.

When fish and seafood are frozen, and subsequently stored and thawed, protein denaturation occurs in muscle tissues. As a result, formation of thaw drip becomes apparent and consequently leads to the leaching out of dissolved materials. Likewise, there is an increase in the release of a watery "cook liquor" when the product is heated. Such water losses result in the loss of water-soluble proteins; however, such losses do not result in any measurable decrease in the nutritive value of the protein [38]. However, such losses lower the proportion of sarcoplasmic proteins in the fish tissue and may also lead to a small loss of water-soluble vitamins and minerals.

Other quality changes, such as lipid oxidation, can also influence the nutritional value of frozen products. Oxidized fish lipids, such as lipid hydroperoxides, may induce oxidative changes in sulfur-containing proteins, producing significant nutritional losses [39].

## VII.  EFFECTS OF FREEZING, FROZEN STORAGE, AND THAWING ON INTRINSIC CHEMICAL REACTIONS

When frozen fish are subjected to excessively prolonged cold storage at temperatures above $-30°C$, a series of intrinsic chemical reactions occurs in fish tissues. These reactions include protein denaturation, breakdown of TMAO and lipid oxidation.

### A.  Protein Denaturation

One of the most prevalent chemical reactions to occur in fish muscle during freezing and frozen storage is the complex phenomenon of protein denaturation. It has been postulated that the rupturing of different bonds in the native conformation of proteins in frozen fish is followed by side-by-side aggregation of myofibrillar proteins, specifically myosin, brought about by the formation of intermolecular cross-linkages [27,40]. It is also believed that the significant decrease observed in the center-to-center distance between the thick filaments of the A-band of the sarcomere after prolonged frozen storage favors the formation of cross-linkages between molecules and stiffens the fibers [41]. Such intermolecular cross-linkages result in aggregation [30], which leads to the formation of high molecular weight polymers [42,43] and subsequent denaturation of myosin during frozen storage.

Several relevant theories on protein denaturation in relation to fish moisture and freezing damage have been formulated. One theory worthy of note is that of protein denaturation being affected by the freezing out of water. The conformation of most native proteins has the hydrophobic side chains buried inside the protein molecule. However, some of these hydrophobic side chains are exposed at the surface of the molecule itself. It has been suggested that the water molecules arrange themselves around these exposed hydrophobic side chain groups so as to minimize the energy of the oil/water interface and, at the same time, act as a highly organized barrier, which mediates the hydrophobic/hydrophilic interactions between protein molecules [44]. These water molecules form a network of hydrogen bonds, which contributes to the stability of the highly organized three-dimensional structure of the proteins. As water molecules freeze out, they migrate to form ice crystals, resulting in the disruption of the organized H-bonding system that stabilizes the protein structure (Fig. 2). As the freezing process continues, the hydrophobic as well as the hydrophilic regions of the protein molecules become exposed to a new environment, which may allow the formation of intermolecular cross-linkages [30], either within the same protein molecule, causing deformation of the protein's three-dimensional structure, or between two adjacent molecules, leading to protein-protein cross-links.

Freezing also concentrates solids, including mineral salts and small organic molecules, within the remaining unfrozen aqueous phase in the cell [45], which results in changes in ionic strength and possibly pH, leading to the denaturation of the protein molecule [46]. Love [47] considered this concentrated salt in the unfrozen phase to be the main protein denaturant in the frozen muscle system. If

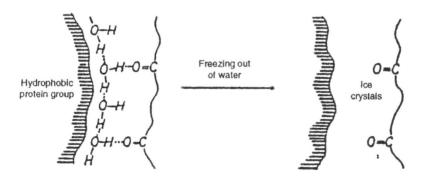

**Figure 2**  Effect of freezing out of water in a protein molecule. (From Ref. 44.)

proteins are denatured over time in the presence of concentrated solutes, it is reasonable to believe that longer exposure of protein molecules to these denaturants (e.g., slow freezing) should be avoided. However, further work must be conducted to determine the effect of the rate of freezing on shelf life of frozen fish as related to the solute concentration effect.

Several methods have been established to determine the extent of protein denaturation during frozen storage of fish and seafood. According to Jiang and Lee [48], protein quality is more sensitively reflected by the enzymatic activities in the muscles than by its extractability, since small microstructural changes in protein molecules can cause more alterations in the enzymatic activities than in extractability. For example, the actomyosin Ca ATPase, which measures the activity of myosin, can be used as an index of protein quality. Since this ATPase is capable of hydrolyzing the terminal end phosphate group of ATP to give ADP [49], this particular enzymatic activity can be determined by measuring changes in the amount of inorganic phosphate present in the muscle. The loss of enzymatic activity reflects the extent of freeze damage and alteration of the protein structure in the muscle system. Connell [40] reported a loss in Ca ATPase activity in muscle during frozen storage. In a more recent study using mackerel, Jiang and Lee [48] observed a loss of ~66% of the original Ca ATPase activity of actomyosin after 6 weeks of storage at –20°C.

Visual examination under a transmission or scanning electron microscope is a powerful technique used in the determination of textural changes in fish muscle due to denaturation. The electron microscopic studies of Matsumoto [30] were able to detect damage to the native structure of the protein: aggregation and an entangled mass was observed. However, results from this technique have to be interpreted cautiously since the fixing processes of tissue or any tissue section may create artifacts by altering the ultrastructural images or by masking the microchanges in the muscle tissue.

The extent of protein denaturation in frozen fish muscle can also be determined by conducting several tests of protein functionality. The physicochemical properties that affect the behavior of protein molecules during processing are defined as the functional properties of the fish myosystem, which include protein homogenate solubility, emulsifying and water-retention properties, gelation, and viscosity [50]. The most popular tests to determine the extent of protein denaturation during frozen storage of fish, in relation to its functionality, are determination of the loss in solubility or extractability of proteins and measurement of the water-retention properties of the fish muscle system.

## B. Breakdown of Trimethylamine Oxide

Quite obviously, protein denaturation during frozen storage produces extensive textural changes and deterioration in fish. These changes are more pronounced in some species of fish, specifically the gadoids, and are related to another intrinsic chemical reaction, the breakdown of TMAO.

TMAO is commonly found in large quantities in marine species of fish. It is believed that these species use TMAO for osmoregulation [51]. Among the marine species, the elasmobranchs contain more TMAO than the teleosts. Among the teleosts, the gadoids have more TMAO than the flatfish. Except for burbot, freshwater species have a negligible amount of TMAO in their muscles, since they do not take in TMAO in their diet beyond their bodies nutritional requirements, and they promptly excrete any excess.

After death, TMAO is readily degraded to DMA and FA through a series of reactions, as shown in Fig. 3. This conversion of TMAO to DMA and FA is typically observed in frozen gadoid species such as cod, hake, haddock, whiting, red hake, and pollock [7].

The presence of air (oxygen) affects DMA and FA formation. It has been suggested that oxygen may actually inhibit the reaction by interacting with metal ions, which otherwise would accelerate the TMAO degradation [52], Lundstrom et al. [53] observed that red hake (*Urophysis chuss*) minces stored in the absence of oxygen showed more rapid DMA and FA formation than red hake fillets stored in air. Likewise, the presence of air (oxygen) in packaged white hake (*Urophysis tenuis*) significantly prolonged the shelf life of the frozen samples [15].

TMAO degradation to DMA and FA was enhanced by the presence of an endogenous enzyme (TMAOase) in the fish tissues, as observed in cod muscles by Amano and Yamada [54]. They also found an enzyme in the pyloric ceca of Allaskan pollock (*Pollachius virens*), which was believed to cause DMA and FA formation in this species [55]. However, evidence also exists which demonstrates that breakdown of TMAO to DMA and FA is nonenzymic in nature [56,57]. The breakdown of TMAO, whether enzymatically or nonenzymatically induced, is believed to produce destabilization and aggregation of proteins.

TMAO has been postulated to be responsible for stabilizing proteins against conformational changes and thermal denaturatrion [58]. However, the conversion of TMAO to DMA and FA has been implicated in gadoid textural problems during frozen storage [12,59,60].

It has been suggested that TMAO's breakdown product, FA, may produce cross-linking of muscle proteins [30] due its high reactivity: FA can covalently bond with various functional groups of proteins, such as the amino, imino, guanido, phenolic, imidazole, and indole residues [61]. This reaction induces both intra- and intermolecular cross-linkages of the molecules, thus producing conformational changes [61].

However, textural changes may also occur during frozen storage of fish species devoid of the TMAO-enzyme system [62]. Such textural changes must then be attributed to another type of mechanism that does not involve the cross-linking of protein molecules, due to the presence of FA. Gill et al. [62] reported that the

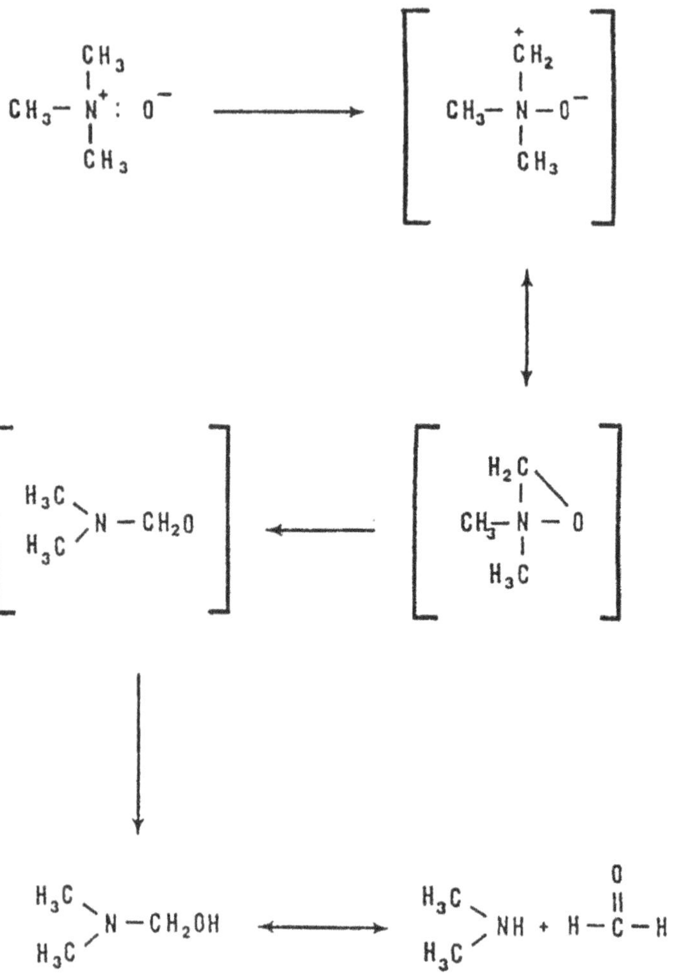

**Figure 3**    Mechanism for the formation of DMA and FA from TMAO. (From Ref. 88.)

presence of FA in red hake resulted in the covalent cross-linking of troponin and myosin light chains, forming high molecular weight aggregates. However, when haddock, a species that does not produce FA, was examined, the same cross-linkages were not found at the molecular level, although textural toughening was observed, which was not as pronounced as that in red hake. Based on these observations, they suggested that textural changes in haddock were probably due to secondary bonds, such as hydrogen or electrostatic bonds, and not due to FA cross-links.

Clearly, the presence of FA is not the only factor involved in textural changes during frozen storage. However, with certain species of fish, it appears to be of primary importance.

To objectively determine the extent of textural deterioration due to DMA and FA formation and subsequent reactions, the measurement of DMA content is recommended. Due to the equimolar formation of DMA and FA in fish muscle and the observed high reactivity of FA, DMA content is routinely used as an index. Consequently, the DMA test indirectly measures the FA value in fish muscle. However, use of this test is limited to those species known to produce DMA and FA during frozen storage.

## C. Lipid Oxidation

Another chemical reaction generally associated with quality changes during freezing, frozen storage, and thawing is lipid oxidation. This phenomenon most commonly occurs in fatty fish and is considered one of the major causes of frozen shelf-life reduction.

Lipid oxidation results in the development of a condition described as "oxidative fat rancidity." The extent of oxidation in fish lipids varies with the quantity and the type of lipids in the fish muscle, i.e., fatty species are more prone to oxidation than lean species, and species with more highly unsaturated fatty acids are less stable than the other species. When oxidative rancidity progresses sufficiently, it leads to the development of obvious off-taste and odor, resulting in reduced shelf life.

Changes in fish lipids may be related to changes in protein during frozen storage. Several reports indicate that the unstable free radical intermediates formed during autoxidation attack the protein molecules, leading to the formation of protein free radicals (Fig. 4) [39]. These protein free radicals may cross-link with other proteins to form protein-protein aggregates and with lipids to form protein-lipid aggregates [7]. Such free radicals may also initiate other reactions, as shown in Figure 4.

Another possible mechanism for reaction between oxidized lipids and proteins occurs through stable oxidation products such as malonaldehyde, propanal, and hexanal [26], which covalently react with specific functional groups on protein side chains, including the -SH group of cysteine, the amino group of lysine, and the N-terminal amino group of aspartic acid, tyrosine, methionine, and arginine [11]. Such interactions increase the hydrophobicity of proteins, making them less water soluble.

Free fatty acids (FFA) formed during autoxidation produce indirect effects on textural degradation by promoting protein denaturation [50]. FFA are believed to bind myofibrillar proteins, specifically actomyosin, rendering it unextractable [26,29]. According to Sikorski et al. [29], when the hydrophobic sites of FFA interact with protein molecules, the protein molecules become surrounded with a more hydrophobic environment, which subsequently results in a decrease in protein

LH + R ———————→ L + LOO + + LOOH + RH + S

PH + L + LOO + S ⇌ P + LH + LOOH + POO + PHS + PHSPH

   LH - LIPID          R - RADICAL

   PH - PROTEIN        S - LIPID SCISSION PRODUCTS

       P + P    ———————→    P P

       P + POO   ———————→    POOP

       P + L     ———————→    PL

       POO + L   ———————→    POOL

       POO + LH  ———————→    POOL + H + POOH + L

       PH + S    ———————→    PHS + PHSPH

**Figure 4** Potential reactions of proteins with lipid radicals and their oxidation products. (From Ref. 27.)

extractability. This interaction may occur through hydrophilic and hydrophobic forces [29].

Several techniques have been developed to assess the extent of lipid oxidation in fish muscle. The most common techniques include (1) the peroxide value (PV) test, which measures the amount of hydroperoxides or peroxides formed during autoxidation (this test provides only a means for predicting the risk of rancidity development) and (2) the thiobarbituric acid (TBA) test, which measures the amount of malonaldehyde formed upon the decomposition of hydroperoxides during the second stage of oxidative rancidity. Other methods are also undoubtedly available. Therefore, the choice of techniques depends on several factors, such as the accuracy required and the availability of equipment.

## VIII. EFFECTS OF FREEZING, FROZEN STORAGE, AND THAWING ON MICROBIOLOGICAL QUALITY AND SAFETY

It is readily apparent that spoilage changes in fresh fish occur most commonly as a result of bacterial activity. The species of bacteria vary according to storage temperature. In fish stored in ice, *Alteromonas, Achromobacter,* and *Flavobacter* spp. predominate. At temperatures between 35 and 55°C, *Micrococcus* and *Bacillus* spp. constitute the main microflora. Some of these microorganisms produce very active proteolytic enzymes, which produce odor, flavor, and textural problems.

When fish and seafood are frozen, the microorganisms present in their tissues are generally inactivated. Thus, during frozen storage, microbiological changes in fish tissue are usually minimal. Microorganisms not destroyed by the freezing process generally do not grow and in some cases die off slowly. Although some microorganisms survive storage at very low temperatures, their activities are suppressed, and bacterial numbers may be considerably reduced if recommended temperatures are maintained [63]. The temperature below which microbial growth is considered minimal ranges from −10 to −12°C [64]. Microorganisms, however, that survive and remain inactivated during frozen storage resume growth when the fish is thawed and may then lead to microbial spoilage of the thawed product.

Frozen fish are far from sterile and cannot therefore be considered a microbiologically safe product. The microbial activities in fish after thawing depend on the degree of freshness of the raw material, the natural microflora in the fish tissues, and the thawing technique utilized.

## IX. STABILITY OF FROZEN PRODUCTS

The effects of various freezing conditions on quality and shelf stability of frozen fish and seafood have received considerable attention recently. Studies have dealt with either the stability of the frozen product as related to storage temperature and fluctuations in storage temperature or the effectiveness of food additives in providing shelf stability to frozen products.

## A. Effects of Storage Temperature

The apparent effects of storage temperature on shelf-life stability of frozen fish are related to protein denaturation and lipid oxidation. The effects of temperature on protein denaturation have been comprehensively studied [65,66]. Maximum denaturation is reported to occur at −4°C in cod muscle [67], while changes in extractable proteins in haddock have been found to be greatest at −2 to −6°C [68].

The rate of lipid oxidation and the accumulation of FFA were observed to increase with temperature [49]. In a study using various species of fish, it was observed that maximum production of FFA due to enzymic activities of lipases occurred at −12 to −14°C [69], while the maximum rate of lipid hydrolysis was detected at temperatures just below freezing [70].

Storage at much lower temperatures can, therefore, prolong the shelf life of frozen fish. For example, cod stored at −160°C showed no detectable deterioration after 6 months of storage [49]. Even at −65 and −50°C, frozen samples exhibited very few changes after 9 months of storage. Such observations also suggest that low storage temperatures limit the problems associated with protein denaturation and lipid oxidation during frozen storage.

Several studies have been conducted in an attempt to determine the shelf life of frozen fish at different temperatures and to establish storage temperatures that can minimize quality deterioration in specific groups of fish. Poulter [71] reported that *Rastrelliger brachysoma* (club mackerel) stored at –10°C remained acceptable until the ninth month of storage, whereas samples kept at –30°C were rejected after 12 months of storage. *Scomber scombrus* (Atlantic mackerel) stored unwrapped at –18°C were rejected after 3 months of storage, while samples at –26°C remained acceptable until the sixth month of storage [72]. Early rejection of fatty species at relatively low temperatures is reportedly due to the development of rancid flavor and odor. Several studies have also reported the same dependence of shelf life for different fish species on temperature [15,73,74].

Clearly, fish composition has an appreciable effect on shelf-life stability of frozen fish. For instance, in a comprehensive study using different fish species, it was found that fatty fish such as mackerel, salmon, herring, sprat, and trout had a shelf life of 2–3 months at –18°C, whereas lean fish such as cod, flounder, haddock, ocean perch, and pollock exhibited storage stability of up to 4 months at the same storage temperature [75].

Based on several studies, it is recommended that those species most susceptible to oxidative rancidity be stored at very low temperatures (at least –29°C) while species less susceptible to rancidity should be stored at temperatures between –18 and –23°C [18]. For species with textural problems due to the TMAO breakdown, the storage temperature must be below –30°C.

## B.   Effects of Fluctuations in Temperature

Fluctuations in storage temperatures affect the shelf-life stability of frozen products due to an increase in the size of the ice crystals formed in fish tissues [26]. With slight increases in temperature, small ice crystals melt faster than larger ones, so that when the temperature drops again, the melted ice refreezes around the large ice crystals, forming larger crystals. These large crystals accelerate freezing damage, leading to shorter storage stability.

## C.   Use of Food-Grade Additives

The effectiveness of different food-grade additives has also received considerable attention recently. The most commonly used types of additives for fish and seafoods function either as antimicrobial agents or as antioxidants.

### 1.   Antimicrobial Agents

Additives are commonly used in the food industry to prevent the growth of bacteria, yeast, and molds. The selection of an antimicrobial agent or any combination of agents is rather complicated, especially when dealing with fish. The effectiveness of an antimicrobial agent depends on several factors, such as the moisture content of the product and the presence of other microbial inhibitors like smoke and salt.

Several antimicrobial agents have been tested for fishery products. For instance, the sorbic acid salt, potassium sorbate (KS), has been found useful in extending the shelf life of fresh fish. Studies have demonstrated that KS, when applied as part of the ice, increased ice storage stability of red hake and salmon up to 28 and 24 days, respectively [76,77]. KS, in combination with modified atmosphere packaging (MAP), was also determined to be an effective method for prolonging the shelf life of fresh whole and filleted haddock on ice [78].

The shelf life of fresh fish may also be extended under refrigerated conditions with the use of Fish Plus, which exhibits its preservative effect due to the combined action of components such as citric acid, polyphophates, and potassium sorbate. Citric acid lowers the muscle pH, which consequently creates an optimum environment for potassium sorbate to exhibit its antimicrobial effects. Dipping in Fish Plus has been found to extend the shelf life of lingcod on ice to as much as a week [79]. Fish Plus may also be used on frozen fish.

## 2. Antioxidants

An antioxidant is a substance capable of delaying or retarding the development of rancidity or other flavor deterioration due to oxidation. It is normally used in conjunction with freezing to reduce the rate of autoxidation during frozen storage. Antioxidants delay the development of rancidity by either interfering with the initiation step of the free radical reaction or by interrupting the propagation of the free radical chain reaction [80].

The first detailed kinetic study of antioxidative action by Bolland and tenHave [81], demonstrated that antioxidants act as hydrogen donors or free radical acceptors. Consequently, they suggested that as free radical acceptors (AH), they react primarily with ROO· and not with the R· radicals:

$$ROO· + AH \rightarrow ROOH + A·$$

A low concentration of this chain-breaking antioxidant (AH) can interfere with either chain or initiation, producing nonradical products:

$$A· + ROO· \rightarrow$$
$$\text{nonradical products}$$
$$A· + A· \rightarrow$$

Different versions of the antioxidant mechanism have been suggested by different authors [63,82]. Although these versions may take place, the original concept of Boland and tenHave [81] is still considered the most important [63].

Other antioxidants may function as metal-complexing agents, which partly deactivate the trace metals, often present as salts of fatty acids [63], which would otherwise promote the oxidative reaction. Citric, phosphoric, ascorbic, and erythorbic (isoascorbic) acids are typical metal-chelating agents.

Among these antioxidants, erythorbic acid was used in studies by Kelleher et al. [83] and Licciardello et al. [37] of shelf life stability of frozen fish. This antioxidant was emphasized due to encouraging results with the use of its salt, sodium erythorbate, in retarding oxidation in whiting, chub mackerel, and white bass fillets [37,84,85]. Licciardello et al. [37] demonstrated the effectiveness of erythorbic acid in the retardation of oxidative rancidity in fillet blocks of Argentine hake stored at −18°C.

However, the use of erythorbic acid is limited to fish species in which rancidity is the main problem. Kelleher et al. [83] demonstrated the effect of this compound on the frozen storage of red hake (*Urophysis chuss*), a gadoid species, in which lipid oxidation is not the limiting factor for shelf-life extension. They found that the rate of DMA formation at −18°C in samples dipped in erythorbate solution was significantly greater than the rate in untreated samples. Such effect of erythorbic acid on DMA formation may be explained by the fact that this acid acts as an alternative and preferred scavenger of oxygen, leaving metal ions, that would otherwise bind to oxygen and be inactivated, available to catalyze the degradation of TMAO to DMA and FA [86].

In summary, several methods have been developed to maintain the quality of fish under frozen conditions. It is evident that conditions during frozen storage must be properly controlled, since they determine the shelf-life stability of frozen fish.

## X. SUMMARY

Freezing, frozen storage, and thawing affect the quality and shelf stability of fish and seafoods. If kept under appropriate conditions, fish and seafood can be stored in the frozen state for several months without appreciable changes in quality.

During frozen storage, microbiological changes in fish and seafoods are very minimal. On the other hand, a series of changes such as protein denaturation, lipid oxidation, texture deterioration, loss of fresh odor and flavor, various enzymatically induced reactions, loss of volatile constituents, nutritional losses, and changes in moisture take place in fish and seafood when subjected to excessively prolonged frozen storage. Likewise, such changes may also occur in frozen and thawed fish and seafood.

Quantitative evaluation of the influence of freezing, frozen storage, and thawing on fish and seafood is rather complex. The many different variables that influence quality, as affected by these treatments, are related to one another. Therefore, it becomes almost impossible to describe some quality changes without actually discussing the other related changes that occur in fish tissues.

# REFERENCES

1. M. Love, Studies on the North Sea cod. I. Muscle cell dimensions, *J. Sci. Food Agric.* 9:609–622 (1958).
2. M. Green-Walker and G. Pull. A survey of red and white muscles in marine fish, *J. Fish Biol.* 7:295–300 (1975).
3. J. George, A histological study of the red and white muscle of mackerel, *Am. Midland Naturalist* 68:487–494 (1962).
4. H. Buttkus and N. Tomlinson, Some aspects of post-mortem changes in fish muscle, *The Physiology and Biochemistry of Muscle as a Food* (E. Briskey, R. Cossers, and T. Trautman, eds.), University of Wisconsin Press, Madison, WI, 1966, pp. 197–203.
5. T. Pitcher and P. Hart, *Fisheries Ecology*, AVI Publishing Co., Westport, CT, 1982.
6. S. Patterson and G. Goldspink. The effect of starvation on the ultrastructure of red and white myotomal muscle of Crucian carp (*Carassius carassius*), *Zellforschung* 146:375–384 (1973).
7. C. Castells, W. Neal, and J. Date, Comparison of changes in TMA, DMA, and extractable protein in iced and frozen stored gadoid fillets, *J. Fish. Res. Bd. Can.* 30:1246–1250 (1973).
8. W. Dyer and D. Hiltz, Sensitivity of hake muscle to frozen storage, *Halifax La. Fish. and Mar. Ser. Nova Scotia Circ. 45*, Halifax, Nova Scotia, 1974.
9. R. Shewfelt, Fish muscle hydrolysis—a review, *J. Food Biochem.* 5:79–100 (1980).
10. R. Boddeke, E. Slijper, and A. Van Der Stelt, Histological characteristics of the body musculature of fishes in connection with their mode of life, *Konik. Ned. Akad. Wetenschappen Ser. C.* 62:576–588 (1959).
11. S. Konosu, K. Watanabe, and T. Shimizu, Distribution of nitrogenous constituents in the muscle extracts of 8 species of fish, *Bull. Jap. Soc. Sci. Fish.* 40:909–915 (1974).
12. D. Crawford, D. Law, J. Babbitt, and L. McGill, Comparative stability and desirability of frozen Pacific hake fillet and minced flesh blocks, *J. Food Sci.* 44(2):363–367 (1979).
13. D. Crawford, D. Law, and J. Babbitt, Yield and acceptability of protein interactions during storage of cod flesh at −14°C, *J. Sci. Food Agric.* 16:769–772 (1972).
14. D. Crawford, D. Law, and J. Babbitt, Shelf-life stability and acceptance of frozen Pacific hake (*Merluccius productus*) fillet portions, *J. Food Sci.* 37:801–802 (1972).
15. E. Santos and J. Regenstein, Effects of vacuum packaging, glazing, and erythorbic acid on the shelf-life of frozen white hake and mackerel, *J. Food Sci.* 55:64–70 (1990).
16. R. Ahvenainen and Y. Malkki, Influence of packaging on shelf-life of frozen foods. II. Baltic herring fillets, *J. Food Sci.* 50:1197–1199 (1985).
17. M. Colokoglu and A. Kundacki, Hydrolytic and oxidative deterioration in lipids of stored frozen mullet (*Mugil cephalus,* L.), *Proceedings of the 6th World Congress of Food Science and Technology*, Dublin, Ireland, 1983.
18. F. Wheaton and A. Lawson, *Processing Aquatic Food Products*, John Wiley and Sons, New York, 1985.
19. M. Jadhav and N. Magar, Presevation of fish by freezing and glazing. II. Keeping quality of fish with particular reference to yellow discoloration and other organoleptic changes during prolonged storage, *Fish Technol*, 7:146–149 (1970).

20.  M. Bito, Studies on the retention of meat color by frozen tuna. I. Absorption spectra of the aquaeous extract of frozen tuna meat, *Bull. Jap. Soc. Sci. Fish.* 30(10):847–857 (1964).

21.  Y. Tauchiya and Y. Tatsukawa, Green meat of swordfish. *Tohotu J. Agric. Res.* 4(2):183 (1954).

22.  P. J. A. Reilly, M. Bernarte, and E. Dangla, Storage stability of brackishwater prawn during processing for export, *Food Technol. Aust.* (May-June): (1984).

23.  C. Ng, C. Tan, S. Nikkoni, Y. Chin, S. Yeap, and M. Bito, Studies on quality assessment of frozen fish. II. K-value, volatile bases, TMA-N as freshness indices and peroxide value and TBA number in rancidity of skin portion. *Refrigeration* 57(662):117–118 (1982).

24.  M. Bito, The observation on blood spots 'shimi' in the frozen-thawed carp meat, *Bull. Tokai Reg. Fish. Res. Lab.* 113:1–5 (1984).

25.  A. Ciarlo, R. Boeri, and D. Giannini, Storage life of frozen blocks of Patagonian hake (*Merluccius hubbis*) filleted and minced, *J. Food Sci.* 50:723–726 (1985).

26.  S. Shenouda, Theories of protein denaturation during frozen storage of fish flesh, *Adv. Food Res.* 26:275–311 (1980).

27.  Z. Sikorski, Protein changes in muscle foods due to freezing and frozen storage, *Int. J. Ref.* 1(3):173–210 (1978).

28.  Z. Sikorski, Structure and protein of fish and shellfish, Part II, in *Advances in Fish Science and Technology* (J. Connell, ed.), Fishing News (Books) Ltd., Surrey, England 1980, pp. 78–81.

29.  Z. Sikorski, S. Kostuch, and J. Lolodziejska, Denaturation of protein in fish flesh, *Nahrung* 19:997–1010 (1975).

30.  J. Matsumoto, Denaturation of fish muscle during frozen storage, *Protein at Low Temperature* (O. Fennema, ed.), ACS Symposium Series (Series #180), ACS, Washington, DC 1979.

31.  S. Jiang, M. Ho, and T. Lee, Optimization of the freezing conditions on mackerel and amberfish for manufacturing minced fish, *J. Food Sci.* 50:727–732 (1985).

32.  A. Borderias, M. Lamua, and M. Tejada, Texture analysis of fish fillets and minced fish by both sensory and instrumental methods, *J. Food Technol.* 18:85–95 (1983).

33.  E. Johnson, M. Peleg, R. Segars, and J. Kapsalis, A generalized phenomenological rheological model for fish flesh, *J. Text. Stud.* 12:413–425 (1981).

34.  M. Bourne, *Food Texture and Viscosity,* Academic Press, New York, 1983.

35.  N. Boyd, *Fish Quality Control,* Part 2, College of Fisheries, University of the Philippines, Quezon City, Philippines, 1981.

36.  M. Estrada, Spoilage pattern of 'alumahan' (*Rastrelliger kanagurta*) stored in ice, M.S. thesis, University of the Philippines, Quezon City, Philippines, 1983.

37.  J. Licciardello, E. Ravesi, and M. Allsup, Extending the shelf-life of frozen Argentine hake, *J. Food Sci.* 45:1312–1517 (1980).

38.  M. Jul, *The Quality of Frozen Foods,* Academic Press, London, 1984.

39.  M. Karel, K. Schaich, and R. Roy, Interaction of peroxidizing methyl linoleate with some proteins and amino acids, *J. Agric. Food Chem.* 23(2):159–163 (1975).

40.  J. Connell, Studies on the protein of fish skeletal muscle. VII. Denaturation and aggregation of cod myosin, *Biochem. J.* 75:530–535 (1960).

41.  L. Jarenback and A. Liljemark, Ultrastructural changes during frozen storage of cod.

III. Effect of linoleic acid and linoleic acid hydroperoxides on myofibrillar proteins, *J. Food Technol.* 10:437–445 (1975).

42.  S. Jiang, M. Ho, T. Lee, and C. Chichester, Interaction between free amino acids and proteins in vitro during frozen storage at $-20°C$, *Proceedings of the 6th World Congress of Food Science and Technology*, Dublin, Ireland, 1983.

43.  E. Childs, Interaction of formaldehyde with fish muscle in vitro, *J. Food Sci.* 38:1009–1011 (1973).

44.  S. Lewin, *Displacement of Water and Its Control of Biochemical Reactions*, Academic Press, London, 1974.

45.  D. Heldman, Food properties during freezing, *Food Technol.* 36(2):92–98 (1982).

46.  F. Ota and T. Tanaka, Some properties of the liquid portion in the frozen fish muscle fluid, *Bull. Jap. Soc. Sci. Fish.* 44:59–62 (1978).

47.  M. Love, Protein denaturation in frozen fish, *J. Sci. Food Agric.* 13:197–200 (1962).

48.  S. Jiang and T. Lee, Changes in free amino acids and protein denaturation of fish muscle during frozen storage, *J. Agric. Food Chem.* 33:839–844 (1985).

49.  W. Dyer and J. Dingle, Fish proteins with special reference to freezing, *Fish as Food*, Vol. I (G. Borgstrom, ed.), Academic Press, New York, 1961, pp. 275–277.

50.  F. Colmonero and A. Borderias, A study of the effects of frozen storage on certain functional properties of meat and fish protein, *J. Food Technol.* 18:731–737 (1983).

51.  K. Yamada, Occurrence and origin of TMAO in fishes and marine invertebrates, *Bull. Jap. Soc. Sci. Fish.* 33(6):591–603 (1967).

52.  National Marine Fisheries Services (NMFS), *Fish News,* New England Fish Development Foundation, Inc., Boston, 1986.

53.  R. Lundstrom, F. Correia, and K. Wilhelm, DMA and formaldehyde production in fresh red hake (*Urophysis chuss*): The effect of packaging materials, oxygen permeability, and cellular damage, *Int. Inst. Ref.* Boston, 1981.

54.  K. Amano and K. Yamada, The biological formation of formaldehyde in cod fish, *The Technology of Fish Utilization* (R. Kreuzer, ed.), Fishing News (Books) Ltd., London, 1965, pp. 73–87.

55.  K. Amano and K. Yamada, Studies on the biological formation of formaldehyde and dimethylamine in fish and shellfish. V. On the enzymatic formation in the pyloric caeca of Allaskan pollack. *Bull. Jap. Soc. Sci. Fish.* 31:60–65 (1965b).

56.  H. Tarr, Biochemistry of fishes, *Ann. Rev. Biochem.* 27:2223–2230, (1958).

57.  J. Spinelli and B. Koury, Non-enzymatic formation of dimethylamine in dried fishery products, *J. Agric. Food Chem.* 27:1104–1110 (1979).

58.  P. Yancey and G. Somero, Counteraction of urea destabilization of protein structure by methylamine regulatory compounds of elasmobrach fishes, *Biochem. J.* 183:317–320 (1979).

59.  K. Yamada, K. Harada, and K. Amano, Biological formation of formaldehyde and DMA in fish and shellfish. VIII. Requirement of cofactor in the enzyme systems, *Bull. Jap. Soc. Sci, Fish.* 35:227–231 (1969).

60.  C. Castells, B. Smith, and W. Neal, Production of dimethylamine in muscle of several species of gadoid fish during frozen storage, especially in relation to presence of dark muscle, *J. Fish. Res. Bd. Can.* 28:1–10 (1971).

61.  J. Walker, *Formaldehyde*, Reinhold, New York, 1964.

62.  T. Gill, R. Keith, and B. Lall, Textural deterioration of red hake and haddock muscle

in frozen storage as related to chemical parameters and changes in the myofibrillar proteins, *J. Food Sci.* 44:661–667 (1979).

63. H. Hultin, Characteristics of muscle tissue, *Food Chemistry,* 2nd ed. (O. Fennema, ed.), Marcel Dekker, New York. 1985, pp. 725–789.

64. C. Dellino, Influence of different freezing techniques freezer types, and storage conditions on frozen fish. *InfoFish Market. Digest* 2:40–44 (1986).

65. S. Hanson and J. Olley, *Technology of Fish Utilization* (R. Kreuzer, ed.), Fishing News (Books) Ltd., London. 1965, pp. 111–115.

66. W. Dyer and M. Morton, Storage of frozen plaice fillets, *J. Fish, Res. Bd. Can.* 13:129–134 (1956).

67. W. Dyer, Protein denaturation in frozen and stored fish, *Food Res.* 16:522–527 (1951).

68. G. Reay, The influence of freezing temperatures on haddock's muscle, Part 2, *J. Soc. Chem. Ind.* 53:265–270 (1933).

69. J. Olley, R. Pirie, and H. Watson, Lipase and phospholipase activity in fish skeletal muscle and its relationship to protein denaturation, *J. Sci. Food Agric.* 13:501–508 (1962).

70. J. Lovern and J. Olley, Inhibition and promotion of post-mortem lipid hydrolysis in the flesh of fish, *J. Food Sci.* 27:551–559 (1962).

71. R. Poulter, Quality changes in fish from the South China Sea. II. Frozen storage of chub mackerel, Paper presented at IPFC/FAO Conference on Fish Utilization, Technology and Marketing. Manila, Philippines, 1978.

72. P. Ke, D. Nash, and R. Ackman, Quality preservation in frozen mackerel, *J. Inst. Can. Sci. Technol. Aliment.* 9(3):135–138 (1976).

73. N. Screenivasan, G. Hiremath, S. Dhananjaya, and H. Shetty, Studies on the changes in Indian mackerel during frozen storage and on the efficacy of protective treatments in inhibiting rancidity, *Myosore J. Agric. Sci.* 10:296–305 (1976).

74. D. King and R. Poulter, Frozen storage of Indian mackerel and big eye, *Trop. Sci.* 25:79–90 (1985).

75. F. Bramnaes, Quality and stability of frozen seafood, *Foods: Time-Temperature Tolerance and its Significance* (W. VanArsdel, M. Copley, and R. Olson, eds.), Wiley Interscience, New York, 1969.

76. M. Fey, Extending the shelf-life of fresh fish by pottasium sorbate and modified atmosphere at 0–1°C. Ph.D. dissertation, Cornell University, Ithaca, NY, 1980.

77. M. Fey and J. Regenstein, Extending the shelf-life of fresh red hake and salmon using a carbon dioxide modified atmosphere and potassium sorbate ice at 1°C, *J. Food Sci.* 47:1048–1054 (1982).

78. J. Regenstein, The shelf-life extension of haddock in carbon dioxide atmosphere with and without potassium sorbate, *J. Food Qual.* 5:285–300 (1982).

79. Seafood Business Report, *Fish Plus:* 58 (1986).

80. W. Nawar, Lipids, *Food Chemistry,* 2nd ed. (O. Fennema, ed.) Marcel Dekker, New York, 1985.

81. J. Bolland and P. tenHave, Kinetic studies in the chemistry of rubber and related materials. IV. The inhibitory effect of hydroquinone on the thermal oxidation of ethyl linoleate, *Trans. Faraday Soc.* 43:201–210 (1947).

82. N. Uri, Mechanism of antioxidation, *Autoxidation and Antioxidants* (W. Wundberg, ed.) Interscience Publishers, New York 1960, pp. 133–169.

83. S. Kelleher, E. Buck, H. Hultin, K. Park, J. Licciardello, and R. Damon, Chemical and physiological changes in red hake blocks during frozen storage, *J. Food Sci.* 46:65–70 (1981).
84. R. Greig, Extending the shelf-life of frozen chub fillets through the use of ascorbic acid dips, *Fish. Ind. Res.* 4:23–29 (1967).
85. R. Greig, Extending the shelf-life of frozen white bass through the use of ascorbic acid dips, *Fish. Ind. Res.* 4:30–35 (1967).
86. J. Regenstein and C. Regenstein, *An Introduction to Fish Science and Technology*, Cornell University, Ithaca, NY, 1985.
87. K. Lagler, J. Bardach, R. Miller, and D. Passino, *Ichthyology*, 2nd ed., John Wiley and Sons, New York, 1974.
88. K. Harada, Studies on enzymes catalizing the formation of formaldehyde and dimethylamine in tissues of fishes and shells, *J. Shimonoseki Univ. Fish.* 23(3):157–163 (1975).

# 5
# Cured and Processed Meats

Roger W. Mandigo and Wesley N. Osburn

*University of Nebraska*
*Lincoln, Nebraska*

## I.  INTRODUCTION

The purpose of this chapter is to review the literature focusing on the effects of freezing, frozen storage, and thawing on the quality of cured and processed meat products. Assuming the reader has an understanding of the identity of various cured meat products, it should be noted that processed meats will refer to any meat product that has received nonmeat ingredients including whole muscle products or meat comminuted in some form to include ground beef and pork.

Cured and processed meats pose a particular challenge in determining the effects of freezing and frozen storage on product attributes since product quality is directly related to the quality of the raw materials used in their manufacture. Often the raw material sources for cured and processed meat products have been previously frozen. Additionally, cured and processed meats will contain ingredients (salt, spices, sodium nitrite, phosphate, smoke), added singly or in combination, that affect the quality, shelf life, and overall acceptability of these products and the intrinsic chemical reactions occurring within these products during frozen storage.

## II.  EFFECTS OF PREPARATION AND PACKAGING

### A.  Raw Material Preparation

Raw meat materials used in the production of cured meats consist of the entire whole muscle portion from primal or subprimal cuts (bone-in) or the muscles from these cuts (boneless). Comminuted processed meats generally consist of the lean and fat trimmings from meat animals. In the production of cured meats, the processing of ham and bacon includes introducing sodium nitrite into the product via injection, immersion, dry rubbing, or a combination thereof. Boneless cured meat products may undergo massaging or tumbling to extract the myofibrillar proteins necessary to bind the muscle pieces together, followed by thermal processing. Comminuted processed meat products are generally formulated by blending the desired amounts of lean and fat trimmings, during which various ingredients may be added. The lean and fat trimmings are reduced in size (by grinding, chopping, flaking, or emulsifying), formed (stuffed into casings or loaf pans or formed into patties), and then either thermally processed or quick frozen fresh prior to or after packaging before being refrigerated or frozen for storage.

A critical point in this process is the quality of the raw materials used to manufacture these products. A large percentage of the raw materials used to manufacture cured and processed meats are either frozen or were frozen and then thawed prior to processing. The functionality of these raw materials (e.g., water-holding capacity, emulsion stability, protein solubility) is affected by freezing rate, length of frozen storage, and packaging.

## 1. Functional Changes Due to Freezing and Thawing

Freezing is the cooling down of moisture-containing foods from a refrigerated temperature ($-2°C$) to a temperature below the freezing point (at least $-15°C$), at which temperature the product can be stored for prolonged periods without significant deterioration in its quality attributes. This freezing process obviously includes the congealing or crystallization of the moisture content and the high refrigeration requirements associated with freezing. The speed at which the zone of maximum crystallization is passed through determines the quality of frozen foods. Slow freezing (0.1–0.2 cm/h) causes very large ice crystals to be produced within the cells, thereby seriously damaging the cell walls. Quick freezing (5.0 cm/h) creates only very fine ice crystals, thereby causing little damage to the frozen tissue [1].

Muscle tissue is detrimentally affected by slow freezing. The moisture in the cell freezes into large crystals mainly in the intercellular spaces, resulting in colloids becoming very dehydrated. Connective tissue and occasionally even the muscle fibers themselves are damaged. These factors must be taken into consideration when utilizing frozen raw materials for the production of further processed meat products.

A major requirement for producing quality frozen products is to freeze rapidly. Most bacteria in food are mesophilic, and consequently growth can be hindered by lowering the temperature quickly. Any refrigerant that comes into direct contact with the surface of the meat improves the heat transfer coefficient, draws the whole surface of the meat into the heat transfer process, and increases the temperature gradient [2].

Foods are largely made up of water. The other components are carbohydrates, fats, proteins, and salts, which either form cell substances or are dissolved in cell moisture. Pure water becomes ice at 0°C, however, solutions do not begin to freeze until the temperature is below 0°C. If freezing takes place very slowly, pure water ice first crystallizes out, and the residual solution becomes increasingly concentrated, freezing at a lower temperature after some delay. This concentration process causes undesirable and irreversible changes in the frozen product [2].

Quick removal of heat results in the rapid formation of many crystallization nuclei, which will allow a solution to freeze without serious shifts in concentration. The more quickly a product is frozen, the more effectively it can be fixed in its original composition and its quality can be maintained, even after thawing [2]. This observation underscores the importance of rate of freezing of raw materials, which will ultimately affect the quality of a further processed meat product.

The functional changes occurring in the storage of frozen meat block components and the relationship of these properties to changes within frankfurters has been investigated [3]. Beef, pork, and pork fat were frozen and stored at −17.8°C from 1 to 37 weeks. Functional and quality tests [pH, water-holding capacity/emulsifying capacity, total extractable protein, 2-thiobarbituric acid (TBA), and peroxide value (for pork fat)] were performed at 6-week intervals on thawed and control meat samples and on frankfurters made from the sample [3].

Frozen storage increased drip loss (beef) and decreased total extractable proteins, water-holding capacity, and emulsifying capacity (beef and pork), supporting the theory that degradation, product formation, and interaction with muscle tissue during frozen storage were responsible for these results. Consequently, short frozen periods were recommended for meat block components intended for use in comminuted meat products [3].

The influence of freezing rate on weight loss during freezing, thawing, and cooking on the water-binding capacity, sensory properties, solubility of myofibrillar proteins, and other physicochemical properties of beef m. longissimus dorsi has also been investigated [4].

The greatest weight losses during freezing, thawing, and cooking were registered at slow freezing temperatures (freezing rates of 0.22 and 0.29 cm/h), resulting in lower water-binding capacity (WBC), decreased protein solubility, and tougher cooked meat. Quickly frozen meat (4.92 and 5.66 cm/h) had higher ratings than slowly frozen samples, but these ratings were lower than for samples with freezing rates of 3.33 and 3.95 cm/h [4].

Freezing rates of 3.33 and 3.95 cm/h exerted the least influence on physicochemical characteristics. Myofibrillar protein solubility, WBC, and tenderness of cooked samples were greatest at these freezing rates. Optimal conditions for meat freezing were concluded to be reached at an average freezing rate of 2–5 cm/h, which is achieved with temperatures somewhat below the eutectic point of meat (−40 to −60°C). Although this study [4] was conducted on a beef subprimal cut not usually cured and/or processed, it is important to note the effect of freezing rates on the functionality of meat tissues. Raw materials used to produce such products can be affected in the same manner, thereby determining the functionality and quality of the final product [4].

Seasoned (salt) and nonseasoned (no salt) pork patties formulated from pre- and postrigor pork were evaluated at 2-week intervals during a 24-week storage period (–20°C) to determine the functionality of prerigor pork on moisture cooking loss [5]. Protein degradation may occur during the processing of prerigor meat, resulting in a reduction of water-holding capacity, which may be negated by the addition of salt. Results of this study demonstrated that unseasoned postrigor-processed patties had lower moisture and cooking losses than unseasoned prerigor patties cooked to 75°C. The addition of seasoning (with salt) decreased moisture cooking loss [5].

Weiners were prepared from hot-boned (prerigor) beef raw materials after storage intervals of 7, 14, and 28 days at –10°C and 21 days at 2°C to determine the effects of storage temperature, length of storage, and levels of added salt on the functional and processing characteristics of hot-boned beef [6]. Level of added salt (0, 3, or 5%) did not affect any chemical, physical, or sensory trait of weiners prepared from prerigor beef raw materials stored at –10°C. All weiner batters exhibited low cookout values in the batter stability tests, and weiners were rated 5.0 or higher for each sensory trait evaluated (8-point scale). Percentage moisture in the finished product was 53.4% for frozen raw materials stored for 28 days, while percentage fat was 27.6% [6].

Batter stability was influenced by the length of frozen storage of the raw material. Values for fat released during cooking increased (0.0 ml at day 7 versus 0.6 ml at day 28) but were still indicative of stable batters. Smokehouse loss (%) decreased over frozen storage time. Although the differences were not significant, smokehouse loss was significantly correlated ($p < 0.05$) with days of frozen storage. The authors [6] suggested that this may partially explain why weiners prepared with prerigor beef stored for 28 days had significantly more moisture and less fat when compared to other storage treatment groups [6]. It was further concluded that the addition of salt may not be necessary to maintain desirable functional properties of prerigor raw materials if stored at –10°C.

## 2. Thawing and Tempering

Frozen meats are either thawed in air, thawed in water, or thawed in a microwave oven. Sausage ingredients in 60- to 100-lb boxes or bags are frequently thawed on pallets with spacer racks between each layer of boxes. This practice is costly in terms of floor space and drip loss, and it downgrades overall plant hygiene [7].

Drip loss as high as 5% (boxed cow or bull meat) is enough to discourage processors from using frozen/thawed meat. This drip or blood loss has a protein content of about 10%. At 5% drip loss, this represents 0.3 lb of protein from a 60-lb block. In a sausage product manufactured to a 4P + 10 formulation, this lost protein translates into a loss of 1.5 lb of finished product [7].

Thawing in water is virtually standard practice for hams and bellies. The cost of thawing in water must be considered from the standpoint of yield loss. Strict control is necessary to avoid excessive bacterial growth and cross-contamination in the thaw tank.

Thawing results in exposure of the meat surface to temperatures above freezing, leaving it susceptible to bacterial growth and other deteriorative changes while the center of the meat is thawing. Slow thawing results in long holding periods at high subfreezing temperatures where enzymatic activity occurs at the maximum rate [7].

Tempering, or partial thawing, can often be employed to reduce the problems of drip loss, bacterial growth, and other deteriorative changes. The use of tempered frozen blocks of meat allows better control over product quality and enables the use of a greater percentage of frozen ingredients in the manufacture of many processed meat products [7]

The effect of thawing rate on the amount of exudate produced by frozen beef has been studied by producing meat cylinders (5 cm in diameter and 8–9 cm in length), which were frozen (–20°C) utilizing slow freezing rates to produce extracellular ice crystals [8]. The frozen cylinders were then sliced at –20°C (0.6 cm thickness and 5 cm diameter for each slice) prior to thawing using a grooved acrylic holder to hold meat samples (in polyethylene bags) during thawing. Thermocouples were placed between the rows of grooves. The acrylic sample holder was connected to a vacuum pump to obtain sufficient adhesion between the meat and the polyethylene bags. Temperature at the center of each slice was recorded by a data logger. Thawing was complete when the sample reached –1°C. Thawed meat slices were placed over metallic grids mounted inside centrifuge tubes and spun for 10 minutes on each side at $2000 \times g$ in a refrigerated centrifuge. Samples were centrifuged after thawing and leaving them at 4°C for different periods of time. Slices of unfrozen meat were analyzed in the same manner. The amount of exudate was obtained by weighing and expressed as weight of exudate per weight of meat. Histological studies and measurement of fiber areas were also conducted.

The results of the study indicated that thawing rate by itself did not modify the potential water-holding capacity of tissues [8]. Thawing conditions affected the amount of exudate by its influence on time allowed for extracellular water resorption. Reduction in water-holding capacity depended on freezing and storage conditions.

A two-step mechanism was also identified for the thawing process [8]. The first step depended on thawing rate (ice melting), which led to water activity increments in the extracellular space. The second step involved water resorption by fibers due to the force generated by incremental water activity. High thawing rates resulted in less water resorption with more water accumulation in the extracellular space, thereby increasing the amount of exudate. Slow thawing rates allowed for water resorption, thereby decreasing the amount of exudate [8].

The amount of exudate was independent of thawing rate when the resorption period was long enough and nonresorbed water accumulated in the extracellular space, suggesting the main resistance to water resorption was the rate of water migration through the sarcolemma. Such findings are important in considering the factors that influence the water-holding capacity of raw materials used for further processed meats since they will directly influence the amount of water retained by that product [8].

## 3. Cryoprotectants

The use of cryoprotectants (5.6%) to stabilize functional properties (gel-forming ability, water-holding capacity, and protein solubility) of prerigor-salted beef and postrigor beef was measured during 6 months of frozen (−28°C) storage [9]. Salt (4% NaCl) accelerated the destabilization of muscle proteins with a subsequent decrease in functional properties, but this effect was reduced by cryoprotectants. The quality of prerigor-salted muscle treated with cryoprotectants and stored for 6 months was approximately equal to that of untreated postrigor meat, prior to freezing.

Cryostabilization of the functional properties of comminuted prerigor and postrigor beef that was vacuum packaged and stored frozen (−28°C) for 5 months was investigated [10]. The addition of 8% polydextrose, alone or in combination with 0.5% phosphates [1:1 mixture of sodium tripolyphosphate (STP) and tetrasodium pyrophosphate (TSPP)], were utilized to determine their effects. Samples were analyzed for pH, cook stability, protein extractability, gel-forming ability, and thermal transition by differential scanning calorimetry.

Polydextrose proved to be an effective cryoprotectant for both pre- and postrigor beef. The functionality of prerigor beef (cook loss, gel-forming ability) could be maintained or improved by addition of 8% polydextrose solution prior to frozen storage. After use of a cryoprotectant, it was no longer required to either salt prerigor beef or avoid thawing of frozen meat. Addition of polydextrose decreased the rate of decline in salt extractability of proteins, but prerigor samples with polydextrose still decreased by 37%. The protective effect of polydextrose on the functionality of frozen meat was greatest with addition prior to freezing, possibly since proteins were maintained in a less denatured state during frozen storage. Phosphates (STP and TSPP) had no cryoprotective effect but did increase pH and enhanced protein extractability, which may enhance gel-forming and water-binding properties. The authors concluded that addition of polydextrose could enhance and maintain the functionality of frozen manufacturing meats (prerigor beef) for processed meat product formulations [10].

Minced beef frozen and stored at −18°C for 6 months without additives, with salt (1.5% NaCl), tripolyphosphate (0.5%), sodium acid pyrophosphate (0.55), and an antioxidant mix (BHA + BHT, 200 ppm) used as cryoprotectants, and the beef batters prepared from this meat were evaluated for gelation and rheological properties [11]. Frozen storage resulted in increased water-holding capacity of the minced beef meat, except with the phosphate treatments. These results suggest that during freezing the number of charged sites in the system increased. Therefore, more water could be bound, when compared to unfrozen meat. The relationships between shear rate and shear stress among treatments resembled Bingham pseudo-plastic behavior. Higher modulus of rigidity values during cooking (0.5°C) revealed higher values for the unfrozen phosphate treatments. Addition of salt lowered the modulus of rigidity values of stored meat compared to the control.

## 4. Shrinkage and Freezer Burn

Weight loss during initial freezing and subsequent resorption of glaze and condensation moisture during storage and thawing were found to be the most important factors influencing total shrinkage of fresh hams [12]. As frozen storage time increased, the biochemical mechanisms that favor resorption of moisture by the muscle deteriorated at a steady rate. Total shrinkage was concluded to be a function of decreasing reabsorption ability of the hams more so than moisture loss during storage. Hams frozen in a blast freezer (–27°C) sustained less shrinkage compared to hams frozen in forced-air (–22°C) and still-air (–20°C) rooms, indicating that faster freezing of meat results in less shrinkage.

Freezer burn is a defect that develops on the surface of exposed frozen meat. It is usually in the form of patches of light-colored tissue, produced by evaporation of ice crystals, which leave air pockets between meat fibers. In hams, rehydration and curing will counteract a small amount of freezer burn at the termination of storage. However, excessive freezer burn results in reduced prices to compensate for trimming losses [13].

Forced-air freezing produces increased amounts of freezer burn on hams between 4 and 6 months of frozen storage compared to still air and blast freezing. However, this difference was negated at the end of 6 months of storage.

Green hams for the production of dry-cured hams were frozen on racks at –29°C, then stored in vats at –18°C for approximately 3 months before being divided into four groups of 15 hams each [14]. Group 1 hams were placed in cure while frozen. Groups 2, 3, and 4 were placed in cure after thawing at 3°C, 15°C, or in running water at 38°C, respectively. Hams were dry-cured for 4 weeks at 3–4°C and salt equalized for 4 weeks at 13°C before smoking for 24 hours at 38°C and aging for 3 months at 24°C. All hams were then sliced and evaluated for weight loss, raw color, aroma, and general appearance. Hams were also weighed fresh after curing, salt equalization, and after 1, 2, and 3 months of aging. At each processing step, the percent weight loss for the frozen group was less than for the other groups, with a significant difference observed between frozen and thawed-in-water hams (Table 1). Lower weight loss was attributed to a lower salt level in the frozen group. Since salt is a dehydrating agent, a lower salt content would be expected to result in a higher moisture content, which, in fact, was observed (54.5%) in the frozen ham group compared to cooler-thawed (49.9%), room-thawed (51.4%), and water-

**Table 1**   Weight Loss by Groups

|                      |          | Weight loss (%)         |                       |                        |
| -------------------- | -------- | ----------------------- | --------------------- | ---------------------- |
| Processing period    | Frozen   | Cooler-thawed, 3°C      | Room-thawed, 15°C     | Water-thawed, 38°C     |
| Cured                | $3.1^w$  | $6.1^x$                 | $5.4^{wx}$            | $5.0^{yz}$             |
| Smoked               | $12.9^x$ | $15.7^{xy}$             | $16.1^{xy}$           | $16.8^y$               |
| Aged 1 month         | $17.5^x$ | $21.0^{xy}$             | $20.6^{xy}$           | $21.4^y$               |
| Aged 2 month         | $22.0^x$ | $25.5^{xy}$             | $25.3^{xy}$           | $26.0^y$               |
| Aged 3 month         | $24.9^x$ | $29.3^{xy}$             | $27.8^{xy}$           | $28.7^y$               |

$^{w,x,y,z}$Means on same line with different superscripts are significantly different ($p < 0.01$).
*Source*: Ref. 14.

thawed (49.2%) hams. The salt content of aged center ham slices for the frozen group was only 3.9%, 5% for cooler and room-thawed hams, and 5% for water-thawed hams. From a processing viewpoint, this indicates that if frozen hams are to be cured, they should either undergo a longer period of curing or have more salt applied. From an economic viewpoint, the researchers indicated that, based on weight loss only, a considerable advantage was apparent in curing frozen hams [14].

## B.  Packaging Effects

Early studies indicated that freezing and storage of frozen meats produced problems concerning wrapping methods and materials, color stability, and package content identity [15]. A moisture-proof, oxygen-impermeable wrapping material is required to inhibit dehydration and flavor changes. Oxygen evacuation before sealing accelerates the color degradation of oxygenated meats, resulting in an unattractive frozen meat color.

Five packaging materials [Saran™ Wrap, Reynold's Heavy Duty Aluminum Foil, Saran™ overwrapped with aluminum foil, Resinite packaging film (polvinylchloride [PVC]), and Cryovac bags] and three storage times (13, 26, and 39 weeks) were used to determine packaging effects on water-holding capacity, thaw and cook losses, TBA values, moisture and fat contents, and sensory characteristics of ground pork [16]. Moisture and juiciness decreased and total moisture losses (thaw + cook) increased over time regardless of packaging. The PVC-packaged pork exhibited the highest decrease in water-holding capacity, while vacuum-packaged pork showed the least. The TBA values were highest for PVC-packaged and lowest for vacuum-packaged pork after 39 weeks of frozen storage. The authors suggested that ground pork purchased at retail in commercial PVC packaging should be repackaged prior to freezing to maintain "fresh pork" characteristics [16]. For consumers, the use of plastic wrap, aluminum foil, or a combination would be effective. Based on sensory data, frozen storage of ground pork should be limited to 26 weeks.

The rate of discoloration of a minced beef/texturized soy protein product (1% added salt) packaged in polyethylene tubes oxygen transmission rate (OTR—1000 $cm^3/m^2/24$ h/atm; 25°C, RH 75%) was analyzed by tristimulus colorimetry to determine color stability for 34 days of frozen (–18°C) storage [17]. Half of the tubes contained an ultraviolet light (UV) barrier (excluded light from 350 to 220 nm). Half of the surface of each tube was covered by black plastic, allowing two sets of treatments to be compared for each tube. Color stability was found to be significantly improved with the UV barrier in the packaging material. The researchers indicated that low-temperature conditions suppressed thermal autooxidation, allowing light-induced discoloration to exceed thermal discoloration below a limiting temperature [17]. Only below 5°C was the effect of thermal oxidation small enough to allow the UV barrier to protect the meat from photooxidation and color deterioration.

An extensive study was conducted to evaluate the effects of freezing, frozen storage, and protective wrap on cooking losses, palatability, and oxidative rancidity development in fresh and cured pork cuts. Hams and bellies were cured, and the loins and shoulders were divided into chops and roasts [18]. One loin chop, shoulder roast, ham steak, and group of bacon slices from each side was randomly assigned to 1 of 7 storage intervals (0, 56, 84, 112, 140, 168, or 196 days), and cuts within each storage interval were equally and randomly divided among four protective-wrap treatments (oxygen-permeable retail wrap, freezer paper, aluminum foil, and polyethylene bags). Cuts assigned to the 0-day storage interval were evaluated fresh, while cuts assigned to other intervals were frozen and stored for the appropriate time. All cuts were frozen at –30°C in still air. Cuts were thawed for 24 hours at 2°C prior to evaluation for proximate composition, cook loss, and sensory (flavor, juiciness, tenderness, overall eating satisfaction, and presence of rancid flavor) traits.

Total cooking losses did not differ among wraps after any storage interval for any of the four cuts or for bacon slices. Fresh ham steaks had lower cooking losses than ham steaks that were frozen and stored for any period of time in any of the four protective wraps evaluated. These results indicated that frozen storage exerted a greater effect on the cooking losses from ham steaks than from other fresh or cured pork cuts evaluated [18].

Acceptability of juiciness of both fresh and cured pork cuts did not differ among wraps at any storage interval. No differences were observed among wraps for bacon slices and ham steaks after any interval of frozen storage. Bacon slices and ham steaks were rated more acceptable in juiciness in the fresh state than bacon slices and ham steaks that had been frozen and stored for any period of time. The juiciness of all four cuts was acceptable throughout 196 days of storage, although ham steaks (any wrap) were lower after 140 vs. 112 days of storage, and bacon slices (aluminum foil or freezer paper) were lower after 84 vs. 56 days [18].

No differences were observed among wraps in acceptability of tenderness of either fresh or cured pork cuts at any storage interval. Ham steaks and bacon slices were rated more acceptable ($p < 0.05$) in tenderness in the fresh state than after frozen storage in any protective wrap. Ham steaks stored for 140 days or longer were less acceptable in tenderness ($p < 0.05$) than those stored similarly for shorter periods, which suggests tenderness deterioration during frozen storage [18].

Acceptability of flavor of both ham steaks and bacon slices was not different among wraps after any storage interval. Ham steaks (any wrap) and bacon slices frozen and stored for any period of time were less acceptable in flavor than their fresh counterparts. Practical storage limits for ham steaks were 140 days (retail wrap, aluminum foil, and freezer paper) and 112 days in polyethylene bags. Bacon slices were rated lower in flavor acceptability after 112 days of storage in all wraps compared to storage for shorter intervals. Bacon slices were generally rated as unacceptable in flavor after 112 days of storage, and ham steaks were approaching the point of being unacceptable after 196 days of storage regardless of protective wrap [18].

Overall eating satisfaction of both fresh and cured pork cuts did not differ significantly ($p > 0.05$) among wraps at any storage interval. Cured cuts were generally rated more acceptable in overall palatability ($p < 0.05$) when in the fresh state than after being frozen and stored. Ham steaks stored for 140 days or more in any wrap were generally rated lower in overall palatability than when stored for shorter periods. Overall eating satisfaction ratings for ham steaks (196 days) and bacon slices (112 days) in all wraps were generally at the point of unacceptability [18].

Taste panel evaluations failed to detect significant differences among wraps in the development of rancid flavors in any of the cuts at any storage interval. The flavor of bacon slices was rated as significantly ($p < 0.05$) more rancid after 112 days of storage (all wraps) than bacon slices stored for shorter periods. Ham steaks were rated more rancid after storage for 196 days than when stored similarly for shorter periods [18]. No consistent significant differences among protective wraps for any measured trait during frozen storage was found, indicating that the palatability of frozen pork was not affected by the type of wrap utilized [18].

Frozen pork patties (seasoned with salt or unseasoned) made from pre- and postrigor pork were subjected to vacuum or oxygen-permeable packaging to determine their effects on frozen storage stability [5]. An interaction between processing (pre- vs. postrigor) and packaging (vacuum vs. oxygen-permeable) was demonstrated, which showed the beneficial effect of hot processing when combined with vacuum packaging to reduce lipid oxidation. It was therefore concluded that salt greatly enhanced lipid oxidation but vacuum packaging greatly inhibited this oxidation [5].

Frozen ($-30°C$) ham steaks and bacon slices were wrapped in oxygen-permeable retail wrap (moisture vapor transmission rate = 22 g/100 sq in./24 h; $O_2$ transmission rate = 505 cc/100 sq in./24 h; $CO_2$ transmission rate = 4720 cc/100 sq in./24 h), freezer paper, polyethylene bags, and aluminum foil to determine the influences of frozen storage and subsequent thawing on the acceptability of cured pork cuts after 24 hours of retail display [19]. Results indicated that regardless of the protective wrap utilized, ham steaks and bacon slices could be stored (still air, no light, −30°C) for at least 140 and 112 days, respectively, and retain an appearance equally acceptable as their fresh counterparts upon thawing (24 h at 2°C) using simulated retail display (on a table in a walk-in cooler, at 2°C under 820 lux of incandescent lighting). No differences were noted for any protective wrap treatment regarding thaw-drip losses, muscle color scores, surface discoloration, psychrotrophic contamination, or off-odors.

The identification and quantification of volatiles generated during lipid oxidation in ground pork (30% fat) packaged in vacuum bags, polyvinyl chloride film, Saran Wrap™, aluminum foil, or Saran™–aluminum foil combination and stored frozen (−17°C) for 13, 26, and 39 weeks was investigated [20]. Samples were analyzed for TBA, pH, sensory traits (odor and flavor), and volatiles measured by gas chromatography/mass spectrometry. Rancid and acid odor and flavor scores and TBA values increased over the duration of storage regardless of packaging treatment. Vacuum-packaged pork had lower scores and PVC-packaged pork had higher scores after all storage periods than pork in any of the other packaging materials used. The exclusion of oxygen during frozen storage appeared to preserve ground pork quality by reducing lipid oxidation [20].

## III. EFFECTS OF FREEZING, FROZEN STORAGE, AND THAWING ON COLOR, APPEARANCE, AND CONSUMER ACCEPTANCE

### A. Color and Appearance

#### 1. Comminuted Products

An experiment to evaluate the effect of frozen storage upon the quality of frozen pork sausage samples and nonfrozen (5°C) samples packaged in modified atmospheres was conducted [21]. Six gas combinations of carbon dioxide and nitrogen gas (0:100, 20:80, 40:60, 60:40, 80:20, and 100:0%), and three control atmospheres (air only, vacuum pack, and a 60:40 $CO_2$:$N_2$ mix) were used. All samples were stored in darkness for 20 days with one half at 4°C and the remainder at −28°C. Hue as an indicator of color deterioration was analyzed, as well as sensory, textural, microbial, and functional properties. The most desirable modified atmospheres to increase the retail shelf life of fresh pork sausages were those with high carbon dioxide concentrations (60–80%), which reduced microbial counts, increased flavor and acceptability scores and water-holding capacity, and improved texture. Color scores decreased with increasing carbon dioxide levels, but this was negated somewhat by freezing sausage prior to retail display. Freezing increased the shelf life of pork sausages.

The effect of UV barriers in packaging materials to reduce discoloration in frozen products was investigated [17]. The color stability of minced beef [ground beef with 70% texturized soy protein (4%), salt (1%), and water (14%)] was packed in polyethylene tubes [oxygen transmission rate approximately 1000 $cm^3/m^2/24$ h/atm (25°C, RH 75%)]. A hydroxybenzophenone UV absorber was added to the polyethylene and used for one half of the minced beef samples, which excluded light in the region of 350 to 220 nm. Color stability was improved by the UV barrier in the packaging material for frozen (–18°C) samples illuminated (520 lux) with fluorescent tubes. In general, a positive effect on color stability of beef with barriers excluding light below 350 nm can be expected for storage temperatures up to approximately 5°C [17].

Mutton, pork, and beef sausages prepared from frozen (–18°C) minced meats stored for up to 52 weeks were subjectively evaluated for color by a sensory panel [22]. Results failed to establish any significant change due to duration of storage with respect to color, since objective measurements did not detect any differences among treatments.

The effects of salt (1%) and light on oxymyoglobin oxidation in frozen pork patties were investigated during freezer storage at –18°C for up to 31 days when placed in the upper layer of an illuminated (700 lux) freezer cabinet or covered with black plastic [23]. The patties (added salt or no salt) were packaged (two per pouch) using two types of polyethylene packaging materials, with and without added hydroxybenzophenone as a UV-light absorber.

Surface color was measured using the Hunter tristimulus colorimeter to monitor pigment oxidation on the product surface. At the end of the experiment, severe discoloration was noted in the pork patties with added salt. Light also affected pigment oxidation, but not as much as salt. The incorporation of a UV barrier in the packaging material gave some protection against light-induced discoloration, thereby prolonging color stability [23].

All beef and soy-extended patties (20% rehydrated soy) were frozen to –18°C for either 24, 48, 72, or 96 hours, stacked 16 patties high, placed in boxes, and stored at –23, –18, or –7°C for 0, 6, 9, 12, 18, or 24 months prior to evaluation. Various quality factors, including sensory evaluation of surface discoloration, freezer burn, and lean color, were measured [24]. Surface discoloration and lean color determinations were made on patties from the interior portions of six patty stacks in each box. Freezer burn was evaluated on the top patties of six patty stacks within a box. Patties stored at –7°C had greater surface discoloration and freezer burn compared to patties stored at –18 or –23°C. Patties stored at –7°C were also rated by panelists as being darker in color and as having less of the light-colored pigments than patties held at –18 or –23°C. Results indicated storage of patties at –7°C should be avoided due to reductions in quality after 6 months of storage. Longer frozen storage produced gradual but continual deteriorations in quality [24].

The sensory and physical characteristics of beef patties containing 20% fat, 8% fat, or 8% fat plus 20% soy protein isolate, soy flour, soy concentrate, or a mixture of carrageenan (0.5%), starch (0.5%), and phosphate (0.2%) were compared after 0, 4, 8, and 12 weeks of storage at –18˚C [25]. Instrumental color determinations were conducted on the surface of the thawed, raw samples. Spectral reflectance was measured every 10 nm over a 400- to 700-nm range under illuminant A with a colorimeter. Percent reflectance at 630 and 580 nm was used as a measure of the red color attributed to oxymyoglobin. Reduction of fat content in all beef patties had no effect on Hunted a* value, but 8% fat patties were darker and less yellow. Patties containing the treatment mixture were lighter, redder, and more yellow than the 8% fat control patties. Differences in lightness were related to fat content and were large enough to possibly influence consumer acceptance of the raw product. Patties stored for 8 or 12 weeks exhibited lower Hunter a* and b* values than patties stored for shorter amounts of time. Red color attributed to oxymyoglobin decreased during the first 4 weeks and again during the second 4 weeks of storage. Pigment oxidation could have been the reason for red color differences attributed to oxymyoglobin, and Hunter a* and L* values. Prolonged storage also decreased Hunter a* and b* values of patties [25].

Ground beef patties containing 15 or 30% fat were packaged in polyethylene bags or vacuum packaged and stored at –12.2, –23.3, or –34.4˚C for 20 weeks to determine the effect of fat, packaging, storage time, and temperature on yields, shear strength, and color (L, a, and b) values [26]. The L and b values and total color difference were affected by fat content and storage time, while storage temperature was insignificant. Total color difference increased and L, a, and b values decreased with both fat content and storage time. Patties containing more fat had higher lightness values than patties containing less fat. Increasing storage time resulted in color loss, and patties tended to become gray in color. The decrease in red color might be explained by the formation of myoglobin and its inability to become oxygenated with time [26].

Physicochemical and sensory effects of 0.4% sodium tripolyphosphate, tetrasodium pyrophosphate, and three commercial phosphate blends were studied in frozen beef patties (–20˚C) over a 90-day storage period. Color changes in phosphate-treated patties were evaluated by determining Hunter L*, a*, and b* values on patties thawed at 24–25˚C for 4 hours [27].

Frozen (–15˚C), salted, ground pork with carnosine (0.5 and 1.5%), sodium tripolyphosphate (0.5%), α-tocopherol (0.02%), or BHT (0.02%) was formulated and stored for up to 6 months in 500-g Whirl-pak bags and evaluated for oxidative changes-TBA-reactive substances (TBARS), color, rancid odor] [28]. Color was evaluated after one month of frozen storage (–15˚C) by an eight-member sensory panel. Color was rated under fluorescent lighting using a 130-mm scale anchored by the descriptive terms "red" and "brown." Samples containing carnosine exhibited greater maintainance of red color than all other samples after one month of storage. The ability of carnosine to prevent metmyoglobin formation could be due to its ability to maintain heme iron in the ferrous state [28].

## 2. Restructured Products

The color and color stability of restructured beef steaks made from semimembranosus muscle with various binders (calcium alginate, crude myosin extract, whey protein concentrate, wheat gluten, soy isolate, or surimi) were compared to controls (intact ribeye muscle, restructured steaks with no additives or with NaCl and sodium tripolyphosphate) to determine effects on initial surface metmyoglobin concentration and sensory color attributes during a 12-week frozen storage period (–23°C) [29]. Binder levels were (1) calcium alginate treatment, (2) calcium carbonate, 0.1%, sodium alginate, 0.5%, and encapsulated lactate, 0.5%, (3) extracted beef myosin treatment, 8.5%, (4) whey protein concentrate, 2.5%, (5) isolated soy protein, 1.5%, (6) wheat gluten, 2.0%, and (7) surimi treatment, 1.5% (Table 2).

Initial surface metmyoglobin (absorbance) was lower ($p < 0.01$) for all treatments. NaCl + STP, crude myosin extract, whey protein, and surimi all had lower ($p < 0.05$) initial surface metmyoglobin concentrations than restructured steaks from calcium alginate or soy protein isolate. Steaks made with various binders showed similar effects on initial surface metmyoglobin concentration and sensory color attributes, except steaks made with calcium alginate or soy isolate protein. During 12 weeks of frozen storage, steaks made with various binders (except soy protein isolate) had similar color stabilities (Table 3).

Surface metmyoglobin concentrations of products increased ($p < 0.01$) with

**Table 2**  Treatment Formulations and Preparation Procedures

| Treatment | Binder (%) | Mixing time (min) | Water (%) | NaCl (%) | STP (%) |
|---|---|---|---|---|---|
| Intact ribeye | — | — | — | — | — |
| Calcium carbonate | 0.1 | | | | |
| Sodium alginate | 0.5 | | | | |
| Encapsulated lactate | 0.5 | 5 | 7.0 | — | — |
| Salt & STP | — | 3 | 7.0 | 1.00 | 0.50 |
| Crude myosin[a] | 8.5 | 5 | (7.0) | 0.13 | 0.05 |
| Whey protein | 2.0 | 5 | 7.0 | 0.13 | 0.05 |
| Wheat gluten | 2.0 | 5 | 7.0 | 0.13 | 0.05 |
| Soy isolate | 1.5 | 5 | 7.0 | 0.13 | 0.05 |
| Surimi | 1.5 | 5 | 7.0 | 0.13 | 0.05 |
| Restructured control | — | 5 | 7.0 | — | — |

[a]Protein content of crude myosin extract was 6%. Final crude myosin content in the product was 0.5%
*Source*: Ref. 29.

**Table 3** Effects of Various Binders on Overall Color Score[a] of Restructured Beef Steaks During Storage

| Treatment | Color score | | | |
|---|---|---|---|---|
| | 0 weeks | 4 weeks | 8 weeks | 12 weeks |
| Intact ribeye | 6.83 | 7.05 | 7.11 | 6.58 |
| Calcium carbonate | 4.53 | 4.77 | 4.54 | 5.10 |
| Salt & STP | 6.25 | 6.27 | 5.19 | 3.83 |
| Crude myosin | 5.37 | 4.63 | 4.52 | 5.03 |
| Whey protein | 5.72 | 5.59 | 4.34 | 5.15 |
| Wheat gluten | 5.50 | 4.63 | 4.23 | 4.82 |
| Soy isolate | 5.38 | 5.18 | 3.76 | 3.85 |
| Surimi | 5.41 | 4.45 | 5.03 | 4.40 |
| Restructured control | 5.93 | 5.20 | 5.34 | 5.42 |

[a]Overall color source by trained sensory panel using 8-point scale (1 = extremely brown; 8 = extremely red).
$LSD_{0.05}$ = 1.088 for comparison of means between treatment for a given storage time.
$LSD_{0.05}$ = 0.371 for comparison of means by storage time within a given treatment.
*Source*: Ref. 29.

storage time. The addition of salt increased the rate of metmyoglobin formation. Restructured calcium alginate steaks had higher ($p < 0.05$) initial surface metmyoglobin concentration but slower ($p < 0.05$) metmyoglobin-formation rates than the other treatments during frozen storage. Calcium carbonate appeared to have a protective effect on myoglobin oxidation rate, which might be due to an increase in pH or to an anion or cation effect [29].

Addition of binders did not effect tristimulus color "L" (lightness), "a" (redness), and "b" (yellowness) color values. Storage time caused "L" values to decrease at the beginning of the storage period and then to increase during the 8- to 12-week storage period. Initial concentration of surface metmyoglobin (reflectance) was lowest ($p < 0.05$) for intact ribeye steaks and highest ($p < 0.05$) for restructured steaks with whey protein. Calcium alginate steaks were lower ($p < 0.05$) in surface metmyoglobin concentration than steaks made with other binders. Panelists also found differences ($p < 0.05$) between treatments in overall color score. Intact ribeye and NaCl + STP steaks had the highest ($p < 0.05$) and calcium alginate steaks the lowest ($p < 0.05$) initial overall color scores (red vs. brown). After 12 weeks of storage, NaCl + STP and soy protein isolate steaks had the lowest scores $p < 0.05$), and intact ribeye scores were the highest ($p < 0.05$).

Color scores were affected by storage time. Color scores decreased for all treatments except intact ribeyes and calcium alginate steaks. Overall color scores of calcium alginate steaks increased as storage time increased. Results from visual surface discoloration appraisal indicated that intact ribeye steaks and steaks containing crude myosin extract, calcium alginate, or wheat gluten had better color stability than steaks from other treatments. Calcium alginate, crude myosin extract, and wheat gluten could be used as effective binders in restructured beef steaks since they did not adversely effect product color or color stability over a 12-week frozen storage period [29].

Restructured beef, pork, and turkey steaks containing salt, sodium pyropolyphosphate, and antioxidants were investigated to determine the role of these nonmeat ingredients on lipid oxidation and discoloration of restructured steaks during frozen storage at $-10°C$ for 16 weeks (beef), 8 weeks (pork), and 8 weeks (turkey) [30].

Restructured beef, pork, and turkey steaks were formulated to contain salt (0%, 1.5%, w/w), salt (1.5%) and sodium pyropolyphosphate (0.3%, w/w), salt (1.5%) and ascorbyl palmitate/$\alpha$-tocopherol (0.02%, based on final fat content), or ice (3%, w/w). Pork and turkey restructured steaks also had additional treatments of salt (1.5%) and tertiary butylhydroquinone (0.02%, based on final fat content). Steaks were wrapped in freezer wrap (1-mm low-density polyethylene).

Lipid oxidation and discoloration occurred simultaneously in pork and turkey, but color loss occurred much earlier than lipid oxidation in beef (Table 4). Phosphates were effective in inhibiting lipid oxidation in pork and turkey steaks, but neither prevented discoloration overall. Results indicate that discoloration and lipid oxidation are interrelated and that pigment oxidation may catalyze lipid oxidation [30].

The effect of mixing and grinding, tempering, addition of NaCl + STP, and processing under a nitrogen atmosphere was investigated on the color of restructured beef steaks [31]. The treatment which included a tempering period [frozen $(-34°C)$ for 48 hours, tempered $(-2°C)$ for 24 hours, refrozen $(-34°C)$ for 12 hours] and the addition of NaCl without STP produced the most rapid increase in rate of discoloration during frozen storage. This rapid rate of discoloration was apparently due to the significant increase in the rate of overall metmyoglobin (myoglobin oxidation) formation, which occurred predominantly during the first month of storage.

The increase in myoglobin oxidation was possibly due to the combination of thawing and fluctuating temperatures resulting in large ice crystal formation [31]. Enhanced ice crystal growth and formation due to temperature fluctuations might enhance shrinkage of muscle fibers and ultimately disrupt muscle cell integrity, resulting in a greater chance of oxidative catalysts coming into contact with myoglobin.

The effect of different gas atmospheres (air, $CO_2$, 8% $O_2$ + 92% $CO_2$, $O_2$) and vacuum-packaging was investigated in another study [32] on the color and color stability of frozen restructured beef steaks during a 3-month storage period at $-23°C$. Initial overall metmyoglobin formation was not different in all restructured steaks from that of the intact control steaks, with the exception of those steaks packaged in $O_2$. The researchers postulated that when equilibrated meat $(2°C)$ was mixed in an $O_2$ atmosphere, the oxygen concentration between the meat particle

**Table 4** Mean Hunter $a_L$ Values for Restructured Pork and Turkey Steaks

| Species | Storage (weeks) | C | S | SP | SAT |
|---|---|---|---|---|---|
| Beef | 0 | 15.71$^{afv}$ | 16.33$^{afv}$ | 15.46$^{afv}$ | 15.97$^{afv}$ |
| Pork | 0 | 14.50$^{amv}$ | 14.38$^{bmv}$ | 14.67$^{amv}$ | 14.85$^{amv}$ |
| Turkey | 0 | 14.69$^{aqv}$ | 15.37$^{aqv}$ | 15.93$^{aqw}$ | 15.06$^{aqv}$ |
| Beef | 1 | 14.30$^{agv}$ | 10.98$^{agw}$ | 11.43$^{agw}$ | 11.79$^{agw}$ |
| Pork | 1 | 13.68$^{amv}$ | 12.44$^{anv}$ | 13.44$^{bmv}$ | 12.12$^{anv}$ |
| Turkey | 1 | 13.25$^{aqv}$ | 8.34$^{brw}$ | 11.58$^{arv}$ | 8.20$^{brw}$ |
| Beef | 2 | 12.91$^{agv}$ | 10.16$^{agw}$ | 10.36$^{agw}$ | 10.43$^{agw}$ |
| Pork | 2 | 10.47$^{bnv}$ | 8.09$^{bow}$ | 10.41$^{anv}$ | 8.24$^{bow}$ |
| Turkey | 2 | 11.37$^{arsv}$ | 8.37$^{brw}$ | 8.48$^{bevn}$ | 7.63$^{brw}$ |
| Beef | 3 | 12.83$^{aghv}$ | 9.88$^{agw}$ | 8.61$^{abhw}$ | 8.58$^{shw}$ |
| Pork | 3 | 12.50$^{anv}$ | 7.33$^{bopw}$ | 10.32$^{anv}$ | 7.18$^{bopw}$ |
| Turkey | 3 | 10.02$^{aghv}$ | 7.70$^{brsw}$ | 7.88$^{buw}$ | 6.86$^{brw}$ |
| Beef | 4 | 12.31$^{aghv}$ | 7.91$^{ahw}$ | 8.06$^{abhw}$ | 8.78$^{ahw}$ |
| Pork | 4 | 11.70$^{anv}$ | 7.16$^{aopw}$ | 9.21$^{anox}$ | 7.00$^{bopw}$ |
| Turkey | 4 | 9.43$^{bsv}$ | 7.02$^{arsw}$ | 6.91$^{buw}$ | 6.88$^{brw}$ |
| Beef | 6 | 11.05$^{ahv}$ | 7.36$^{ahw}$ | 7.22$^{shw}$ | 7.90$^{ahw}$ |
| Pork | 6 | 11.17$^{anv}$ | 5.69$^{apw}$ | 8.22$^{sox}$ | 6.06$^{bpw}$ |
| Turkey | 6 | 9.58$^{asv}$ | 6.35$^{asw}$ | 7.07$^{sux}$ | 6.54$^{brx}$ |

Each value is the mean of three samples/treatment/species/week. Control (C): salt—1.5% (S); salt—1.5% + phosphate—0.3% (SP); salt—1.5% + alpha-tocopherol/ascorbyl palmitate—0.02% (SAT).
[a-c]Means with different superscripts in the same column, within a storage time, are significantly different ($p < 0.05$).
[f-u]Means with different superscripts in the same column for beef (f–h), pork (m–p), and turkey (q–u) are significantly different ($p < 0.05$).
[v-y]Means with different superscripts in the same row for the same species are significantly different ($p < 0.05$).
*Source*: Ref. 30.

interface reached a critical level, which resulted in greater metmyoglobin formation [32]. This would indicate that the amount of oxygen diffused into meat particles pre- and postmixing was a possible decisive factor in formation of initial metmyoglobin formation.

Restructured steaks in a pure oxygen atmosphere had the greatest rate of overall metmyoglobin formation, occurring mainly during the first month of storage, possibly due to a chloride anion–promoted myoglobin autoxidation accelerated by high oxygen concentration in close proximity to the meat particles. Surface metmyoglobin formation rates were greater than overall metmyoglobin formation rates during frozen storage, especially during the first month of storage. This may be due to oxygen absorbed onto the surface of the steaks after slicing, which remained during vacuum packaging, or to not enough vacuum, resulting in discoloration due to metmyoglobin formation. Sensory color scores reinforced the results obtained by chemical analyses [32].

Carbon dioxide gas treatments did not affect pH, and overall metmyoglobin concentrations compared favorably with intact steaks. As the oxygen content of the gas atmosphere increased, rates of overall metmyoglobin content increased [32].

Flaked, cured pork was manufactured using 25 combinations of NaCl and STP [33]. The effect of frozen storage was evaluated at 3-week intervals from time of production through 18 weeks at –23°C. Color was evaluated by a laboratory sensory panel of 10–12 people using a scale of 1 to 9, with 9 being the most acceptable. A color-difference meter was used to objectively evaluate the color of each product. Measurements were taken directly through the vacuum packaging. As time of frozen storage increased, the color of the products faded and became less acceptable. However, all products were found to be aceptable initially and after 18 weeks of storage [33].

## 3.  Whole Muscle Products

Frozen green hams (stored in vats at –18°C for 3 months) were divided into groups and cured either frozen or thawed (3°C, 15°C, or in running water at 38°C). They were used to produce dry-cured hams. After curing, salt equalization, smoking, and aging, the hams were sliced and the exposed surface of a slice was subjectively evaluated for raw color [13]. No significant differences existed in subjective evaluation criteria of the color of the cut surface of any group, and all slices were rated as being red in color (approximately 3.0 average on a 4.0 scale, with 4.0 being dark red).

Ham steaks and bacon slices were frozen and stored at –30°C in still air and in four protective wraps (oxygen-permeable, polyethylene, aluminum foil, and freezer paper) to determine the effects of frozen storage and thawing on the acceptability of cured pork after retail display [19]. After 24 hours of simulated retail display, each cut was evaluated for muscle color and extent of surface discoloration. In general, muscle color scores and amount of surface discoloration did not differ among wrap treatments within storage intervals for any cut. However, ham steaks and bacon slices could be stored for 140 and 112 days, respectively, and still be considered acceptable.

## B.  Consumer Acceptance

Frozen (–23°C) restructured pork patties were evaluated over a 6-week period to determine the effects of cooking method, reheating method, and frozen storage on physical, chemical, and palatability factors [34]. An untrained panel evaluated the pork patties for color, visual texture, flavor, eating texture, juiciness, and aroma using a 9-point scale, with 1 being unacceptable and 9 being excellent. Juiciness and aroma were evaluated on a scale of 1 to 9, with 5 being the most acceptable.

Panel evaluations as affected by storage time indicated a higher mean color score immediately after cooking and reheating than when reheated following 6 weeks of postcooking storage. No differences between storage times were found for any other trait. It was concluded that the storage times used in the study were adequate for product protection. Mean panel evaluation scores for all traits were acceptable initially and after 6 weeks of storage.

Ground beef patties (20% fat) were frozen in liquid nitrogen or liquid carbon dioxide in a cryogenic tunnel at $-74°C$ or in a walk-in mechanical freezer at $-29°C$ and then stored in plastic-lined boxes for 120 days at $-29°C$ [35]. A consumer sensory panel (approximately 70 university students) evaluated samples from the three freezing treatments using a 9-point hedonic scale for four quality attributes: flavor, texture, juiciness, and overall acceptability. The consumer panel also evaluated samples from the freezing treatments as a complete sandwich including a bun and condiments added free choice.

Consumer panel scores for overall acceptability of ground beef patties and complete sandwiches with condiments were lower for the mechanically frozen treatment than carbon dioxide and nitrogen treatments. Carbon dioxide and nitrogen freezing in a cryogenic tunnel produced ground beef samples higher in overall acceptability than the same freezing treatments when samples were consumed with a bun. This observation was attributed to the faster rate of freezing in the cryogenic freezing treatment, which produced greater quality retention than mechanical freezing [35].

The effects of temperature, light, and storage period on the physicochemical and sensory (appearance and off-odor) changes in vacuum- or nitrogen-packed meat loaves were evaluated after 1, 3, 7, 14, 21, 28, 35, and 49 days of storage at $-4$, 0, 3, and $7°C$ under dark and lighted display [36]. At the end of each storage period, samples were subjectively evaluated under incandescent light (920 lux) by four panelists using a 7-point hedonic scale for appearance and odor. Slight or advanced green discoloration terms were also available to describe surface color attributes of the loaves and exudate.

Overall appearance ratings were based mainly on surface discoloration and appearance of product resulting from microbial growth or pigment color alteration. No significant change was observed in the overall appearance of either vacuum or nitrogen-packed samples at $-4°C$. After 14 days of storage, the appearance of meat loaves stored at 0, 3, and $7°C$ had deteriorated, mainly due to microbial growth. Light accelerated the dissociation of nitric oxide from the heme and then the heme reacted further to produce a greenish porphyrin compound in vacuum-packaged samples which was most likely due to the high numbers of lactobacilli found in this study [36].

Seasoned (salt) and nonseasoned (no salt) pork patties formulated from pre- and postrigor pork were evaluated at 2-week intervals during a 24-week storage period ($-20°C$) to determine the functionality of prerigor pork on moisture cooking loss

[5]. Taste panel results, based on a hedonic scale of 1 to 7, indicated that hot processing had little effect on rancidity development. Seasoning (salt) added prior to frozen storage reduced flavor scores significantly, while vacuum packaging improved the flavor of seasoned patties to a level similar to that of nonseasoned patties stored in an air-permeable packaging. The latter was rated higher for flavor than all other treatments.

An untrained, 50-member consumer panel evaluated weiners prepared from hot-boned (prerigor) beef raw material after storage intervals of 7, 14, and 28 days at −10°C and 21 days at 2°C to determine the effects of storage temperature, length of storage, and levels of added salt on the sensory attributes of hot-boned beef [6].

Sensory ratings for appearance, juiciness, flavor, tenderness, and overall desirability were affected by length of raw material storage. Raw materials stored for 14 days produced weiners with higher appearance scores than those stored for other time periods. Ratings for flavor and tenderness were lower for weiners made from raw materials stored 28 days. Overall desirability was higher for weiners manufactured after the 14-day storage period when compared to weiners made after 7 or 28 days of storage.

Flaked, cured pork was manufactured using 25 combinations of NaCl and STP [33]. The effects of frozen storage were evaluated at 3-week intervals from time of production through 18 weeks of storage at −23°. Panel members evaluated general acceptability on either eight or nine cold (4°C) half slices of meat using a scale of 1 to 9, with 9 being the most acceptable. As salt content increased to 1.5%, taste panel scores also increased, indicating a more palatable product. However, beyond 1.5% the panel scores tended to decrease. In general, all organoleptic evaluations decreased as frozen storage time increased, but products found to be acceptable initially were still acceptable after 18 weeks of storage.

Frozen green hams for the production of dry-cured hams were frozen on racks at −29°C and then stored in vats at −18°C for approximately 3 months before being cured while frozen or being cured after thawing at 3°C, 15°C, or in running water at 38°C [14]. After 5 months of processing (dry-curing, salt equalization, smoking, and aging), the hams were sliced and evaluated by an experienced taste panel for various attributes, including overall satisfaction. No differences were observed for overall satisfaction among curing treatments. Mean overall satisfaction was approximately 6.8 on a 9-point hedonic scale.

The effects of freezing, frozen storage, and protective wrap on cooking losses, palatability, and oxidative rancidity development in fresh and cured pork cuts (loins and shoulders—fresh; hams and bellies—cured) were investigated [18]. The acceptability of flavor, juiciness, tenderness, and overall eating satisfaction was evaluated by a 6-member taste panel using a 9-point scale. No differences were observed among wraps for acceptability of juiciness, tenderness, flavor, and overall eating satisfaction for either fresh or cured pork cuts at any storage interval. Consistent significant differences for measured traits were not observed during frozen storage.

Frozen (–30°C) ham steaks and bacon slices were wrapped in oxygen-permeable retail wrap, freezer paper, polyethylene bags, and aluminum foil to determine the influences of frozen storage and subsequent thawing on the acceptability of cured pork cuts after 24 hours of retail display [19]. No differences in retail acceptability scores for bacon slices and ham steaks attributed to type of protective wrap were found after any storage interval. Ham steaks stored for 168 days or longer were rated less acceptable after thawing and 24 hours of retail display than those stored for shorter periods. Storage of ham steaks up to 140 days could be accomplished using the protective wraps in this study without loss of acceptable appearance upon thawing and retail display.

Physicochemical and sensory effects of 0.4% sodium tripolyphosphate, tetrasodium pyrophosphate, and three commercial phosphate blends were studied in frozen beef patties (–20°C) over a 90-day storage period (27). Sensory evaluation of the treatments included juiciness, texture, flavor, and overall acceptability on a 7-point hedonic scale. Overall, treating patties with phosphate resulted in significant improvements in overall acceptability when compared to control patties (no phosphate) when stored at –20°C for up to 90 days.

# IV. EFFECTS OF FREEZING, FROZEN STORAGE, AND THAWING ON PALATABILITY ATTRIBUTES

## A. Comminuted Products

Long-term frozen storage (52 weeks at –18°C) of meats (mutton, pork, and beef) used to manufacture "English"-type fresh skinless sausages was investigated to determine its effect on sausage quality and palatability [22]. A significant decrease in Warner-Bratzler shear force values for all sausages was observed with prolonged storage of the meat. However, only a slight decrease in the texture of mutton sausages was observed with increases in storage period ($p < 0.05$). Objective measurements showed no significant changes in color, juiciness, and shrinkage. The TBA number and pH of all meats increased during frozen storage. None of the meats became obviously rancid. The high unsaturated fatty acid content of pork did not lead to an appreciable increase in the extent of lipid oxidation. Subjective evaluation by a taste panel did not establish any significant change with respect to prolonged storage of meat in the texture, juiciness, color, flavor, or general acceptability of beef and pork sausages ($p > 0.05$). All sausages were organoleptically acceptable even when made from meats stored for 52 weeks at –18°C [22].

Pre- and postrigor goat meat processed as is or after freezing in chunks or mince form was analyzed to determine its effects on the physicochemical and organoleptic properties of patties [37]. Patties made from chilled, postrigor goat meat in frozen chunk or mince forms produced significantly higher ($p < 0.05$) cook yields with less diameter shrinkage than patties made from prerigor meat frozen in similar forms. This observation might indicate the possibility of partial "thaw rigor" in hot frozen meat.

The organoleptic scores (appearance, flavor, texture, juiciness, and overall acceptability) for all patties that were precooked (72°C), frozen (–10°C, 10 days), thawed (4°C, 16 h), and reheated (100°C, 20 min, internal 60°C) were generally lower than for fresh-cooked patties. However, these decreases could not be easily explained since the trends were not systematic [37].

Whole hog sausage patties (approximately 33% fat) made with 0% phosphate, 0.375% STP, or 0.441% Lem-O-Fos[R] (STP with natural antioxidant) in combination with 0, 0.5, 1.0, and 1.5% NaCl were stuffed into fibrous casings, frozen (–34°C) for 4 hours, held overnight (–9°C), tempered (–3°C), sliced into 9.5-mm patties, placed in plastic foam boat trays (8 patties/tray), packaged using oxygen-permeable PVC film overwrap, or vacuum packaged in oxygen-impermeable Saran™-polyethylene bags and assigned to frozen (–9°C) storage periods of 0, 2, 4, or 8 weeks [38]. Extended periods of storage (4–8 weeks) reduced cooking yields (2–6% compared to 0–2 weeks of frozen storage). Juiciness and texture scores were also reduced, but the development of off-flavor and rancidity increased. TBA values were not affected ($p > 0.05$) by NaCl level, possibly due the antioxidant effect of phosphates or vacuum packaging during extended periods of frozen storage [38].

A study was conducted to determine the effects of reduced NaCl levels, reduced NaCl levels with STP, and reduced NaCl levels with Lem-O-Fos[R] on properties of precooked (72°C) pork sausage patties during frozen storage (–9°C for 0, 2, 4, or 8 weeks) [39]. Precooked patties were reheated from the frozen state to approximately 60°C and served to taste panelists who evaluated the patties for juiciness, saltiness, texture and off-flavor. Results indicated alkaline phosphates and vacuum packaging increased pH values, cooking yield, saltiness and juiciness scores while protecting against off-flavor and rancidity development. A minimum level of 1.0% NaCl in precooked sausage patties was desirable to maintain yield and sensory properties over extended periods of frozen storage.

Beef and soy-extended patties were frozen to –18°C within 24, 48, 72, or 96 hours and stored at –23, –18, or –7°C for 0, 6, 12, 18, or 24 months prior to evaluation for aerobic plate counts, TBA values, freezer burn, color, odor, flavor, and juiciness [40]. The 96-hour freezing rate for both products resulted in greater rancid flavor and higher TBA values (9 months of storage) and reduced beef flavor and juiciness when compared to all other freezing rates immediately postfreezing. Storage at –7°C for 6 months produced significant reductions in quality attributes with the exception of juiciness. Longer frozen storage periods produced gradual but continual deteriorations in color, flavor, odor, and TBA values. No distinct advantages were apparent in either short- or long-term storage for utilizing –23°C over –18°C as a storage temperature. Substitution of 20% rehydrated soy was found to reduce the rate of quality deterioration during frozen storage [40].

The effect of prechilling (air at $-10°C$ or $CO_2$ snow), freezing method, or rate (liquid $N_2$ immersion, $CO_2$ snow and air blast at $-25$ or $-14°C$) and frozen storage at $-17°C$, for 6–9, 24–30, 56–63, or 88–98 days on ground beef quality was determined by evaluating color, WHC, pH, shrinkage, TBA, and textural properties [41]. Liquid $N_2$ immersion freezing produced higher cooking and total shrink losses, a paler color, and lower WHC. Patty quality during frozen storage ($-17 \pm 2°C$) was significantly reduced, as indicated by increases in storage loss, shear values, hardness, gumminess, chewiness, surface reflectance, and TBA values. Carbon dioxide prechilling resulted in lower freezing, storage, and total losses and an overall redder surface color [41].

The effects of two fat levels (6 or 20%) and two freezing temperatures ($-43$ or $-20°C$) on sensory, shear, cooking, and compositional properties of beef patties made from U.S. Select beef rounds and U.S. Choice plates was determined [42]. Patties formulated with 20% fat had higher beef flavor and initial tenderness ($-20°C$ frozen patties only) scores, but they exhibited higher shear values than the 6% fat beef patties. Freezing patties (especially 6% fat patties) at $-43°C$ improved juiciness, beef flavor intensity, and detectable connective tissue scores and reduced the amount of energy (kg/cm) required to shear the samples. It was suggested that U.S. "select" beef rounds in conjunction with rapid freezing would produce acceptable low-fat ground beef patties [42].

The physicochemical and sensory effects of 0.4% sodium tripolyphosphate (STP), tetrasodium pyrophosphate (TSPP), and three commercial phosphate blends were studied in frozen beef patties over a 90-day storage period (samples were tested at 0, 2, 7, 30, 60, and 90 days) at $-20°C$ [27]. Addition of phosphates significantly ($p < 0.05$) increased pH, soluble orthophosphates, Hunter color a* values, cook yield, and overall acceptability scores. Proximate analysis, texture and flavor scores, and Hunter L* and b* values were unaffected by phosphate addition. Phosphates improved overall patty quality by increasing cook yields and reducing the development of oxidative rancidity. However, it was noted that phosphates may have interfered with TBA analysis [27].

The effects of combinations of boning temperature (hot or cold), time the patties were processed (immediately after grinding and before freezing; after freezing and thawing of bulk ground beef), and use of texturized soy concentrate (0 or 20%, rehydration ratio of 2.6:1 or 1.5:1) on sensory, Instron, and cooking properties of ground beef patties were evaluated [43]. With the exception of hot-boned, 0% soy patties, both sensory and instrumental measures of tenderness indicated that patties made from frozen and thawed bulk ground beef were more tender than patties processed immediately after grinding and before freezing. Patties processed after grinding and prior to freezing had higher juiciness scores and higher cooking yields than patties made from frozen/thawed bulk ground beef [43].

The manufacture of hot-processed all-beef patties coupled with immediate freezing produced products with higher tenderness ratings. However, incorporation of 20% rehydrated soy concentrate reduced shear values, juiciness scores, cooking loss, patty shrinkage during cooking, and beef flavor intensity scores [43].

The effects of temperature, light, and time of storage on the physicochemical (gas composition), pH, extract release volume [(ERV), an indicator of water-holding capacity], exudate, TBA, and sensory (appearance and off-odor) changes of vacuum- or nitrogen-packed, cooked (70°C) meat loaves were evaluated at 7-day intervals over a 49-day storage period at four storage temperatures (–4, 0, 3, and 7°C) under dark and lighted display environments [36]. The $CO_2$ concentration of vacuum- and nitrogen-packed meat loaves stored at –4°C did not change during 49 days of storage. No significant changes were noted for pH, ERV, or overall appearance. Off-odor was suppressed significantly by storage of samples at –4°C, but in general, off-odor increased in both types of packages as storage temperature or time increased [36].

Cooked meat loaves with 0–5% added blood were vacuum packaged and stored for 1 year at –20°C prior to being evaluated for rancidity and off-flavor by a descriptive sensory panel [44]. Small but significantly higher rancidity and off-flavor were found in meat loaves with 1–2% added blood. Higher concentrations of blood resulted in no further increase in rancidity and off-flavor, possibly due to the ability of heme compounds to act as an antioxidant when present in high concentrations relative to that of polyunsaturated fatty acids (Table 5).

Pre- and postrigor beef semimembranosus muscles were ground with salt (0, 2, and 4%), divided into two treatment groups (frozen or freeze-dried) and evaluated during storage (0, 5, 10, and 15 weeks) for composition, TBA value (lipid oxidation), sensory traits, and rehydration ratio [45]. Pre- and postrigor beef patties selected for freezing were stored at –29°C, while the patties selected for freeze-drying were subsequently stored at 25°C during the storage period.

At each salt level, prerigor frozen samples contained less moisture than postrigor frozen samples due to quicker evaporation of moisture from prerigor samples compared to postrigor samples. Rehydration ratios of prerigor, freeze-dried beef were higher than those of postrigor, freeze-dried beef, probably due to the higher pH values of the former. No differences in TBA values were noted between pre- and postrigor frozen beef treatments at any salt level. In general, TBA values for all samples studied significantly increased with salt level [45]. Prerigor, frozen beef patties were more tender than postrigor frozen samples, due probably to higher pH, resulting in higher subsequent water-holding capacity and lower cooking loss for the prerigor samples. The same results were observed for the freeze-dried samples. In both pre- and postrigor beef samples, frozen meat was more tender than freeze-dried beef. Prerigor frozen beef patties were more cohesive than postrigor beef patties, again probably due to higher pH, which allowed greater amounts of salt-soluble proteins to be extracted [45].

Differences in TBA values were not significant between pre- and postrigor frozen beef samples at any salt level. Frozen beef (pre- and postrigor) was rated as

**Table 5** Calculated Molar Ratio of Polyunsaturated Fatty Acids (PUFA) to Heme in Meat Loaves with 0–5% Blood Added

| Percent blood in meat mince | μmol PUFA/g[a] in standard recipe | HPLC[b] mg heme protein/g in minces | μmol heme/g[c] in standard mince | μmol heme/g after heating (30% reduction) | Approximate molar ratio |
|---|---|---|---|---|---|
| 0 | 32.9 | 3.0 | 0.16 | 0.11 | 299:1 |
| 1 | 32.5 | 5.0 | 0.27 | 0.19 | 171:1 |
| 2 | 32.3 | 7.0 | 0.38 | 0.27 | 120:1 |
| 3 | 32.0 | 9.0 | 0.49 | 0.34 | 94:1 |
| 4 | 31.6 | 10.0 | 0.54 | 0.38 | 83:1 |
| 5 | 31.3 | 12.0 | 0.65 | 0.46 | 68:1 |

[a]MW linoleate = 280.
[b]Data are means of three measurements for each blood percent; standard deviation = 7%.
[c]MW myoglobin = 18,500, MW heme = 616.
*Source*: Ref. 44.

being less rancid and exhibiting less off-flavor than freeze-dried beef. Freeze-dried beef samples (pre- and postrigor) were rated inferior to frozen beef (pre- and postrigor) in all sensory traits studied. Based upon energy usage and meat quality, freeze-dried ground beef should not be used as a raw material for the manufacture of hamburger or sausages [45].

Ground beef patties (20% fat) frozen by liquid nitrogen or liquid carbon dioxide in a cryogenic tunnel at –74°C or in a mechanical freezer at –29°C and stored in plastic-lined boxes for 120 days at –29°C were analyzed for sensory attributes by an experienced taste panel to determine the effect of fat, packaging, storage time, and temperature on flavor, texture, juiciness, and overall acceptability [35].

Panel scores for flavor, texture, juiciness, and overall acceptability of ground beef patties were lower for the mechanical freezing treatment than the carbon dioxide and nitrogen (cryogenic methods) treatments, but no differences were noted between method of cryogenic freezing (carbon dioxide vs. nitrogen). These observations were attributed to the faster rate of freezing with the cryogenic freezing treatments, which resulted in increased quality retention over the mechanically frozen patties, rather than any inherent advantages in the cryogenic method itself [35].

Chub-packed ground beef containing one of three levels of mechanically deboned beef [(MDB) 0, 10, or 20%] were stored at –12, –18, or –23°C and analyzed for sensory attributes initially and at 2-week intervals up to 24 weeks [46]. Level of MDB and storage time significantly affected all palatability traits (tenderness, juiciness, connective tissue residue, appearance, aroma, flavor, and overall desirability). Frozen storage temperature did not affect palatability traits. Level of MDB increased panel ratings for tenderness and detectable connective tissue residue, did not affect juiciness, appearance, aroma, or rate of decline for overall desirability with storage time, but did decrease flavor. Ground beef containing 20% MDB was unacceptable in flavor after 10 weeks of storage, while ground beef containing 10 or 0% MDB were acceptable in flavor after 24 weeks of storage. Frozen storage time affected the rate of deterioration of palatability traits, whereas rates of deterioration were not affected by level of MDB [46].

The identification and quantification of volatiles generated during lipid oxidation in ground pork (30% fat) packaged in vacuum bags, polyvinyl chloride film, Saran Wrap™, aluminum foil, or Saran™–aluminum foil combination and stored frozen (–17°C) for 13, 26, and 39 weeks was investigated [20]. A trained sensory panel analyzed samples for the intensity of rancid odor and flavor using a 7-point semi-unstructured line scale with end anchors of 1 = none and 7 = intense. In general, rancid odor, acid odor, rancid flavor, and acid flavor intensity scores of cooked fresh ground pork increased over time in frozen storage. Intensity scores for rancidity, acid odor and flavor, and TBA values increased over time of storage regardless of packaging treatment. Vacuum-packaged pork had lower intensity scores, and PVC-packaged pork had higher intensity scores during all storage periods than any of the other packaging materials used. The exclusion of oxygen by vacuum packaging during frozen storage appeared to preserve ground pork quality by reducing lipid oxidation [20].

Beef and soy-extended patties (20% rehydrated soy) were frozen to $-18°C$ in either 24, 48, 72, or 96 hours and stored at $-23$, $-18$, or $-7°C$ for 0, 6, 9, 12, 18, or 24 months prior to evaluating various quality factors, including sensory evaluation of ground beef flavor intensity and juiciness [24]. Ground beef flavor and juiciness were influenced by freezing rate. Slower freezing resulted in lower ground beef flavor scores. Panel scores indicated that a slower freezing rate resulted in a more rancid flavor in 96-hour patties than the faster freezing rates. Juiciness ratings were also affected by freezing rate in a similar manner, particularly for all-beef patties, as evidenced by the 96-hour product receiving lower juiciness scores immediately after freezing. It was concluded that storage of patties at $-7°C$ should be avoided due to reductions in quality after 6 months of storage. Longer frozen storage produced gradual but continual deteriorations in quality [34].

The sensory and physical characteristics of beef patties containing 20% fat, 8% fat, or 8% fat plus 20% soy protein isolate, soy flour, soy concentrate, or a mixture of carrageenan (0.5%), starch (0.5%), and phosphate (0.2%) were compared after 0, 4, 8, or 12 weeks of storage at $-18°C$ [25]. An eight-member trained sensory panel evaluated samples for intensity of beef flavor, off-flavor, rubbery texture, cohesiveness, and juiciness. Increased time in frozen storage increased off-flavor and rubbery texture. Unstored patties had lower off-flavor scores than patties stored for 4, 8, and 12 weeks. Patties stored for 8 or 12 weeks had higher rubbery texture scores than patties stored 0 or 4 weeks. Detrimental changes in flavor occurred during the first 4 weeks of frozen storage, while changes in lipid oxidation and texture occurred between 4 and 8 weeks of storage. Therefore, quality was lowered in frozen-stored high-fat or low-fat soy-extended beef patties [25].

## B. Restructured Products

Frozen restructured pork patties (25% fat) were utilized to evaluate the effects of cooking method (convection oven or electric grill), reheating method (microwave oven, convection oven, or infrared oven), and frozen storage (2, 4, or 6 weeks) on physical, chemical, and palatability factors [34]. Results revealed no differences for chemical composition, shear force values, percent cooking loss, or percent change in cooked patty thickness due to frozen storage. TBA values increased between 0 and 2 weeks and remained relatively stable between weeks 2 and 6. Taste panel evaluations showed mean color score was higher ($p < 0.05$) immediately after cooking and reheating than when the samples were reheated following 6 weeks of postcooking storage ($-23°C$). No other differences due to length of frozen storage were found for any other taste panel trait.

Restructured beef and pork steaks made from frozen (−34.5°C) and flaked (11 different flake sizes) beef chucks and pork shoulders were manufactured and stored 3 weeks at −34.5°C prior to evaluation for textural properties by a texture profile panel and Instron measurements [47]. Results showed as flake size increased, visually detected fibers, first bite hardness, cohesiveness of the chewed mass, number of chews required for swallowing, detectable amount of connective tissue, and shear force increased. It was concluded that considerable attention must be given to flake size to achieve the desired textural properties [47].

Mechanical tenderization and acid-neutralized liquid smoke were utilized in a fully cooked restructured pork item made from hot-processed sow meat to determine their effects on sensory and physical attributes [48]. Four treatment groups were formulated: (1) nontenderized, no liquid smoke (NTNS), (2) nontenderized, liquid smoke (NTS), (3) tenderized, no liquid smoke (TNS), and (4) tenderized with liquid smoke (TS). All treatments were cooked (68°C), packaged, and stored at −23°C and evaluated for proximate composition, cooking loss, sensory attributes, Kramer shear determinations, and Instron analysis. TBA values were determined for each of the fully cooked products from each treatment after 0, 30, 60, and 90 days of frozen storage (−23°C).

Differences were not found among treatments for proximate composition; sensory juiciness, cohesiveness, connective tissue, or flavor; Kramer shear value; or percent cooking loss between tenderized and nontenderized or for mechanical tenderization with liquid smoke treatments. Tensile strength was greater for tenderized than nontenderized treatments. Initial TBA values for fresh-frozen and cooked-frozen restructured pork samples showed no practical differences between the two products. After 30 days of storage (−23°C), the cooked-frozen product had significantly higher TBA values than the fresh-frozen product [48].

Restructured pork roasts made from pork shoulders were manufactured by grinding or flaking, with salt treatments (0.40% NaCl or 0.51% KCl, equivalent ionic strength) and stored frozen (−24.4°C) for 1 or 4 weeks. Frozen storage period (1 or 4 weeks) did not affect sensory properties (moistness, off-flavor intensity, mouthfeel), but cooking shrinkages were higher for roasts stored 4 weeks than for those stored 1 week (31.08% vs. 27.12%, respectively) [49].

## C. Whole Muscle Products

A two-phase experiment was conducted to evaluate cook-in-the-bag processing and the effect of tripolyphosphates on sensory properties and shelf life of fresh-frozen and precooked pork roasts [50]. In Phase I, pairs of boneless pork loins ($n = 10$) were pumped to contain 0.5% tripolyphosphate in water or water only and allotted to three treatments: PVC-wrapped, convection oven cookery; vacuum packaged, precooked, and reheated in a water bath; and vacuum packaged and cooked in a water bath. Loins were evaluated for sensory juiciness, tenderness, pork flavor intensity and off-flavor, total cooking loss, Warner-Bratzler shear, proximate analysis, pH, and percent free water. In Phase II, pairs of boneless pork loins ($n = 30$) were pumped (10%) with 5% tripolyphosphate solution and fabricated into roasts (two per loin). Four roasts from each pair were allotted to one of the following treatments: PVC-wrapped, stored at $-20°C$ then conventionally cooked; precooked in a vacuum cooking bag, stored at $4°C$, and then warmed in the bag; stored at $-20°C$ in a vacuum cooking bag and cooked in the same bag; or precooked in a vacuum cooking bag, stored at $-20°C$, and then warmed in the bag. Storage times were 0, 14, or 28 days. Samples were evaluated for the same parameters as in Phase I.

After 14 days of storage, vacuum-packaged and frozen roasts were juicier ($p < 0.05$) than conventional and precooked frozen roasts (Table 6). After 28 days, the

**Table 6** Palatability Attributes of Pork Roasts Cooked Using Different Methods and Stored for 0, 14, and 28 days

| | Day | CO[a] | PCRF[a] | PCFR[a] | VPFR[a] |
|---|---|---|---|---|---|
| | | | Cooking treatment | | |
| Juiciness[b] | 0 | 9.44 | 8.45 | 7.98 | 9.05 |
| | 14 | 8.36[ef] | 9.04[de] | 6.99[f] | 10.04[d] |
| | 28 | 8.64[d] | 8.63[d] | 6.64[e] | 8.61[d] |
| Tenderness[b] | 0 | 10.78 | 11.75 | [x]11.63 | 10.98 |
| | 14 | 10.04[e] | 11.69[d] | [xy]10.40[de] | 11.42[de] |
| | 28 | 10.86 | 10.61 | [y]9.94 | 10.22 |
| Pork flavor intensity[b] | 0 | 11.75 | [x]11.81 | 11.13 | [x]11.80 |
| | 14 | 11.32 | [xy]11.20 | 10.70 | [xy]11.16 |
| | 28 | 11.12 | [y]10.76 | 10.10 | [y]10.76 |
| Off-flavor intensity[c] | 0 | 14.02 | 14.06 | 14.17 | 13.96 |
| | 14 | 13.80 | 13.55 | 14.37 | 13.64 |
| | 28 | 13.50 | 13.82 | 13.52 | 13.69 |
| WBS (kg) | 0 | 2.53 | 2.39 | 2.32 | 2.34 |
| | 14 | 2.40 | 2.32 | 2.27 | 2.06 |
| | 28 | 2.52 | 2.83 | 2.53 | 2.32 |

[a]CO = PVC wrapped, $-20°C$ storage, convection oven cookery. PCRF = Precooked in vacuum bag, $4°C$ storage, reheated in the bag. PCFR = Precooked in vacuum bag, $-20°C$ storage, reheated in the bag. VPFR = Fresh roast vacuum bag, $-20°C$ storage, cooked in the bag.
[b]Means derived from sensory panel scores with possible range 0–15, where 0 = extremely tough, dry, or bland and 15 = extremely tender, juicy, or flavorful.
[c]Means derived from sensory panel scores with possible range 0–15, where 0 = extreme off-flavor and 15 = no off-flavor.
[def]Mean values in same row bearing unlike superscripts differ significantly ($p < 0.05$).
[xy]Mean values in same column bearing unlike superscripts differ significantly ($p < 0.05$).
*Source*: Ref. 50.

precooked and frozen roasts were less juicy and tenderness decreased from day 0 to day 28, but at 14 days these roasts were more tender than control roasts. Pork flavor intensity decreased during 28 days of storage in precooked and frozen roasts and fresh, vacuum-packaged frozen roasts. However, no differences in off-flavor or WBS values were observed [50]. Cooking loss was higher in control roasts stored for 28 days than those stored for 0 days. Cooking losses of precooked roasts were similar to control roasts, while fresh, vacuum-packaged, and frozen roasts sustained less cooking loss. TBA values did not change during storage, suggesting that oxidative rancidity was not a problem under these storage conditions [50].

Flaked, cured pork was manufactured using 25 combinations of salt (NaCl) and sodium tripolyphosphate (STP) stored frozen (−23°C) and evaluated at 3-week intervals from the time of production through 18 weeks [33]. As time of frozen storage increased, freezer yields, cooking yields, product color, and nitrite levels decreased, while shear values and rancidity increased. Organoleptic scores decreased as storage time increased, but products were found to be acceptable initially and after 18 weeks of storage for juiciness, cured pork flavor, and general acceptability.

A study to compare freezing and cold storage effects on the quality of precooked sliced beef from the posterior end of *latissimus dorsi* muscles removed from 2-year-old steer carcasses aged for 2 weeks at +3°C was conducted [51]. The muscles were sliced across the fibers (1.5 cm thick), packaged in laminated pouches, frozen in a freezing tunnel (−35°C), and stored at −90°C for 2 days or 2 months, respectively, for samples to be refrozen or cold stored after thawing and cooking, respectively.

Frozen, raw meat slices were thawed (0°C) and fried in margarine (175°C) with butylated hydroxyanisole (BHA) until an internal temperature of 70–75°C was reached. Fried meat slices undergoing pasteurization with cold storage were packed in laminated pouches and then given a pasteurizing treatment for 9–12 minutes in a convection oven (105°C) until an internal temperature of 80°C was reached. The slices were cooled (10°C) and stored in the dark at 3 or 8°C from 1 day to 3 weeks (−90°C), then reheated in a convection oven (110°C) until an internal temperature of 65°C was reached (approximately 10 minutes) before sensory analysis. Fried meat slices undergoing the frozen storage treatment were immediately frozen in the freezing tunnel (−35°C, approximately 45 minutes) and packaged in heat-sealable nylon/polyethylene laminated pouches with 0.5 or 9 ml of headspace. After frozen storage at −20°C for 2 months, the meat slices were reheated in the package directly from the frozen state, under the same conditions as the refrigerated samples for sensory analysis. An experienced six-member taste panel evaluated each sample twice. Deterioration of flavor with time of refrigerated storage was evident. A clear advantage in flavor was obtained for frozen, precooked beef after 2 months of storage at −20°C over the pasteurized, refrigerated product stored at 3°C for a few days, with a headspace volume of approximately 10 ml per 75-g pouch. Higher juiciness scores were also noted for the frozen product [51].

The effects of precooking, curing, and frozen storage were compared as processing methods to reduce warmed-over flavor in pork chops [52]. Sixty-four pork loins were divided into two weight groups (light, 6.3–7.7 kg; heavy, 9.9–11.3 kg). Eight frozen pork loins (stored at –18°C for 7–28 days) from each weight group were randomly selected and assigned to each of four treatments: (1) frozen loins cut into chops and oven-broiled (fresh), (2) loins precooked (internal temperature of 66°C) with no added seasonings and cut into chops, (3) loins cured with 0.5% salt and 40 ppm $NaNO_2$ and precooked (66°C), then cut into chops, and (4) loins cured with 2% salt and 120 ppm $NaNO_2$ and precooked (66°C), then cut into chops. Twenty center loin chops were removed from each loin for analysis. Four chops from each group were individually packaged in mylar pouches, frozen at –26°C, and stored at –18°C for 0, 28, 56, and 84 days from the initiation of the study. Chops assigned to the oven-broiled treatment were cooked to an internal temperature of 74°C. All other cuts were oven-broiled to an internal temperature of 43°C for sensory analysis. Sensory analysis was conducted by a six-member experienced panel for juiciness, tenderness, flavor, and acceptability using an 8-point hedonic scale (1 = extremely undesirable; 8 = extremely desirable). Warmed-over flavor was evaluated on a 5-point intensity scale (1 = extremely detectable; 5 = none detected) [52].

Flavor scores for uncured, oven-broiled and cured (120 ppm nitrite), precooked samples changed very little during frozen storage. Flavor scores for cured (40 ppm nitrite), precooked samples decreased slightly during frozen storage, while scores for uncured, precooked samples decreased markedly after 28 days of storage. In general, an overall increase in warmed-over flavor was observed in samples from all processing treatments during the 84-day storage period, with the greatest increase seen in uncured, precooked samples [52]. Results indicated that precooking of previously frozen pork loins to an internal temperature of 66°C, fabricating the loins into chops, and subsequently oven-broiling them to an internal temperature of 43°C improved tenderness and juiciness scores when compared to non-precooked chops cooked to 74°C internal temperature. It was concluded that successful freezing and storage of precooked pork chops depended upon protection from the development of warmed-over flavor, and nitrite effectively inhibited warmed-over flavor development in precooked frozen chops (52).

Green hams for the production of dry-cured hams were frozen on racks at –29°C, then stored in vats at –18°C for approximately 3 months before being divided into four groups of 15 hams each [14]. Group 1 hams were placed in cure while frozen. Groups 2, 3, and 4, respectively were placed in cure after thawing at 3°C, 15°C, or in running water at 38°C. Hams were dry-cured for 4 weeks at 3–4°C and salt equalized for 4 weeks at 13°C prior to being smoked for 24 hours at 38°C and then aged for 3 months at 24°C. Hams were then sliced (2.5 cm), broiled (internal temperature of 77°C), and evaluated by an experienced palatability panel for tenderness, flavor, saltiness, and overall satisfaction. Tenderness, flavor, and overall satisfaction scores were similar for all ham groups. The water-thawed group exhibited higher ($p < 0.05$) saltiness scores than the frozen group. This was verified when the slices were analyzed for salt content. It was concluded that frozen hams absorbed salt more slowly, and water-thawed hams, which were warmer and softer when placed in the cure, readily absorbed the salt [14].

## V. EFFECTS OF FREEZING, FROZEN STORAGE, AND THAWING ON NUTRITIONAL VALUE

The stability of vitamins in foods is generally influenced by pH and the presence of oxygen, light, metals, reducing agents, and heat. Vitamins are almost invariably destroyed to some extent during food processing, with the extent of destruction being a key factor in balancing the economics of processing against food quality [53].

Thiamine in meat is partially decomposed by curing, smoking, cooking, canning, dehydration, and ionizing radiation [53]. Refrigeration and frozen storage has little effect on the vitamin levels of meat [54]. In a review of vitamins, research investigating the effects of freezing and storage on the content and retention of vitamins was cited indicating that vitamin retention depended upon storage temperature and other factors [53]. Thiamine losses were found to be 22–23% in ground meat and 13–21% in salami stored at −12 to −24°C for one year, while losses of riboflavin in ground meat was 15–16% and in sausages 7–34%. Vitamin $B_6$ (pyridoxine) losses were small or virtually nonexistent for ground meat sausages and Vienna sausages [53]. However, up to one half of the vitamin $B_6$ may be lost in the drip during thawing of frozen meat (20–50%) [53]. During a year of frozen storage of tinned meat, pantothenic acid losses over 0–22% and pyridoxine (vitamin $B_6$) losses were 10–50% [53].

Seven types of beef products—loin (strip), steak (strip loin), trimmings (70% lean, 30% fat), ground beef patties (bulk), ground beef patties (vacuum-packaged), ground beef (chub), and mechanically processed beef (MPB)—were commercially packaged (vacuum-packaged or placed in a polyethylene-lined box), frozen, and stored at top corner pallet positions in three identical freezer rooms for one year [54]. The storage temperatures were (1) −23°C constant temperature, (2) −23 and −21°C, and (3) −21 and −18°C. Freezer rooms with the dithermal storage regime were controlled at the lower temperatures during one 12-hour period and the higher temperature during the other 12-hour period by two thermostats and a time clock. Bulk-packed hamburger patties exhibited a weight loss in excess of 1%. Results of sensory, nutritional, and other analyses of frozen boxed beef products indicate that mechanically processed beef and hamburger patties should not be frozen and stored for longer than 6 months.

Nutritional analysis indicated that thiamine content was reduced with increasing storage time. The greatest loss of thiamine occurred during the first 6 months of frozen storage. Mechanically processed beef sustained the greatest thiamine loss (66.2%). Thiamine losses of 55.4 and 48.52% were found in bulk patties and vacuum-packaged patties, respectively, which may indicate that vacuum packaging offers some protection against thiamine losses. There were no differences in thiamine, riboflavin, and niacin contents as a result of storage temperature variation [54].

Some changes in the mineral contents of frozen foods may occur due to diffusion into the drip at the thawing stage and to moisture loss. Means for mineral levels in all boxed beef products (except mechanically processed beef) tended to increase at 12 months, which may be attributed to a corresponding loss in moisture. Objective, sensory, and nutritional quality of beef was not seriously affected by temperature fluctuations, provided the maximum range did not exceed a 3°C difference between the controlled dithermal temperature levels up to 6 months for some products (MPB) and 12 months for other products (strip loin and strip loin steak) [54].

Thiamine, riboflavin, pantothenic acid, and nicotinic acid contents in pork loins aged 1, 3, or 7 days at 1.1°C and 7 days at 4.4°C and kept in frozen storage (−17.8°C) for 48 weeks were investigated [55]. Loins from pork carcasses aged 3 days at 1.1°C showed the greatest storage losses during storage for up to 48 weeks, losing 20% of the thiamine after 8 weeks of storage, which increased to 33% after 24 weeks of storage. Pork loins from carcasses aged for 7 days at 4.4°C after 24 weeks of storage, and throughout the remainder of the storage period at 4.4°C, exhibited greater retention of thiamine than at 8 or 16 weeks. Enzyme activity during the frozen state may help to increase the thiamine content, with longer aging periods aiding this reaction [55]. Riboflavin losses in these same pork loins were 18% after 8 weeks and 20% after 32 weeks. Pantothenic and nicotinic acids showed the greatest losses during frozen storage in pork loins from carcasses aged 1 day at 1.1°C. The pantothenic acid losses were 18% at 8 weeks and 24% at 16 weeks of frozen storage. Nicotinic acid losses were observed to be the least of any of the vitamins; 9% was lost after 8 weeks and 18% was lost after 32 weeks of frozen storage. Although the pork loins were not processed (cured and smoked) in this study, results indicate the possibility of additional losses of nutrients if raw materials (pork loins) stored frozen for prolonged periods (up to 48 weeks) were to be subsequently cured and thermally processed.

A series of studies investigating the effects of thawing and cooking on the nutritive value of frozen ground meat (pork, beef, and lamb) have been conducted [56–58]. Ground meat (1/8-in. grind) was made into loaves (pork, approximately 907 g), molded into metal pans or patties (beef, lamb, 112 g), frozen, wrapped in aluminum foil and stored at −24.4°C from 6 to 18 weeks. Thawing methods used were thawing during cooking (cooking from the frozen state) and thawing at room temperature (18.3–23.8°C) to an internal temperature of 2.8°C.

The patties (beef and lamb) were cooked by four methods: panbroiling, oven-broiling, dielectrically without prebrowning, and dielectrically with prebrowning. Loaves were baked in a household gas oven at 176.6°C to an internal temperature of 87.8°C or dielectrically cooked to an internal temperature of 85–87.8°C. Samples (pork loaves) were analyzed for thiamine, riboflavin, and niacin. Beef and lamb patties were also analyzed for lysine.

The average vitamin content of frozen raw pork was 0.50 mg thiamine, 0.19 mg riboflavin, and 2.64 mg niacin per 100 g. No appreciable losses occurred during thawing. Loaves thawed at room temperature and baked dielectrically retained 10% more thiamine in the drained cooked meat than loaves cooked by oven-baking. From 2 to 7% of the thiamine was found in the cooked drip of the loaves, and the loaves retained approximately 66–74%. Riboflavin rententions ranged from 77 to 89% in the drained meat, 4 to 5% in the drip, or 81 to 94% in total. Loaves cooked dielectrically from the frozen state contained 8% of the riboflavin in the drip. Drained cooked loaves retained 88–100% of the niacin, and the drip contained 6–12%. The total was 100–107%. Dielectrically cooked loaves thawed at room temperature had a higher apparent retention of niacin. However, this finding could not be explained by the researchers [56].

The average vitamin contents of frozen raw beef were 0.08 mg of thiamine and 0.20 mg of riboflavin per 10 grams. The average thiamine retention in drained cooked patties was 84%. Significant differences were observed for thiamine retention. Thawed, unbrowned, dielectrically cooked patties had lower retention, and thawed, prebrowned, dielectrically cooked patties had higher retention. Dielectrical cooking resulted in more thiamine in the drip. Riboflavin retentions averaged 79%. The only significant difference observed in riboflavin retention was increased retention in the drip of patties thawed and dielectrically cooked without prebrowning. The average lysine content of frozen raw beef was found to be 3.08 g/100 g. An average of 89% lysine was retained in the drained cooked patties [57].

The average vitamin content of frozen raw lamb was 0.07 mg/100 g thiamine and 0.18 mg/100 g riboflavin [58]. The average thiamine retention in drained cooked patties was 86%. Thawed oven-broiled patties retained slightly less thiamine than patties cooked by other methods. More thiamine was found in the drip of dielectrically cooked meat than in meat cooked by other methods. Riboflavin retention in cooked drained patties was 55%, in drip it was 7%, and total retention averaged 62%. The average lysine content of frozen raw lamb was 2.3 g/100 g, with an average of 92% retained in the cooked patties [58].

Commercial raw and char-broiled frozen beef-soy patties were heated by various methods, and vitamin $B_1$ and $B_2$ retentions were determined [59]. Heating methods were (1) hot air convection followed by holding patties hot (internal product temperature ranged from 71.1 to 82.2°C) for 0, 30, 90, or 180 minutes and infrared, high-pressure steam or (2) microwave followed by holding for 30 minutes. Raw patties were analyzed for thiamine and riboflavin retention while frozen, immediately after cooking, and after 30, 90, and 180 minutes of holding. Char-broiled patties were analyzed for thiamine and riboflavin retention while frozen and after being heated and held in a food warmer for 30 minutes. Riboflavin retention was high (96.2–99.1%) in beef patties except for infrared- or steam-heated patties (80.0 and 89.5%, respectively). Infrared heating was destructive to riboflavin due to absorption of radiation, which is deterimental to light-sensitive riboflavin. Lower riboflavin retention in steam-heated patties might be due to leaching losses. Raw patties displayed acceptable thiamine retention when cooked by convection oven and held up to 90 minutes. Convection heating of char-broiled patties followed by a 3-hour hold time and infrared heating resulted in the lowest thiamine retention, possibly due to the length of heating and intensity of the heat energy. It was noted that the dry ingredient mix for the beef-soy patties contained very high levels of thiamine. Consequently, all beef-soy patties contained high levels of this vitamin. This study indicated that for thin, flat products, the retention of heat-labile nutrients is not increased. Steam or microwave heating caused a significant loss of thiamine compared to raw patties but was not significantly different from that of convection heated patties [59].

The thiamine content of freshly prepared ham loaves and similar loaves after 2 and 4 months of frozen storage was investigated [60]. Ham loaves were prepared by mixing raw ground pork and smoked ham with appropriate seasonings and placing 475 g of this ham loaf mixture into loaf pans lined with aluminum foil. The loaves were cooked immediately (85°C internal temperature), cooked frozen (2 or 4 months' storage) and reheated (85°C internal), or frozen (2 or 4 months' storage) and then cooked. Loaves undergoing frozen storage were wrapped in aluminum foil and stored for the appropriate time (approximately –32°C for 24 hours, then approximately 18°C for 2 or 4 months). After cooking, freezing, and frozen storage of the raw or cooked loaves and subsequent cooking or reheating, the mean concentration of thiamine (6.00–6.10 μg/g) was slightly lower than for the raw mixture (approximately 6.34 μg/g). However, when the percent retention after cooking was considered, significant amounts were lost (25–30%). The retentions were of the same magnitude regardless of treatment [60].

Frozen, raw meat slices (1.5 cm thick) from latissimus dorsi muscles were fried in heated margarine (175°C with BHA) and underwent pasteurization with cold storage or frozen storage (–20°C for 2 months) prior to reheating in the package directly from the frozen or refrigerated state. Freezing and cold storage had no significant effect on retention of vitamins $B_1$ and $B_2$ [51].

## VI. EFFECTS OF FREEZING, FROZEN STORAGE, AND THAWING ON INTRINSIC CHEMICAL REACTIONS

Food quality can be affected greatly by reactions such as oxidation, insolubilization of proteins, and glycolysis, which frequently accelerate during freezing (–1 to

–15°C) [61]. Since proper frozen storage is conducted at temperatures below –18°C (0°F) the greatest potential for damage arising from increased rates of reactions would be during freezing and during thawing.

Investigation of possible relationships between discoloration and lipid oxidation of frozen (–18°C) pork patties (1% added salt) stored for up to 31 days indicated that lipid oxidation underwent a lag period during frozen storage, while pigment oxidation was dependent upon storage conditions [23]. A constant lag period for lipid oxidation was observed, followed by a constant rate of lipid oxidation independent of light. A coupling mechanism was suggested to exist between photooxidation of the meat pigment and lipids, since oxymyoglobin was transformed to metmyoglobin and was activated into a superoxide radical ion, which then was able to produce hydroperoxide byproducts. The superoxide radical ion and hydroperoxide were capable of transforming iron into a highly reactive state or may have reacted with lipid compounds directly [23].

Although the results of this study did not prove that pigment oxidation initiated lipid oxidation, it did indicate that two completely independent mechanisms for lipid oxidation (light-catalyzed and iron-catalyzed) could occur. Light-dependent lipid oxidation was preceded by light-induced discoloration, suggesting modification of myoglobin by UV light, which may subsequently promote lipid oxidation in the presence of salt [23].

A study was conducted to determine the stability of phosphatase activity as an indicator of thermal processing in domestically prepared cured/canned picnics, which had been heat-processed (61, 62.7, 65.5, 68.3, or 70.7°C), during frozen storage (15 or 36 months at –34°C). The effect of frozen storage time on residual phosphatase activity (based on calculated internal temperature) demonstrated that the greatest loss of residual phosphatase activity, as evidenced by an increase in calculated internal temperature, occurred during the first 15 months of storage, with the least amount of loss occurring during the next 21 months of storage [62].

Effects of freezing rates on oxidation of commercial fresh pork sausage chubs were evaluated in association with location extremes during production [63]. Three treatments requiring 9 hours (chubs placed individually in suspended wire-mesh freezer baskets), 2.4 days (chubs placed in boxes 2 rows deep × 6 chubs per row placed into suspended wire-mesh freezer baskets), or 6.8 days (chubs placed into boxes, the boxes stacked, and styrofoam insulation placed around outside of boxes) to achieve temperature declines from 7 to –15°C were tested. Sausage chubs were analyzed after 4, 8, 12, 16, or 20 weeks of frozen (–15°C) storage and 3 weeks of postfrozen refrigerated (1°C) storage. The TBA values did not increase with slower freezing ($p > 0.05$), but Hunter a values were greater ($p > 0.05$) for the two slower rates. Overall, pallet location of the boxed chubs only minimally affected oxidative stability. Significant correlations ($p < 0.05$) indicated a positive relationship between lipid and myoglobin oxidation [63].

Fresh and frozen beef were investigated to determine the effects of $K_2HPO_4$

(0.00, 0.50, and 0.75%), NaCl (2.5 and 3.0%), and corn oil temperature (5, 11, and 21°C) on emulsion capacity. The emulsion capacity of frozen (–24°C for 20 days) beef was 6.4% higher than that of fresh meat [64].

A rapid increase in rate of discoloration during frozen storage [frozen (–34°C) for 48 hours] of restructured beef steaks undergoing a tempering period (–2°C for 24 hours) and then refrozen (–34°C for 12 hours) with added sodium chloride and with no sodium tripolyphosphate was observed [31]. This rapid rate of discoloration was due to an increase in the rate of overall metmyoglobin (myoglobin oxidation) formation, which occurred predominantly during the first month of storage. This increase in myoglobin oxidation was believed to be due to the combination of thawing and fluctuating temperatures, which resulted in large ice crystal formation and subsequently enhanced the disruption of muscle cell integrity, leading to a greater exposure to oxidative catalysts [31].

A subsequent study reported no difference in initial overall metmyoglobin formation in frozen restructured beef steaks from that of intact control steaks, with the exception of steaks packaged in $O_2$[32]. It was postulated that oxygen concentration between the meat particle interface reached a critical level, which resulted in greater metmyoglobin formation, thereby indicating oxygen diffusion into meat particles pre- and postmixing, which was concluded to be a possible decisive factor in initial metmyoglobin formation [32].

Restructured steaks in pure oxygen atmosphere had the greatest rate of overall metmyoglobin formation, possibly due to a chloride anion–promoted myoglobin autoxidation accelerated by a high oxygen concentration within the meat particles [32]. Surface metmyoglobin formation rates were greater than overall metmyoglobin formation rates during frozen storage, possibly due to oxygen absorbed on the surface of the steaks after slicing and remaining during vacuum-packaging or to insufficient vacuum, resulting in discoloration due to metmyoglobin formation [32].

## VII. EFFECTS OF FREEZING, FROZEN STORAGE, AND THAWING ON MICROBIOLOGICAL QUALITY AND SAFETY

Meats preserved by freezing are usually maintained at temperatures that prevent microbial growth. Freezing and subsequent thawing kills some microorganisms. Those that survive freezing usually die slowly during frozen storage. However, it is impractical to consider freezing and frozen storage as a means for markedly reducing bacterial loads. Bacteria that survive frozen storage will continue to grow on thawed meat [65].

Microbiological activity is controlled by two conditions present in frozen foods: water activity ($a_w$) is limited and the temperature is usually sufficiently low to prevent microbial growth. Although some microorganisms may be killed during freezing, many survive and exist in various states of viability during frozen storage. Freezing is more lethal to microorganisms if the food has also been subjected to salting, heating, acidification, or refrigeration treatments normally associated with processed meats [66].

## A.  Comminuted Products

The microbiological effects of 0.4% sodium tripolyphosphate, tetrasodium pyrophosphate (TSPP), and three commercial phosphate blends (Brifisol 414B, 414P, and 614P), added in solution or in powder form, were studied in frozen (90 days, $-20°C$) and subsequently temperature-abused (24–25°C, 24 hours) beef patties [67]. Results indicated that the addition of 0.4% levels of pure or blended phosphates did not significantly ($p > 0.05$) reduce numbers of misophiles, psychrophiles, lactic acid bacteria, presumptive *Staphylococcus aureus*, or viable anaerobic spores in beef patties stored up to 90 days at $-20°C$. Consequently, spoilage was not prevented in temperature-abused patties by incorporation of phosphates. The most effective phosphate tested in terms of antimicrobial activity was a solution of Brifisol 414B. Addition of phosphates increased the pH of the patties. Lactic acid bacteria numbers were lower when Brifisol 414B and TSPP were added to the meat in liquid form [67].

The effects of storage temperature, light, and time on the microflora of vacuum- or nitrogen-packed cooked (internal 70°C) meat loaves were evaluated at 7-day intervals during a total storage time of 49 days at $-4$, 0, 3, or 7°C under dark or lighted display environments [68]. Storage of cooked meat loaves at $-4°C$ for 49 days produced little increase in lactobacilli numbers (from $\log_{10}$ 5.2 to 6.3) but resulted in significant ($p < 0.05$) increases in psychrotrophs ($\log_{10}$ 3.9 to 5.9) and anerobes ($\log_{10}$ 5.1 to 6.3) under vacuum. This study suggested that both vacuum and nitrogen packaging stimulated the development of lactobacilli except when the products were stored at freezing temperatures ($-4°C$), which would be expected in view of the optimal temperature for growth of these microorganisms on meat [68].

A study to determine the microbiological quality of frozen Canadian meat pies was conducted [69]. The types of meat pies examined were: meat pies, filler only precooked; meat pies, both filler and pastry precooked; pot pies, filler only precooked; pot pies, both filler and pastry precooked; and shepherd pies, with no pastry enclosure.

One hundred and twenty samples of these various frozen meat pies were collected from 20 manufacturers across Canada and analyzed for aerobes, coliforms, Salmonella, *Clostridium perfringens*, *S. aureus*, yeast, and mold counts. Salmonella was not isolated from any of the pies, and only low numbers of *C. perfringens* and *S. aureus* were found. The highest aerobic, coliform, yeast, and mold counts were observed in pies with uncooked pastry. The degree of precooking was observed to be the main factor associated with the microbial content of the pies. The researchers concluded that the degree of contamination in these pies was not high enough to warrant establishment of microbiological standards or guidelines for these products [69].

Beef and soy-extended patties (20% rehydrated soy) were frozen to –18°C in 24, 48, 72, or 96 hours and stored at –23, –18, or –7°C for 0, 6, 9, 12, 18, or 24 months prior to evaluation of aerobic plate counts [24]. After 6 months of frozen storage, patties with or without soy held at –7°C had lower aerobic plate counts, perhaps due to greater dehydration (freezer burn), than patties stored at –23 or –18°C. The freezing process was found to produce decreases ($p < 0.05$) in aerobic plate counts in patties manufactured without soy [24].

The effects of combining fresh beef, blast-frozen beef, or a combination of fresh and frozen beef on the microbial flora of frozen (liquid nitrogen or liquid carbon dioxide) beef patties stored for 6 months at –20°C was investigated [70]. Freezing reduced bacterial counts before frozen meat was mixed with fresh beef. Combining frozen meat with fresh meat increased bacterial counts. Counts of *E. coli* and *C. perfringens* were reduced considerably by freezing and frozen storage. The researchers suggested that the "mechanical" effects of freezing on meat tissue may be important in making nutrients available for the uninjured microorganisms present in fresh meat. As a result, the potential for microbial spoilage may increase significantly when frozen meat is used in combination with fresh meat, especially when mixed in equal proportions [70].

A series of studies investigating the effect of thawing and cooking on the microbial counts of frozen ground meat (pork, beef, and lamb) were conducted [56–58]. In each study it was observed that thawing always increased plate counts, while cooking decreased plate counts regardless of the thawing and cooking methods.

The microbial counts of freshly prepared ham loaves after 2 and 4 months of frozen storage was investigated [60]. Ham loaves were cooked immediately (85°C internal temperature), cooked after frozen storage (2 or 4 months) and reheated (85°C internal), or stored frozen (2 or 4 months) and cooked. Loaves undergoing frozen storage were wrapped in aluminum foil and stored at approximately –32°C for 24 hours before being stored at approximately –18°C for 2 or 4 months. All samples had large reductions in microbial counts [60].

## B. Emulsion-Type Products

Weiners prepared from hot-boned (prerigor) beef after storage intervals of 7, 14, or 28 days at –10°C and 21 days at 2°C were analyzed to determine the effects of storage temperature, length of storage, and levels of added salt on microbial characteristics [6]. Microbial levels of the raw materials for all storage groups were low, with day 28 exhibiting the lowest levels (log 3.0), followed by day 14 (log

4.0), and day 7 having the highest microbial levels (log 5.0). These counts indicated that hot-boned beef would be acceptable for weiner production when stored frozen for up to 28 days. Microbial counts for the weiners remained low throughout a 21-day retail case display period [6].

Frankfurters were vacuum- or nitrogen-packaged and stored at –4, 0, 3, or 7°C for 49 days under light or dark display conditions to determine the effects of temperature and storage duration on microbial growth [71]. The data indicated that temperature was the most important factor for the growth of lactobacilli and psychrotrophic and anaerobic bacteria at freezing temperatures (–4 and 0°C). Microbial counts were relatively low and stable for up to 49 days of storage [71].

## C.   Whole and Restructured Products

Green hams for the production of dry-cured hams were frozen (–29°C) and then stored (–18°C for 3 months) before being divided into four groups of 15 hams each [14]. Group 1 hams were placed in the cure while frozen. Groups 2, 3, and 4, respectively, were placed in the cure after being thawed at 3°C, 15°C, or in running water at 38°C. Hams were dry-cured for 4 weeks at 3–4°C and salt equalized for 4 weeks at 13°C before being smoked for 24 hours at 38°C and then being aged for 3 months at 24°C. Hams were then sliced and evaluated for microbial growth.

No bacteria of public health significance were detected in or on hams before curing or after 3 months of aging. Freezing and storage appeared to injure some microorganisms, which resulted in the failure to recover selected pathogens and some nonpathogens from thawed hams. Higher initial surface counts were obtained from hams thawed before curing. Hams cured frozen had higher surface counts after curing. The investigators suggested that the results may have been related to lower salt and higher moisture levels in hams cured frozen, as well as to the method used to thaw hams [14].

Hams frozen and stored (12 months at –17.8°C) before being thawed (7 days at 3°C) and then dry-cured were sampled microbiologically after thawing, after curing, and after 30 days aging at 34.5°C and 62.5% relative humidity [72]. Surface samples and intramuscular shank samples contained the greatest total number of microorganisms. Moisture content appeared to be the major limiting factor to bacterial growth. However, the salt-tolerant microorganisms were the predominant flora. No bacteria of public health significance were isolated on or within ham tissues [72].

Frozen (–30°C) ham steaks and bacon slices were wrapped in oxygen-permeable retail wrap, freezer paper, polyethylene bags, and aluminum foil to determine the influence of frozen storage (0–196 days) and subsequent thawing on the psychrotrophic counts of cured pork cuts after 24 hours of retail display [19]. Length of frozen storage had no effect on $log_{10}$ psychrotrophic counts from either ham or bacon slices. It was concluded that frozen storage did not improve the bacteriological quality of the meat [19].

# VIII.   STABILITY OF FROZEN PRODUCTS

## A.   Comminuted Products

An investigation of the color quality of frozen minced beef during storage and the combined effects of light and added salt on prerigor beef was compared to postrigor beef [72]. Surface discoloration was related to lipid oxidation by determining surface color and thiobarbituric acid reactive substances (TBARS) on the product surface.

The minced hot or cold deboned beef product (pure beef; beef plus 1% NaCl; and beef plus 1% NaCl, 8 ppm $K_4[Fe(CN_6)]$), was packed in 450-g portions in polyethylene tubes (oxygen transmission rate = 1000 $cm^3/m^2/24$ h/atm at 25°C, RH 75%) and frozen to an internal temperature of –10°C (hot deboned product reached –10°C less than 6 hours postmortem). The tubes were placed in a freezer cabinet display [–18°C, fluorescent light (300–400 nm UV), 30 days storage], and partially covered with black plastic [72]. Results indicated that a better general oxidative stability and color quality (initial color and color stability during storage) may be expected during frozen storage for products made from prerigor compared to postrigor meat [72].

Pork backfat with varying levels of polyenoic fatty acids (up to approximately 15%) was obtained from pigs fed diets containing varying amounts of an ensiled corn cob mix [73]. The diet containing the highest amount of corn cob mix was supplemented with all-$\alpha$-tocopherol acetate. Pork backfat was removed from the carcasses and vacuum packaged or packaged in polyethylene freezer bags before being stored in the dark for 3, 6, or 9 months at –20°C. The backfat was analyzed for fat content, fatty acid composition, oxidation stability, peroxide and TBA values, and iodine and $\alpha$-tocopherol levels [73]. No rancidity was observed in backfat samples after 9 months of storage. No significant advantage was apparent for vacuum packaging over oxygen-permeable packaging for reducing lipid oxidation during frozen storage. Backfat produced from pigs fed the supplemented diet appeared to be more stable with regard to lipid oxidation, however, all backfat samples were far from being oxidized after 9 months of storage [73].

The influence of polyphosphate (0.0, 0.25, or 0.50% of an 85% STP and 15% sodium hexametaphosphate mixture) on the sensory (cohesiveness, juiciness, off-flavors) and chemical (proximate composition, lipid oxidation) characteristics of battered and breaded, cooked, restructured beef and pork nuggets (20% fat) was evaluated every 4 weeks over a 20-week frozen (–23°C) storage period [74]. No differences were observed between treatments for proximate composition and expressible moisture of cooked beef or pork nuggets. Sensory panelists detected no differences in off-flavors of beef nuggets between treatments or across storage times, but they did detect differences in off-flavors for pork nuggets. Addition of polyphosphate increased sensory scores for cohesiveness in both beef and pork nuggets, probably due to enhancement of protein extraction and increased bind. No differences across storage times were found for cohesiveness and juiciness. Lower

TBA values were observed for beef nuggets made with polyphosphates, but no additional protection from lipid oxidation was detected when polyphosphate level was increased from 0.25 to 0.50%. The TBA values for polyphosphate treated pork nuggets were lower initially than the control and remained lower throughout storage. Results indicated that polyphosphates protected lipids from oxidation and concluded that polyphosphates may protect the flavor of such products if they were stored frozen for long periods of time [74].

The effects of meat grinder wear on lipid and myoglobin oxidation in commercial fresh pork sausage chubs (with antioxidants) after frozen storage (–15°C for 4, 8, 12, and 16 weeks) and subsequent postfrozen refrigeration (1 day and 2 or 3 weeks at 1°C) were investigated (75). Lipid oxidation (TBA values) was not affected by length of frozen storage. Length of frozen storage significantly affected meat color as evidenced by a general decrease in Hunter a and b values during storage. However, an increase in these values was observed between 12 and 16 weeks of storage, which could not be explained. During frozen storage, the largest single increase in internal metmyoglobin "ring" formation occurred after the first 4 weeks of storage. After 3 weeks of postfrozen refrigeration, the internal color of the sausage improved, regardless of the length of frozen storage. Lipid oxidation and metmyoglobin formation were inversely related, supporting the concept that in spite of correlations between the two types of oxidation, a consistent interdependence may not exist [75].

Cooked meat loaves with 0–5% added blood were vacuum packaged and stored for 1 year at –20°C before being evaluated for rancidity and off-flavor development by a descriptive sensory panel [44]. Small but significantly higher rancidity and off-flavor values were detected in meat loaves with 1–2% added blood. Higher concentrations of blood resulted in no further increase in rancidity or off-flavor, possibly due to the ability of heme compounds to act as an antioxidant when present in high concentrations relative to that of polyunsaturated fatty acids. Approximate molar ratio values (polyunsaturated fatty acids to heme) were 94:1, 83:1, and 68:1 for 3, 4, and 5% added blood, respectively. Since the meat loaves were produced from homogeneous batches of pork fat, ground beef, and blood, the experimental conditions guaranteed a decrease in this molar ratio. Previous research had reported that heme compounds increasingly act as antioxidants beginning at 350:1, with complete inhibition at approximately 90:1 [44].

## B.  Restructured Products

The effectiveness of rosemary oleoresin (0.05 or 0.10%) with or without STP (0.3%), and tertiary butylhydroquinone (0.02%) with STP (0.3%) was determined by evaluating their antioxidant activities in raw restructured beef steaks stored at –20°C for 6 months (samples tested at 0, 3, or 6 months) and in cooked steaks stored at 4°C for 6 days [76]. During refrigerated storage, TBARS and sensory scores indicated that STP provided protection against the development of warmed-over flavor. A linear concentration effect of rosemary oleoresin observed for both TBARS values and sensory scores (intensity of warmed-over flavor, presence or absence of nonmeat flavors, and perceived flavor) indicated rosemary oleoresin to be an effective antioxidant, which in combination with STP produced an additive protective effect during frozen storage. Hexanal content and phospholipid fatty acid profiles generally increased during frozen storage, most noticeably with the salt-only control [76].

The binding properties and shelf-life characteristics of solid muscle restructured beef produced utilizing meat blocks bound together with algin/calcium/adipic acid binding gel were evaluated [77]. Steakettes were allocated to four treatment groups: Group 1 steakettes were evaluated the day they were cut (fresh, day 1 evaluation); Group 2 steakettes (fresh, day 6 evaluation) were placed on styrofoam trays, overwrapped with permeable polyvinylchloride film, and stored at 4°C (7, 21, or 35 days) before being subjected to continuous fluorescent display lighting; Group 3 steakettes were packaged in the same manner as Group 2 steakettes, held at 4°C under continuous fluorescent display lighting for 24 hours, then overwrapped with plastic-coated freezer paper, frozen, and stored at –20°C for 7, 21, and 35 days; Group 4 steakettes were treated the same as Group 3 steakettes but were vacuum packaged. Steakettes were analyzed for surface discoloration, steakette juncture integrity (subjective), objective binding strength, proximate composition, and TBA values [77]. Results of this study indicated that refrigerated storage time had no effect on surface discoloration, juncture integrity, binding strength, or proximate composition of raw or cooked steakettes. Binding strength values were lower for frozen than for fresh groups for both raw and cooked steakettes, which the authors suggested might be due to the additional handling of steakettes in preparation for freezing and/or the freezing/thawing process, which may have disrupted interactions between gel constituents and muscle components [77]. Fresh steakettes displayed higher percent moisture and lower cook yield than frozen steakettes. The lower moisture content of frozen steakettes was postulated to have been due to additional storage and purge losses during thawing prior to analyses. Rancidity development was greatest in steakettes stored in oxygen-permeable packages. Algin/calcium/adipic gel maintained the integrity of restructured meat products following extended refrigerated and/or frozen storage [77].

## IX.  SUMMARY

Freezing and thawing affects the retention of nutrients, color stability, lipid oxidation, shelf life, and overall acceptability of cured and processed meat products. It is important to consider the pre- and postprocessing effects of freezing and thawing on overall product quality of cured and processed meats. The rate of freezing, frozen storage conditions (temperature, time, packaging), and subsequent method of thawing affect the microbial load, functionality, and other physicochemical properties of the product raw materials, which ultimately affects the attributes of the final cured or processed meat product. Freezing will not improve the quality of the raw materials or of the product postprocessing. Freezing results in gradual deterioration of raw product attributes as frozen storage time increases, regardless of the method or technique incorporated to stop its advance. The challenge is to reduce the rate of raw product deterioration through improved packaging materials, freezing systems, and other novel approaches. The most efficient way to reduce the deterioration of product attributes due to freezing is to not freeze the raw product materials in the first place. Since this is not a viable option in a majority of instances, it is important to understand the losses that may occur to raw products during freezing. The quality of the final cured or processed meat product is directly related to the quality of the raw materials used in their manufacture. Postprocessing freezing does not improve final product quality, though it may extend the shelf life of these products with minimal additional losses in quality during frozen storage.

Cured or processed meat products intended for retail sale distributed in the frozen state face additional concerns with regard to transportation time and temperature, retail storage conditions, and display lighting. Efforts must be continued to explore the effects of freezing and frozen storage on the mechanisms responsible for desirable color and sensory attributes of these products. When freezing is a part of the total process, understanding the consequences of freezing and taking appropriate steps to minimize the deterioration caused by freezing is vitally important.

## REFERENCES

1.  J. Buchmüller, Chilling and freezing meat products with liquid nitrogen, *Fleischwirtsch Int.* 3:25 (1987).
2.  J. Buchmüller, Chilling and freezing meat products with cryogenic liquids, *Fleischwirtsch Int.* 2:27 (1986).
3.  A. J. Miller, S. A. Ackerman, and S. A. Palumbo, Effects of frozen storage on functionality of meat for processing, *J. Food Sci.* 45:1466 (1980).
4.  L. Petrović, R. Grujić, and M. Petrović, Definition of the optimal freezing rate—2. Investigation of the physico-chemical properties of beef m. longissimus dorsi frozen at different freezing rates, *Meat Sci.* 33:319 (1993).
5.  B. H. Chiang, H. W. Norton, and D. B. Anderson, The effect of hot-processing, seasoning and vacuum packaging on the storage stability of frozen pork patties, *J. Food Proc. Pres.* 5:161 (1981).
6.  J. O. Reagan, S. L. Pirkle, D. R. Campion, and J. A. Carpenter, Processing, microbial and sensory characteristics of cooler and freezer stored hot-boned beef, *J. Food Sci.* 46:838 (1981).
7.  A. Bezanson, Thawing and tempering of frozen meat, *Proceedings of the Meat Industry Res. Conf.*, Am. Meat Inst. Found., Chicago, IL, 1975, p. 51.
8.  S. Gonzalez-Sanguinetti, M. C. Anon, and A. Calvelo, Effect of thawing rate on the exudate production of frozen beef, *J. Food Sci.* 50:697 (1985).

9.  J. W. Park, T. C. Lanier, J. T. Keeton, and D. D. Haman, Use of cryoprotectants to stabilize functional properties of prerigor salted beef during frozen storage, *J. Food Sci.* 52(3):537 (1987).

10. J. W. Park, T. C. Lanier, and D. H. Pilkington, Cryostabilization of functional properties of pre-rigor and post-rigor beef by dextrose polymer and/or phosphates, *J. Food Sci.* 58(3):467 (1993).

11. S. Barbut and G. S. Mittal, Phosphates and antioxidants as cryoprotectants in meat batters, *Meat Sci.* 30:279 (1991).

12. B. H. Ashby and G. M. James, Effects of freezing and packaging methods on shrinkage of hams in frozen storage, *J. Food Sci.* 38:254 (1978).

13. B. H. Ashby and G. M. James, Effects of freezing and packaging methods on freezer burn of hams in frozen storage, *J. Food Sci.* 38:258 (1978).

14. J. D. Kemp, B. E. Langlois, and J. D. Fox, Composition, quality and microbiology of dry-cured hams produced from previously frozen green hams, *J. Food Sci.* 43:860 (1978).

15. L. J. Bratzler, Technical problems in prepackaged fresh and frozen meats, *Proceedings of the 7th Res. Conf.*, Am. Meat Inst. Found., Chicago, 1955, p. 2.

16. M. S. Brewer and C. A. Z. Harbers, Effect of packaging on physical and sensory characteristics of ground pork in long-term frozen storage, *J. Food Sci.* 56(3):627 (1991).

17. H. J. Andersen, G. Bertelsen, and L. H. Skibsted, Colour stability of minced beef. Ultraviolet barrier packaging material reduces light-induced discoloration of frozen products during display, *Meat Sci.* 25:155 (1989).

18. L. E. Jeremiah, The effect of frozen storage and protective wrap upon the cooking losses, palatability and rancidity of fresh and cured pork cuts, *J. Food Sci.* 45:187 (1980).

19. L. E. Jeremiah, The effects of frozen storage and thawing on the retail acceptability of ham steaks and bacon slices, *J. Food Qual.* 5:43 (1981).

20. M. S. Brewer, W. G. Ikins, and C. A. Z. Harbers, TBA values, sensory characteristics, and volatiles in ground pork during long-term frozen storage: Effects of packaging, *J. Food Sci.* 57(3):558 (1992).

21. I. G. Legarreta, W. R. Usborne, and G. C. Ashton, Extending the retail storage time of pork sausage using modified atmospheres and freezing, *Meat Sci.* 23:21 (1988).

22. M. M. Verma, A. D. Alarcon Rojo, D. A. Ledward, and R. A. Lawrie, Effect of frozen storage of minced meats on the quality of sausages prepared from them, *Meat Sci.* 12:125 (1985).

23. H. J. Andersen and L. H. Skibsted, Oxidative stability of frozen pork patties. Effect of light and added salt, *J. Food Sci.* 56(5):1182 (1991).

24. B. W. Berry, Changes in quality of all-beef and soy-extended patties as influenced by freezing rate, frozen storage temperature, and storage time, *J. Food Sci.* 55(4):893 (1990).

25. M. S. Brewer, F. K. McKeith, and K. Britt, Fat, soy and carrageenan effects on sensory and physical characteristics of ground beef patties, *J. Food Sci.* 57(5):1051 (1992).

26. M. Bhattacharya, M. A. Hanna, and R. W. Mandigo, Effect of frozen storage conditions on yields, shear strength and color of ground beef patties, *J. Food Sci.* 53(3):696 (1988).

27.  R. A. Molins, A. A. Kraft, H. W. Walker, R. E. Rust, D. G. Olsen, and K. Merkenich, Effect of inorganic polyphosphates on ground beef characteristics: Some chemical, physical and sensory effects on frozen beef patties, *J. Food Sci.* 52(1):50 (1987).

28.  E. A. Decker and A. D. Crum, Inhibition of oxidative rancidity in salted ground pork by carnosine, *J. Food Sci.* 56(5):1179 (1991).

29.  C. M. Chen and G. R. Schmidt, Color and its stability in restructured beef steaks during frozen storage: Effects of various binders, *J. Food Sci.* 56(6):1461 (1991).

30.  J. G. Akamittath, C. J. Brekke, and E. G. Schanus, Lipid oxidation and color stability in restructured meat systems during frozen storage, *J. Food Sci.* 55(6):1513 (1990).

31.  Y. H. Chu, D. L. Huffman, G. R. Trout, and W. R. Egbert, Color and color stability of frozen restructured beef steaks: Effect of sodium chloride, tripolyphosphate, nitrogen atmosphere, and processing procedures, *J. Food Sci.* 52(4):869 (1987).

32.  Y. H. Chu, D. L. Huffman, G. R. Trout, and W. R. Egbert, Color and color stability of frozen restructured beef steaks: Effect of processing under gas atmospheres with differing oxygen concentration, *J. Food Sci.* 53(3):705 (1988).

33.  K. L. Neer and R. W. Mandigo, Effects of salt, sodium tripolyphosphate and frozen storage time on properties of a flaked, cured pork product, *J. Food Sci.* 42(3):738 (1977).

34.  J. F. Campbell and R. W. Mandigo, Properties of restructured pork patties as affected by cooking method, frozen storage and reheating method, *J. Food Sci.* 43:1648 (1978).

35.  J. G. Sebranek, P. N. Sang, R. E. Rust, D. G. Topel, and A. A. Kraft, Influence of liquid nitrogen, liquid carbon dioxide and mechanical freezing on sensory properties of ground beef patties, *J. Food Sci.* 43:842 (1978).

36.  B. H. Lee, R. E. Simard, C. L. Laleye, and R. A. Holley, Shelf-life of meat loaves packaged in vacuum or nitrogen gas 2. Effect of storage temperature, light and time on physicochemical and sensory changes, *J. Food Prot.* 47(2):134 (1984).

37.  G. S. Padda, R. C. Keshri, B. D. Sharma, and T. R. K. Murthy, Physico-chemical and organoleptic properties of patties from hot, chilled and frozen goat meat, *Meat Sci.* 22:245 (1988).

38.  R. G. Matlock, R. N. Terrell, J. W. Savell, K. S. Rhee, and T. R. Dutson, Factors affecting properties of raw-frozen pork sausage patties made with various NaCl/phosphate combinations, *J. Food Sci.* 49(5):1363 (1984).

39.  R. G. Matlock, R. N. Terrell, J. W. Savell, K. S. Rhee, and T. R. Dutson, Factors affecting properties of precooked-frozen pork sausage patties made with various NaCl/phosphate combinations, *J. Food Sci.* 49(5):1372 (1984).

40.  B. W. Berry, Changes in quality of all-beef and soy-extended patties as influenced by freezing rate, frozen storage temperature, and storage time, *J. Food Sci.* 55(4):893 (1990).

41.  R. Hanenian, G. S. Mittal and W. R. Usborne, Effects of pre-chilling, freezing rate, and storage time on beef patty quality, *J. Food Sci.* 54(3):532 (1989).

42.  B. W. Berry, Fat level and freezing temperature affect sensory, shear, cooking and compositional properties of ground beef patties, *J. Food Sci.* 58(1):34 (1993).

43.  B. W. Berry and K. F. Leddy, Effects of hot processing, patty formation before or after freezing-thawing and soy usage on various properties of low-fat ground beef, *J. Food Qual.* 11(2):159 (1988).

44.  I. M. Oellingrath and E. Slinde, Sensory evaluation of rancidity and off-flavor in frozen stored meat loaves fortified with blood, *J. Food Sci.* 53(3):967 (1988).

45. J. C. Kuo and H. W. Ockerman, Effects of rigor state, salt level and storage time on chemical and sensory traits of frozen and freeze-dried ground beef, *J. Food Prot.* 48(2):142 (1985).

46. H. R. Cross, A. W. Kotula, and T. W. Noland, Stability of frozen ground beef containing mechanically deboned beef, *J. Food Sci.* 43(2):281 (1978).

47. B. W. Berry, J. J. Smith, and J. L. Secrist, Effects of flake size on textural and cooking properties of restructured beef and pork steaks, *J. Food Sci.* 52(3):558 (1987).

48. J. C. Cordray, D. L. Huffman and W. R. Jones, Restructured pork from hot processed sow meat: Effect of mechanical tenderization and liquid smoke, *J. Food Prot.* 49(8):639 (1986).

49. L. W. Hand, R. N. Terrell, and G. C. Smith, Effects of chloride salt, method of manufacturing and frozen storage on sensory properties of restructured pork roasts, *J. Food Sci.* 47(6):1771 (1982).

50. S. L. Jones, T. R. Carr, and F. K. McKeith, Palatability and storage characteristics of precooked pork roasts, *J. Food Sci.* 52(2):279 (1987).

51. B. Jakobsson and N. Bengtsson, A quality comparison of frozen and refrigerated cooked sliced beef, *J. Food Sci.* 37:230 (1972).

52. L. F. Miller, H. B. Hedrick, and M. E. Bailey, Sensory and chemical characteristics of pork chops as affected by precooking, curing and frozen storage, *J. Food Sci.* 50: 478 (1985).

53. Vitamins, *Developments in Food Science*, Vol. 21 (J. Davidek, J. Velisek, and J. Pokorny, eds.), Elsevier, New York, 1990, pp. 230–293.

54. W. Moleeratanond, B. H. Ashby, A. Kramer, B. W. Berry, and W. Lee, Effect of a di-thermal storage regime on quality and nutritional changes and energy consumption of frozen boxed beef, *J. Food Sci.* 46:829 (1981).

55. B. D. Westerman, B. Oliver, and D. L. Mackintosh, Influence of chilling rate and frozen storage on B-complex vitamin content of pork, *J. Agric. Food Chem.* 3(7):603 (1955).

56. K. Causey, E. G., Andreassen, M. E. Hausrath, C. Along, P. E. Ramstad, and F. Fenton, Effect of thawing and cooking methods on the palatability and nutritive value of frozen ground meat. I. Pork, *Food Res.* 15:237 (1950).

57. K. Causey, E. G. Andreassen, M. E. Hausrath, C. Along, P. E. Ramstad, and F. Fenton, Effect of thawing and cooking methods on the palatability and nutritive value of frozen ground meat. II. Beef, *Food Res.* 15:249 (1950)

58. K. Causey, E. G. Andreassen, M. E. Hausrath, C. Along, P. E. Ramstad, and F. Fenton, Effect of thawing and cooking methods on the palatability and nutritive value of frozen ground meat. III. Lamb, *Food Res.* 15:256 (1950)

59. C. Y. W. Ang, L. A. Basillo, B. A. Cato, and G. E. Livingston, Riboflavin and thiamine retention in frozen beef-soy patties and frozen fried chicken heated by methods used in food service operations, *J. Food Sci.* 43:1024 (1978).

60. L. C. West, M. C. Titus, and F. O. Van Duyne, Effect of freezer storage and variations in preparation on bacterial count, palatability and thiamine content of ham loaf, Italian rice, and chicken, *Food Technol.* 13(6):323 (1959).

61. O. R. Fennema, Rates of chemical deterioration in frozen foods, *Proceedings of the Meat Indus. Res. Conf.*, American Meat Sci. Association, American Meat Institute Found., Chicago, 1971, p. 35.

62. W. E. Townsend, Stability of residual acid phosphatase activity in cured/canned picnic samples stored at –34°C for 15 and 36 months, *J. Food Sci.* 54(3):752 (1989).
63. M. P. Wanous, D. G. Olson, and A. A. Kraft, Pallet location and freezing rate effects on the oxidation of lipids and myoglobin in commercial fresh pork sausage, *J. Food Sci.* 54(3):549 (1989).
64. O. Zorba, H. Y. Gokalp, H. Yetim, and H. W. Ockerman, Salt, phosphate and oil temperature effects on emulsion capacity of fresh or frozen meat and sheep tail fat, *J. Food Sci.* 58(3):492 (1993).
65. R. V. Lechowich, Microbiology of meat, *The Science of Meat and Meat Products*, (J. F. Price and B. S. Schweigert, eds.), Food and Nutrition Press, Inc., Westport, CT, 1978, pp. 411–412.
66. M. L. Speck and B. Ray, Effects of freezing and storage on microorganisms in frozen foods: A review, *J. Food Prot.* 40(5):333 (1977).
67. R. A. Molins, A. A. Kraft, H. W. Walker, R. E. Rust, D. G. Olsen, and K. Merkenich, Effect of inorganic polyphosphates on ground beef characteristics: Microbiological effects on frozen beef patties, *J. Food Sci.* 52(1):46 (1987).
68. B. H. Lee, R. E. Simarrd, C. L. Laleye, and R. A. Holley, Shelf-life of meat loaves packaged in vacuum or nitrogen gas 1. Effect of storage temperature, light and time on the microflora change, *J. Food Prot.* 47(2):128 (1984).
69. K. Rayman, K. F. Weiss, G. Riedel, and G. Jarvis, Microbiological quality of Canadian frozen meat pies, *J. Food Prot.* 49(8):634 (1986)
70. A. A. Kraft, K. V. Reddy, J. G. Sebranek, R. E. Rust, and D. K. Hotchkiss, Effect of combinations on microbial flora of ground beef patties, *J. Food Prot.* 44(11):870 (1981).
71. R. E. Simard, B. H. Lee, C. L. Laleye and R. A. Holley, Effects of temperature, light and storage time on the microflora of vacuum- or nitrogen-packed frankfurters, *J. Food Prot.* 46(3):199 (1983).
72. H. J. Andersen, G. Bertelsen, and L. H. Skibsted, Colour and colour stability of hot processed frozen minced beef. Results from chemical model experiments tested under storage conditions, *Meat Sci.* 28:87 (1990).
73. J. H. Houben, and B. Krol, Effect of frozen storage and protective packaging on lipid oxidation in pork backfat with slightly increased levels of polyenoic fatty acids, *Meat Sci.* 13(4):193 (1985)
74. D. L. Huffman, C. F. Ande, J. C. Cordray, M. H. Stanley, and W. R. Egbert, Influence of polyphosphate on storage stability of restructured beef and pork nuggets, *J. Food Sci.* 52(2):275 (1987).
75. M. P. Wanous, D. G. Olson, and A. A. Kraft, Oxidative effects of meat grinder wear on lipids and myoglobin in commercial fresh pork sausage, *J. Food Sci.* 54(3):545 (1989).
76. S. M. Stoick, J. I. Gray, A. M. Booren, and D. J. Buckley, Oxidative stability of restructured beef steaks processed with oleoresin rosemary, tertiary butylhydroquinone, and sodium tripolyphosphate, *J. Food Sci.* 56(3):597 (1991).
77. T. S. Muller, R. C. Johnson, W. J. Costello, J. R. Romans and K. W. Jones, Storage of structured beef steakettes produced with algin/calcium/adipic acid gel, *J. Food Sci.* 56(3):604 (1991).

# 6
# Fruits

**Grete Skrede**

*MATFORSK—Norwegian Food Research Institute*
*Ås, Norway*

## I. INTRODUCTION

The influence of freezing, frozen storage, and thawing on fruit quality have been extensively investigated through several decades [1–13]. These studies have included investigations into the freezing process itself, the influence of temperature on ice crystal formation, the effects of cell rupture on tissue texture, and the effects of enzymatic reactions on odor and flavor as well as on ascorbic acid and color deterioration. Fruits have been subjected to a variety of storage conditions and times to gain a better understanding of the reactions taking place and to minimize the extent of damage to susceptible fruit tissues. Since some types of fruit are better suited to freezing, varietal characteristics are important, prompting a constant search for new types of frozen fruit and a better supply of suitable fruit cultivars.

Frozen fruits constitute a large and important food group in modern society. Fruit may be more extensively used if available during the off-season. In addition, frozen fruit can be transported to remote markets that could not be accessed with fresh fruit. Freezing also makes year-round further processing of fruit products, such as jams, juice, and syrups from frozen whole fruit, slices, or pulps possible.

The quality demanded of frozen fruits or frozen fruit products depends upon the intended use of the product. If the fruit will be eaten without any further processing after thawing, texture characteristics are more important than if the fruit will be made into juice.

In this chapter recent literature dealing with freezing, frozen storage, and thawing are presented. The review concentrates on the physical and chemical changes taking place in fruits during freezing and storage under controlled laboratory conditions, since little information is available about results obtained under authentic industrial conditions. Moreover, most studies cited deal with frozen fruit and fruit products intended for further processing, as this currently is the area of most commercial interest. The text of the chapter is based upon research findings mainly since 1975, with no effort made to cover completely existing knowledge. Consequently, complete coverage of the state of the art is not provided directly, but may be obtained from the cited references.

## II. EFFECTS OF PREPARATION AND PACKAGING

Most fruits are soft in texture even before freezing and thawing. Freezing tends to disrupt the structure and destroy the turgidity of the living cells in fruit tissues. Preparation methods for fruits intended for freezing are influenced by the fragility of fruit tissues and must be chosen carefully. This is in contrast to vegetables, where the more fibrous compounds tend to preserve the structure after freezing and thawing.

### A. Effect of Maturity Level and Storage Prior to Freezing

Harvesting of fruits at optimum maturity level for commercial use is difficult. Objective methods for judging maturity are scarce, and the need for efficient production often implies the use of mechanical harvesting at a time when most of the crop has reached an acceptable maturity level. Several studies demonstrate the influence of maturity on freezing performance. Attention should be given to this aspect of quality when the suitability of fruits for freezing and frozen storage is considered. Also, the importance of freezing fruits immediately after harvesting has been evaluated.

Mango is an easily perishable fruit with a short harvesting season. According to Singh [14], ripe mango fruit must be frozen without delay the day of harvest. Marín et al. [15] reported studies where mangoes were harvested at the preclimacteric hard-green stage. The fruits were stored at 12°C and 85–90% relative humidity within 24 hours of harvest until they reached proper maturity levels (about 10 days). The fruits were reported to perform well when peeled, cut into slices, frozen in an air-blast freezer at –40°C, and stored for 4 months at –18°C. Polyak-Feher and Szabone-Kismarton [16] reported lower amounts of drip from unripe frozen apricots, cherries, sour cherries, and plums compared with ripe fruits.

Urbànyi and Horti [17] studied color (CIELAB) and carotenoid content during frozen storage of tomatoes with varying degrees of maturation. Increased maturation caused a darker color (lower L* values) and increased redness (higher a* values) in tomatoes. Yellowness was only minimally influenced by stage of ripeness. During frozen storage color gradually became lighter, redness decreased, and yellowness increased. The ripest samples underwent the greatest change and were a deeper red and had a higher saturation level than the least ripe samples at the end of the storage period. Total carotenoid content also decreased with storage time. Carotenoid retention was highest in the most mature samples.

Cano et al. [18] studied the influence of maturity level on the freezing performance of sliced bananas. Fruits were processed during a ripening period of up to 23 days. Freezing caused about 60% inactivation of peroxidase in bananas at all maturity levels. Only postclimacteric fruits showed similar inactivation of polyphenoloxidase. No inactivation was observed in preclimacteric or climacteric bananas.

Plocharski [19] compared quality of strawberries frozen within 2 hours of picking with fruits frozen after 5–6 hours. After 3 months of storage, significant differences in sensory properties were observed between fruits frozen immediately and those frozen after the delay. Fruits frozen immediately in two out of three cultivars were rated higher in texture, color, and flavor intensity than fruits frozen after the delay. However, after 6 and 12 months of storage, the positive effects of immediate freezing were not detectable. Differences in composition between strawberries were also observed. Fruits frozen immediately after picking revealed tendencies for higher titratable acidity, lower pH, higher content of ascorbic acid and anthocyanins, and higher drained weight than fruits frozen after the 5- to 6-hour delay, and these differences remained unchanged during the 12-month storage period at $-20°C$.

Plocharski [19] also evaluated the effect of storing strawberries for 6 days at $6°C$ under normal and controlled atmospheres. The main conclusion drawn from this experiment was that storage of strawberries before freezing, regardless of the atmosphere, had little negative effect on the sensory quality of frozen fruits when compared with fruits stored for 6–8 hours after harvesting. Storage beyond 6 days, however, resulted in a severe decrease in frozen fruit flavor and a decrease in fruit texture desirability. Storage of fruits for longer periods prior to freezing resulted in losses due to rotting and desiccation of the fresh fruits.

## B.  Effects of Peeling, Slicing, and Cutting

Fruits used for freezing must be prepared beforehand, because it is not usually possible to wash and cut them after thawing without difficulties and losses. The cleaning, rinsing, sorting, and cutting of fruits intended for freezing is simlar to the preparation of fruits for other types of conservation [20]. These unit operations, therefore, are not specific to frozen fruits.

Stone fruits may be pitted either by halving the fruit prior to pitting or by punching the pits out of whole fruit [21]. These processes clearly influence fruit performance during the freezing process. Polyak-Feher and Szabone-Kismarton [16] compared the amount of drip from ripe and unripe, whole and stoned apricots, cherries, sour cherries, and plums frozen at $-20°C$ and stored for 3 and 6 months.

The amount of juice tended to be greatest with stoned fruits and least with underripe whole fruits.

Fruits requiring peeling before consumption or further processing must be peeled prior to freezing. This may be done by scalding in hot water, steam, or hot lye solutions [21]. Peeled fruits studied prior to freezing include kiwi [22], bananas [18], and mango [15].

Gradziel [23] studied freezing performance of peeled and unpeeled tomatoes. Tomatoes were peeled by immersion in boiling water for 30 seconds. After freezing at −10 or −40°C and storage for periods up to 330 days, no effect of peeling was reported for the quality factors evaluated.

Freezing rate has been known to influence freezing performance for quite some time [10,24,25]. If freezing is fast, many small ice crystals form. Slow freezing permits early-forming ice crystals to grow. A slow freezing rate, therefore, results in fewer but larger ice crystals in the tissue. The presence of large crystals causes more structural damage within the cell.

Fast freezing of fruits may be achieved by decreasing the size of the product to be frozen. Consequently, it has been common practice to cut larger fruits, such as tomatoes [17], bananas [18], breadfruits [26], mangoes [14,15,27,28] and kiwi [29,59], into cubes or slices prior to freezing. Trials have also been conducted on sliced strawberries [20,30].

Morris et al. [30] demonstrated that slicing of strawberries prior to dipping in calcium chloride or low-methoxyl pectin solutions resulted in firmer fruit, when compared to whole berries. In addition, when calcium solutions were used, sliced fruits were more red and more yellow than whole fruits, indicating better color retention in the sliced fruits.

## C.  Blanching

Fruits lose their characteristics of freshness if they are heat-treated for enzyme inactivation [4]. Therefore, only a few types of fruits are blanched before freezing. Losses of soluble substances, minerals, and water-soluble vitamins to the blanching water may also occur, and efforts should be made to keep these losses to a minimum [20].

The main advantage of blanching is inactivation of enzymes. Enzyme inactivation prevents detrimental changes in color, odor, flavor, and nutritive value. A well-established procedure is to inactivate catalase and nearly all peroxidase in a blanching procedure, which keeps the blancing time as short as possible at a carefully chosen temperature. Blanching is hot water also removes tissue air and thereby reduces the occurrence of undesirable oxidation reactions during freezing, frozen storage, and thawing of the product. Although blanching has limited use as a pretreatment when freezing fruits, some applications are of interest.

Bananas undergo rapid browning due to cellular disruption and exposure to oxygen during the peeling and slicing operations. Cano et al. [18] examined the ability of thermal treatment to improve freezing performance of bananas. Whole peeled bananas were blanched in boiling water for 11 minutes and cooled in ice water for 5 minutes prior to slicing, freezing, and storage (24 hours) at -24°C. Blanching of bananas prior to freezing produced almost complete inactivation of polyphenoloxidase and peroxidase. In bananas frozen without heat treatment, varying degrees of enzyme inactivation occurred, depending upon the maturity level of the banana. Water blanching produced a product with acceptable sensory quality—bright color, firm texture, and desirable flavor with no darkening—even when the product was thawed in air for a relatively long period of time. Microwave treatment of bananas prior to freezing also inhibited enzyme activities. However, the fruits became stale and showed significant darkening in the central part of the slices, due to nonenzymatic Maillard reactions. Microwave heating was, therefore, found not to be a suitable pretreatment for freezing banana slices.

Passam et al. [26] proposed a method for freezing easily perishable breadfruit by partially cooking peeled segments prior to freezing. Fully mature fruits were peeled, cut into slices, precooked for 0–10 minutes, cooled, packaged, and frozen (-15°C). After 10 weeks of frozen storage, the fruit segments were placed directly into boiling water and cooked to an appropiate firmness (the time required depended upon the initial blanching period). Cooked breadfruits were then assessed for flavor, texture, and color by persons familiar with local fresh breadfruits. Samples frozen without preboiling discolored during cooking, and the flavor was impaired. Cooking for only 1 minute prior to freezing tended to produce an inconsistent texture, with some segments remaining hard longer than others. With cooking times of 2 or 5 minutes, the frozen breadfruit segments were scored just as high as the control (cooked, nonfrozen breadfruit) for flavor, color, and texture. Breadfruits precooked for periods longer than 5 minutes appeared to be less acceptable after storage. Often samples had desirable flavor but had a soft, sometimes crumbly, texture.

As a substitute for thermal blanching, Serratos et al. [31] evaluated the possibility of applying sulfur dioxide ($SO_2$) and carbon monoxide (CO) gases as enzyme inhibitors in apple segments prior to freezing. The apple segments were deaerated under vacuum prior to the gas treatment. Treatment with CO was found to be ineffective for inhibition of catalase and polyphenolixidase. Treatment with $SO_2$, however, appeared to inhibit enzyme activity as effectively as thermal blanching, and the inhibition was maintained during a storage period of 195 days at both -8 and -18°C.

To avoid undesired $SO_2$ odor and taste remaining in samples without further heat treatment, $SO_2$ was diluted with nitrogen ($N_2$) gas and the effects of various gas mixtures on the inactivation of both enzymes were investigated. Samples were subjected to gas treatments of 75, 50, and 25% $SO_2$ for 15 seconds. After 10 days of storage -18°C, both enzymes showed 100% inactivation at all $SO_2$ levels.

Residual $SO_2$ was 36–72 ppm, with the lowest level in apple segments treated with the lowest $SO_2$ levels. When samples treated with an atmosphere containing 25% $SO_2$ were submitted to vacuum for a period of 4 hours, residual $SO_2$ dropped to nondetectable levels and no regeneration of either catalase or polyphenoloxidase was observed.

Serratos et al. [31] determined Hunter Lab color values, browning, pH, residual $SO_2$, and soluble solids in $SO_2$-treated apple segments during the 195-day storage period at –8 or –18°C. The results were compared with those of thermally blanched apples stored under the same conditions. Hunter L (lightness) and b (yellowness) values remained nearly constant at about 37 and 10 days, respectively, during storage at both –8 and –18°C for both gas-treated and thermally blanched samples when made into apple puree. Intensity of red (a) values increased slightly from about 0 at the beginning of the storage period to about 3 after 195 days of storage at –8 and –18°C. In heat-treated apples a tendency towards a green color (a from 0 to –3) was observed. Browning, as measured by absorbance at 420 nm of a clarified ethanol extract, showed no significant changes during storage at either temperature, concurring with the Hunter L values observed. The pH of the $SO_2$-treated apples decreased to an average of 3.39, while that of the thermally treated apples rose from 3.65 to 3.83. These changes were attributed to the dissolution of $SO_2$ in the gas-treated tissues and the dilution effect in the samples blanched in water. Baking the samples after thawing resulted in similar pH values for the two treatments. Soluble solids were higher in the gas-treated samples, indicating loss of soluble solids during the blanching process. The $SO_2$ levels of gas-treated apples initially were 700–800 ppm. During frozen storage the levels decreased to about 450 ppm. After baking, the $SO_2$ levels in the frozen/thawed apples decreased to between 50 and 100 ppm. No differences between treatment at storage temperatures of either –8 or –18°C were detected in taste or texture of baked apples.

Lee et al. [32] reported nonsignificant changes in color hue (a/b) and only minor changes in pH, amino-N, and total and reducing sugars when pasteurized and fresh passion fruit juice were stored at –20°C for up to 12 months. Frozen storage reduced the flavor scores of heated and unheated juice. After 12 months of storage, the heated juice had a decreased but acceptable flavor, while the unheated juice was regarded as being unacceptable.

## D.  Production of Pulp, Puree, Juice, and Nectar

Fruits are often subjected to various processing operations prior to freezing. Common products include pulps, purees, juices, and nectars. These products are of particular interest for fruit that does not meet the standards required for fresh fruit and, therefore, is not suitable for sale on the fresh market [33]. Such products have the advantage of being less susceptible to oxidation from the surrounding air but often have the disadvantage of more comprehensive treatments during the production processes [34,33]. Fruit juices and nectars, which represent a large proportion of drinks in the international market, have been studied thoroughly in recent years. However, few studies deal with the actual process of making these products.

## E. Effect of Addition of Ingredients

### 1. Sugars and Syrups

An extremely important pretreatment involves the addition of sugars [4,14,20,24, 28]. The addition of sugar and syrup has the effect of excluding oxygen from the fruit and thus helps to preserve color and appearance. However, sugar syrup with a 30–60% sugar content must cover the fruit completely to act as a barrier to oxygen transmission and thereby prevent browning [24]. The use of sweeteners is intended to enhance the natural fruit flavor and protect the fruit against the action of enzymes.

Sugars act by withdrawing water from cells by osmosis, leaving very concentrated solutions inside the cells [20,25]. The high concentration of solutes depresses the freezing point and therefore reduces freezing within the cells. If the cells do not actually freeze, no ice crystals will form to cause structural damage [25]. With only limited structural damage during freezing, improved texture, reduced damage to membranes, and consequently less mixing of enzymes and substrates will result.

The major problem when using sugars as cryoprotectants is getting sufficient amounts of sugar into the fruit so that the internal tissues are protected [25]. Syrup is generally considered a better protecting agent than dry sugar. If dry sugar is used, it should be added 1 or 2 hours before freezing to allow the sugar to dissolve in the fruit juice [24]. With use of dry sugar, often the surface layers of a fruit are protected while central tissues are damaged as a result of the sugar not reaching them [25]. This problem is exacerbated by the fact that the tissue in the middle of a fruit freezes more slowly than the surface layers and thereby is more prone to ice crystal formation. The transport of liquid between the cells and their surroundings can only take place as long as the liquids are not frozen, i.e., during the pretreatment of fruits and during the thawing process. Loss of color into the surrounding medium occurs in frozen berries where anthocyanins pass into the sugar syrup [24]. This change occurs very slowly at –18°C or colder temperatures but is greatly accelerated at higher temperatures and on thawing.

Clear-cut evidence that sugar has a protective effect on flavor, odor, and color during freezing, storage, and thawing has been difficult to find [4]. Kulisiewicz and Kolasa [35] reported that addition of 20% sugar enhanced the color, flavor, and consistency of frozen strawberries. More recently, Fraczak and Zalewska-Korona [36] found that strawberries frozen in sugar with a fruit: sugar ratio of 1:4 had better ascorbic acid retention during frozen storage than strawberries frozen without sugar. Vitamin C losses just after freezing were 2.5%, while after 6 months of storage at –18°C they were 25.0%.

A slight but significant protective effect of sugar addition on anthocyanin pigment stability in frozen strawberries has been shown [37]. Individually quick-frozen strawberries packed with 10, 20, or 40% added sucrose were stored at –25°C for 3 years. Analyses of total monomeric anthocyanin pigment, polymeric color, browning, and color density were performed on acetone extracts obtained directly from fruits homogenized in liquid nitrogen. Sugar addition had a significant stabilizing effect on the total monomeric anthocyanin pigment content. Also, polymeric color and browning were significantly influenced by the addition of sugar to the fruits prior to freezing. However, the significant differences occurring were not always in increasing or decreasing order of sugar level. Color density was not influenced by sugar addition. The mechanisms of the protective effect of sugar addition was suggested to be inhibition of pigment-degradative enzymes, i.e., β-glucanase and polyphenoloxidase; steric interference with condensation reactions, i.e., anthocyanin-phenolic polymerization; anthocyanin-ascorbate interaction; or provision of a partial oxygen barrier.

## 2.  Ingredients in Addition to Sugars

Fruits may be frozen in syrup with the addition of small amounts of ascorbic acid to prevent changes in color and flavor arising from the action of oxidative enzymes after destruction of cell tissues [4]. The decrease in enzymatic browning results primarily from the action of syrup as a barrier to oxygen diffusion. Ascorbic acid is used as a reducing agent to keep the phenolic substances in a reduced and colorless state.

Gorgatti Netto et al. [28] evaluated freezing ability of mango slices packed in syrup. Slices were rinsed in 0.05% ascorbic acid solution prior to immersion in syrup. Various syrups were tested: sucrose (25–40°Brix), glucose, and a combination of both. Storage conditions were –20°C for periods up to 4 months. No significant changes in pH, °Brix, or total and reducing sugar content occurred during frozen storage. However, the total acidity slightly decreased during storage. Ascorbic acid declined only slightly during storage in sucrose syrup but was reduced by about 30% in combination with sucrose and glucose (50:50). Polyphenolase was only active during the first 30 days of storage. At the end of the storage period, mango pieces were pulped and color was evaluated by a Munsell disk colorimeter. No changes in color of pulped mango slices were observed during storage.

Kozup et al. [38] examined the effect of certain chemical treatments applied to fresh strawberries to determine the color stability of frozen berries and products made from them. The experiments were performed with whole berries and with puree, and the treatments consisted of two levels of each of the following: sodium bisulfite, citric acid, stannous chloride, EDTA, acetaldehyde and pasteurization. Berries and puree were treated, frozen, and stored (–18°C). After 2 months of storage all samples were made into jam, sealed in cans, and stored at 24°C. Analyses were performed at the beginning of this storage period and after 3 months. Color was evaluated using a Color Difference Meter (CDM), by browning index (absorbance 430 nm/500 nm), and by sensory evaluation. Sodium bisulfite, stannous chloride, and EDTA provided the greatest stabilizing effect on strawberry color, while citric acid caused a small amount of color stabilization in the jam. Pasteurization did not show any effect, and acetaldehyde had a detrimental effect on color as evaluated in this study. The authors suggested that acetaldehyde concentration in this study may have been too high, as previous research showed that low levels of acetaldehyde improved strawberry color.

Fuster and Prestamo [39] studied the effect of various pretreatments in combination with sugar syrup on texture and flavor of strawberries. The fruits were immersed either in an aqueous solution of sugar (30%), starch (0.3%) and agar (0.3%) for 15 seconds or in an aqueous solution of $CaCl_2$ (0.28%) and NaCl (0.5%) for 15 seconds followed by 15 seconds in 30% sugar syrup. Fruits were frozen and stored for 11 months. Texture deteriorated in all cases compared with fresh strawberries and was not improved by any of the treatments. Samples immersed in the sugar, starch, and agar solution had a better flavor than samples treated with $CaCl_2$ and NaCl prior to sugar syrup.

Abufom and Olaeta [40] investigated the stability of chirimoya fruit pulp during frozen storage. Variables studied included pasteurization (75°C, 10 min) and addition of ascorbic acid (0.15%), citric acid (0.2%) and EDTA (0.02%). The fruit pulp was frozen at −38°C and stored at −18°C for up to 120 days. Pasteurization was detrimental to sensory quality. The additive mixture provided good control of enzymatic browning and sensory quality was better than the control without pasteurization or any additives.

Morris et al. [30] also performed preprocessing experiments with strawberries to investigate the effect of dipping in calcium chloride (0.18% Ca), low-methoxyl pectin (0.3% LMP), or sucrose (40 °Brix) solutions on the quality of frozen-and-thawed fruits. The greatest firming effect was achieved using Ca and pectin. Drained weight loss was reduced by pectin, Ca, and sucrose. Dipping of whole and sliced strawberries in Ca or water, followed by freezing with and without pectin, had little effect on color. However, red color intensity (a) as measured on a Gardner Color Difference Meter, was greater when fruits were dipped in a Ca solution. In a second experimental series, calcium and sucrose dip did not affect the color of thawed whole fruits. Treatments with vacuum (172 mm Hg) during the dipping period resulted in darker fruits (lower CDM L) and lower red and yellow values compared with samples where reduced pressure was not applied. Heating fruits to 70°C prior to freezing produced a lighter color, with higher red and yellow values.

## 3. Effect of Osmotic Dehydration

Pinnavaia et al. [41] described a method for the removal of water from fruit prior to freezing by direct osmosis (the osmo-dehydro-freezing process). In this process, fruits were placed in a hypertonic solution, such as a concentrated sugar solution [42–44]. The osmotic pressure buildup causes water to diffuse out of the tissue cells while sugar moves into the cells. As a result, dehydration of the fruit occurs and an increased sugar content is obtained. Strawberries, raspberries, apricots, and cherries were tested by immersion of whole or halved fruits in a high-fructose corn syrup with 1% NaCl added [41]. The treatment lasted for 8 and 16 hours at room temperature. After treatment, fruits were allowed to drip for 5 minutes to drain excess syrup and were subsequently frozen in an air-blast freezer at –40°C. Frozen products were stored at –20°C. Thawing was carried out at ambient temperature, and weight loss, firmness, and drip loss were determined. Whole strawberries lost about 15% water during the 8 hours in the hypertonic solution. When the treatment was expanded to 16 hours, about 40% of the initial water content was removed. For strawberry pieces the amounts were 40 and 65%, respectively. Similar results were obtained for the other types of fruit tested. In addition to the water losses, the gain in solutes (mainly mono-, di-, and polysaccharides) in the fruit increased. The osmo-treated fruits exhibited a higher freezing rate than untreated fruits and a freezing point depression of about 3°C compared with fresh fruits. In the frozen state, osmo-dehydro-frozen fruit was described as having a pleasant, firm texture compared with frozen fresh fruit, which had a very hard texture. At thawing the osmo-dehydro-frozen fruits showed a reduced thawing time and a small drip loss. Tomasicchio et al. [45] compared dehydro freezing of pineapples and strawberries with ordinary freezing at –20°C. Pineapples exhibited better quality after dehydro freezing, whereas strawberries so treated were very similar to conventionally frozen berries.

## F.  Packaging

Fruits exposed to oxygen are susceptible to oxidative degradation, resulting in browning and reduced storage life of the products [24,46]. Packaging of frozen fruits therefore to a large extent is designed to exclude air from the fruit tissue. Replacement of oxygen with a sugar solution or inert gas, consuming the oxygen by glucose-oxidase, and/or the use of vacuum and oxygen-impermeable films for packaging prevent and retard browning and other oxidative color changes.

In most studies dealing with frozen storage of fruits, only minor reference is made to the packaging material. Fruits are packed with or without previous evacuation in plastic bags, plastic pots, paper bags, cans, or more specifically in polyethylene bags. Information about penetration properties and thickness of the plastic film are rarely given [23,33].

Bisset and Berry [47] demonstrated the importance of packaging material to the stability of frozen concentrated orange juice. Containers tested were cans of paper laminated with polyethylene having aluminum ends and cartons of paper, aluminum foil, and polyethylene laminate. There was little difference in ascorbic acid retention between package types when juice was stored at –20°C. When stored at –6.7°C, the foil-barrier package resulted in better ascorbic acid retention than the fiber/polyethylene package. After 8 months of storage, ascorbic acid retention was 79 and 43% for the foil and the fiber/polyethene containers, respectively.

Blonski et al. [48] studied the effect of packaging material on freezing performance of strawberries. The fruits were packed in paper bags with a polyethylene lining or in polyethylene pouches held in outer cartons. The product was stored at –25 to –30°C for 1 year and analyzed at 3-month intervals for dry matter, vitamin C, pigments, color, and consistency. The polyethylene/carton combinations yielded superior frozen storage characteristics.

Gradziel [23] investigated the effect of thickness of packaging material on ascorbic acid retention in tomatoes during frozen storage. Thickness of the plastic containers did not significantly influence ascorbic acid content of the frozen tomatoes.

Venning et al. [33] studied chlorophyll and ascorbic acid stability in kiwi (cv. Hayward) pulp stored in two different packaging materials. One type was polyethylene (70 μm, a film used for commercial frozen kiwi pulp in New Zealand). The other type was a foil laminate [12 μm PET (polyethylene terephthalate), 98 μm aluminum foil, 65 μm LLDPE (linear low-density polyethylene)]. Extreme care was taken during freezing and storage to prevent the occurrence of pinholes in the packages. Prior to analysis, samples were thawed in a water bath at 15°C for 15 minutes. During this period the pulp reached –2.5 to –1.5°C. The type of packaging had no effect on the extent of chlorophyll degradation or color changes in the frozen pulp during the 1-year storage period at –9, –18, or –25°C. Packaging material also appeared to exert no influence on ascorbic acid levels when pulp samples were stored at –18 and –25°C (Fig. 1). When stored at –9°C, however, packaging type had a substantial effect on ascorbic acid stability (Fig. 2). Pulp packed in foil laminate lost about 26% of its ascorbic acid after 55 weeks of storage. Pulp packed in polyethylene lost 88% of it original ascorbic acid after 55 weeks of storage at –9°C and exhibited a correspondingly large increase in dehydroascorbic acid.

## III. EFFECTS OF FREEZING, FROZEN STORAGE, AND THAWING ON COLOR, APPEARANCE, AND CONSUMER ACCEPTANCE

Appearance is the main quality factor that attracts the consumers' attention. Visual judgment is made before a product is tasted or even purchased and will automatically determine if the product is purchased and ultimately consumed. Consequently, substantial interest has developed in objective scientific evaluation and measurement of the appearance of fruit and fruit products [6,49].

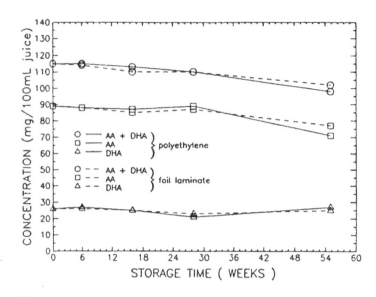

**Figure 1**   Changes in total ascorbic acid (AA + DHA), ascorbic acid (AA) and dehydro-ascorbic acid (DHA) of frozen kiwi pulp with seeds packed in polyethylene or foil laminate and stored at –18°C. (From Ref. 33.)

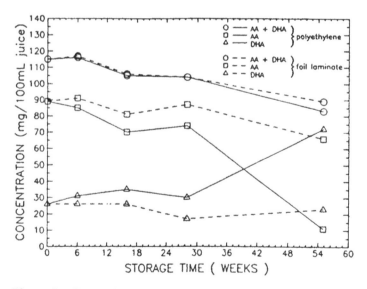

**Figure 2**   Changes in total ascorbic acid (AA + DHA), ascorbic acid (AA) and dehydro-ascorbic acid (DHA) of frozen kiwi pulp with seeds packed in polyethylene or foil laminate and stored at –9°C. (From Ref. 33.)

Visual perception is activated when light reflected from an object reaches the eye and is transformed into an image. The glossiness of the surface together with spectral characteristics of the light contribute to the perception [50]. Sensory evaluation is widely used to assess the appearance of fruits [6]. Major emphasis has been placed on instrumental analysis for assessing appearance, mainly based on color. Methods have evolved for analyses using various color scales based upon how color is perceived by the human eye [51] and for analysis of the pigments contributing to color [52]. Limited information is available about factors such as size, glossiness, and freshness, which contribute significantly to the overall visual appearance of fruits and their products.

CIELAB and the Hunter Lab system, L, a, and b represent the three axes of the color space: L for lightness, a for red-green chromaticity, and b for yellow-blue chromaticity [51]. L values vary between zero for black and 100 for white, while positive a values represent intensity in red and negative a values represent intensity in green. Positive b values represent intensity in yellow and negative values represent intensity in blue. The term "hue angle," defined as $h_{ab} = \tan^{-1} b/a$, is a measure of the color hue. Hue angles between 0° and 90° represent red-yellow color, between 90° and 180° green-yellow, between 180° and 270° blue-green, and between 270° and 360° red-blue color. The term chroma is defined as $C = (a^2 + b^2)^{1/2}$ and is used as a measure of chromaticity of samples and does not include the lightness of the color. Total color difference, $\Delta E = (\Delta L^2 + \Delta a^2 + \Delta b^2)^{1/2}$, includes all color parameters and defines color differences between two samples.

Pigments in fruits consist of anthocyanins [53,54,66,108], chlorophylls [29,33, 55,59] and carotenoids [15,17,56]. Together these chemical compounds create the enormous variations in color that are found in fruits. The pigments are often more abundant in the outer layers of the fruit. However, some fruits are more evenly pigmented through the entire tissue. For fruits consumed directly without further processing (cutting, peeling, homogenizing, or juice extraction), the surface color is often the most important quality characteristic. In contrast, the entire pigment content is more critical in fruits undergoing further processing, since the final product must have an acceptable color.

## A. Anthocyanines

The red pigments in fruits are basically flavenoids, occurring as water-soluble glycosides, mainly formed from glucose, rhamnose, arabinose, galactose, and xylose, either as mono- or disaccharides [57,58,60]. The main group of compounds are anthocyanins. The amount of anthocyanins varies widely between types of fruit. The level of anthocyanins rises sharply as the fruits ripen. An amount of pelargonidine-3-glucoside of 0.96 mg/100 g has been reported for unripe Senga Sengana strawberries, while 12.30 mg/100 g was reported for ripe fruits. Correspondingly, black currants have been shown to contain levels of between 110 and 430 mg/100 g [61].

Recurring problems of discoloration and loss of color during processing have been reported by Holdsworth [57]. Considerable research has been conducted over the years to identify the pigments present in red fruits. Carrick [62], as early as in 1930, described color changes during freezing of vinifera grapes. More recent studies of freezing, frozen storage, and thawing of anthocyanin-pigmented fruits and color stability have been conducted on cherries, strawberries, blueberries, blackberries, cranberries, and raspberries.

## 1.  Cherries

Several studies have been conducted to investigate the freezing performance of cherries. Cherries vary in pigmentation from lightly colored varieties and cross-varieties with a clear syrup to heavily pigmented varieties with dark-colored pulp [63]. In an investigation of nine types of cherries, the most heavily pigmented varieties were found to be best suited for quick freezing, because they maintained the fresh fruit color when both pigment separations and Hunter color analyses were used to evaluate fruit color.

Polesello and Bonzini [53] evaluated pigment stability during frozen storage of deeply red-colored, sweet cherries from three cultivars. In these experiments, sweet cherries were frozen in a forced-air tunnel and stored in polyethylene bags at –20°C. Anthocyanins in the fresh fruits were identified by relative retention ($R_F$) values from thin-layer chromatography (TLC) and subsequent evaluation of spectral performance. No new chromatographic bands were observed after 4 months of storage, but changes in color intensity were observed, especially in the free cyanidin band. In fresh fruits this band had a weak color intensity, which became more intense after frozen storage, indicating accentuated hydrolysis. Total anthocyanin degradation for the three cultivars evaluated varied from 33.6 to 73.1% after 4 months of frozen storage at –20°C. Pigment loss was not related to initial anthocyanin concentration of the cherries. Objective color measurements of the cherry surface using a HunterLab D25L Color Difference Meter before and after frozen storage did not produce differences, probably due to the strong pigmentation of the product.

Urbányi [64] investigated the effect of temperature during frozen storage on the color of freeze-dried sour cherries. Cherries were freeze-dried at 40, 50, or 60°C followed by frozen storage at either –6 or –20°C in air or under nitrogen, with and without light for periods of up to 120 days. Results revealed no effect of freeze-drying on either anthocyanin content of cherries or CIELAB color values of the fruit surfaces. However, pigment content decreased more during storage at –6°C than at –20°C. Oxidation played an important role in the breakdown of the anthocyanins and was catalyzed by light. Significant correlations between anthocyanin concentration and each of the CIELAB L*, a*, b*, and C* values were obtained.

More recently Urbányi and Horti [65] used three varieties of sour cherries to study changes in hue and coloring during frozen storage. Whole fruits packed in polyethylene pouches were frozen and stored at –20°C for periods up to one year.

Prior to analyses, fruits were thawed in a microwave oven. Measurements were carried out with a Momcolor-D tristimulus colorimeter, which determines the surface color through reflectance.

Surface CIELAB color revealed similar changes during frozen storage for all varieties. The fresh cherry surface color range was 19.85–23.02 for L*, 20.75–30.52 for a*, and 3.71–5.77 for b*. During the first 242–298 days of the storage period, surface color became more yellow. This change in hue appeared to be caused mainly by decreased redness (a* values). After prolonged storage, the direction of the shift in hue changed, so hue became redder. This change in color was especially apparent in the Újfehértói fürtös cultivar (Fig. 3). After one year of frozen storage, the corresponding values were L* 19.39–21.39, a* 16.28–20.98, and b* 6.03–7.41. Average hue values increased for the three varieties from 10.04 to 16.39. Most correlations between CIELAB color values and storage time were not statistically significant, due to the diverse changes in the parameters during storage.

Anthocyanin content of cherries was also determined in this study from absorbance (530 nm) measured in HCl-ethanol extracts [65]. Significant and linear increases in anthocyanin content were observed throughout the one-year storage period. Values ranged from 54.14 to 201.62 mg/100 g during frozen storage of

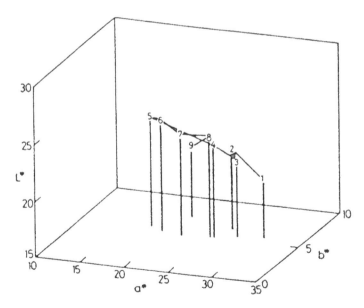

**Figure 3** Color point in CIELAB space of frozen cherries in various phases of storage: 1 = day of freezing, 2 = 109 days, 3 = 188 days, 4 = 209 days, 5 = 242 days, 6 = 272 days, 7 = 298 days, 8 = 355 days. (From Ref. 65.)

fruits from the Érdi variety. In a 196-day storage study, this change was reversed and anthocyanin content increased. This deviation was related to the maturity level of the fruits according to the authors. In the less ripe fruits of the first experiment, biosynthesis of anthocyanins may have continued during frozen storage. In the second experiment, the continous formation of anthocyanins may have been slower, since the fruits were more ripe.

Both a* and b* values have been shown to increase with decreasing anthocyanin concentration in the very dark syrup of black currants [61]. Decreasing a* values were not observed until a 1:10 dilution was reached of syrup made from black currants with an anthocyanin content of about 250 mg/100 g. Decreases in b* values were detected in syrup dilutions of 3:10 or greater. These differences in the sensitivity in a* and b* produced a change in hue angle ($\tan^{-1}$ b/a) toward a more red color when syrups were diluted more than 2:10. Low sensitivity at high pigment concentration may also have contributed to the somewhat inconsistent results observed in the surface color of sour cherries [65]. In ligher-colored fruits, such as tomatoes, significant correlations have been reported between carotenoid content and CIELAB parameters [17].

## 2.  Blueberries

The effects of freezing and thawing on the appearance of highbush blueberries have been thoroughly studied [54,66,108]. Factors determining the appearance of these fresh and processed small fruits are surface pigmentation, structural features on the surface imparting glossiness (glaucousness), the presence of pigmented exudate, syrup, or water on the surface, and shriveling, skin splitting, or other defects. Changes in these factors can occur during frozen storage, thawing, or processing as a result of enzymatic browning, enzymatic or thermal degradation of pigments, anthocyanin complex formation, pigment formation from leukoanthocyanins, leakage of pigments intracellularly or into the cooking water or syrup, loss of the waxy bloom, or physical disruption of the fruit tissue.

Fruits from 11 highbush blueberry cultivars were refrigerated within a few hours of harvest, packed in polyethylene freezing containers, and frozen at –13°C. After 2–3 months of frozen storage at this temperature, berry samples were thawed overnight in a refrigerator. Fruits were then visually evaluated for bloom, and tristimulus reflectance measurements were obtained on intact berries in the thawed samples with a Gardner XL-23 tristimulus colorimeter. Droplets of condensate or exudate adhering to the surface were removed prior to analysis. The thawing process produced darkening of the blueberry fruits (decreased L* values) for most cultivars compared with raw, unfrozen controls. Waxy bloom was partly retained during freezing and thawing. Decreases in L* values were accompanied by increases in a* (redness) and b* (yellowness). Thus, the hue angle ($\tan^{-1}$ b/a) increased, indicating a shift in berry color from blue toward red. Differences in changes in hue among cultivars were observed. No relationships between tristimulus parameters and total anthocyanin content of the raw or thawed samples were detected, probably because of the high pigment content of the epidermal cells in blueberries.

Microscopic observations of berry samples are sometimes carried out to assess differences in fruit appearance and reflectance properties. Light microscopy revealed loss of pigments from epidermal and subepidermal cells of the thawed berries in this experiment. Some variations among cultivars were reported, particularly in the Elliot and Burlington cultivars. Fruits from Elliott lost juice and pigment directly through the skin after thawing, while juice was exuded through the skin of Burlington. Investigations of the ultrastructure of the cuticular region and the morphology of the surface by transmission electron microscopy (TEM) revealed no obvious cultivar differences in cuticle structure either before or after freezing and thawing, possibly as a result of artifacts arising during sample preparation, since various solvents, which may influence wax structure, were present. Scanning electron microscopy (SEM) revealed surface wax differences between the two cultivars. The wax of unfrozen Elliott consisted of discrete plates, while unfrozen Burlington had a more uniform and continuous wax coating. Freezing and thawing had little or no effect on the Burlington wax coating but caused the surface of Elliott to appear softened or partially annealed. In several locations, the thawed Elliott fruit surface had no visible wax. The authors suggested the wax coating on Elliott blueberries may be a less effective barrier to diffusion than Burlington wax. This would explain the apparent ease with which pigment-bearing celluar fluids leaked through the skin of thawed Elliott berries.

Yang and Yang [67] studied color changes in lowbush blueberry puree stored at $-20°C$ for periods up to 7 weeks. Initial color values were 6.59, 5.94, $-0.05$, and $-0.48$ for Hunter L, a, b, and hue ($\tan^{-1}$ b/a), respectively, corresponding to the dark bluish-red color of the blueberry puree. After 4 weeks of storage, lightness (L) decreased slightly to 5.64. Six weeks of storage produced increased b values, indicating that a more yellow hue had begun to develop in the blueberry puree. Hue values increased to 0.66 by the end of the 7-week storage period.

Lenartowicz et al. [68] studied the quality of six highbush blueberry cultivars and their suitability for freezing. Fruits were frozen to $-20°C$ in a plate freezer or by the use of liquid nitrogen in an experimental spray tunnel. Berries were packed in polyethylene bags and stored at $-20°C$ after freezing. Anthocyanin content was measured in both fresh and frozen fruits after 3 months of frozen storage. No significant differences between anthocyanin content of fresh and frozen berries were observed for all of the cultivars tested, and there was no effect of freezing method. Consequently, all varieties were suitable for freezing.

## 3. Blackberries

Sapers et al. [54,66,108] evaluated differences in the color and processability of ripe fruits and juice from 40 thornless blackberry cultivars and selections after freezing and thawing. The berries were iced immediately after harvest, stored at

1°C overnight, and packed in polyethylene freezer containers prior to freezing and storage at –13°C. Samples were thawed for 24 hours in a refrigerator or for 3–4 hours at room temperature for analysis.

Samples were harvested when ripe, based on their dull black color and ease of separation from the plant. Tristimulus reflectance measurements on intact fruits indicated small but significant differences in Hunter L, a, and b values among cultivars. After 6 months of storage, many fruits in each sample had turned red while still in the frozen state. The proportion of red fruit in frozen samples varied with successive harvest of the same cultivar during one season and between the two consecutive years. Therefore, no blackberry cultivar tendencies to turn red when frozen could be confirmed. The tendency of blackberries to turn red during frozen storage had previously been related to incomplete ripeness of the fruits, which could not be recognized visually in the berries [69]. This change constitutes a serious quality problem, because red-colored blackberries are not attractive to the consumer. The color of frozen individual blackberries was retained during thawing [54,66,108]. The number of red berries increased steadily during each of eight freeze/thaw cycles [69].

Total anthocyanin values of blackberry fruit are repored to be higher for black than for red berries [54,66,70,108]. Juice obtained from black-colored blackberries contained 20–30% less anthocyanin when the berries were thawed slowly in a refrigerator than when they were thawed rapidly at room temperature. Red-colored blackberries behaved similarly, and the thawing effect was smaller. Tristimulus measurements in the transmission mode indicated that undiluted juice from slowly thawed fruit was significantly lighter (higher L value) and more orange (higher Θ-value) than corresponding values for rapidly thawed fruit. Also, these differences were more pronounced for the ripest blackberries. Enzymatic anthocyanin degradation may have been more comprehensive, according to the authors, due to the longer thawing period. Differences due to thawing were generally smaller than differences among cultivars or between red and black subsamples and may well not be of practical importance.

Polesello et al. [70] using spectral and TLC analyses revealed similar anthocyanins in black and red berries. The spots on the TLC plate for each anthocyanin were smaller and less intense for red fruits when compared with those of black fruits. A significant increase in anthocyanin content as determined by absorbance in 1% HCl/methanol extract was observed during frozen storage. Anthocyanin content of black fruits on the average increased from 253.1 to 297.3 mg/100 g fresh weight, while that of red fruits increased from 187.9 to 258.9 mg/100 g fresh weight. This color change was due to several factors, including pigment composition and the physical and chemical state of the fruit, according to the authors.

## 4. Cranberries

The anthocyanins of cranberries are located in the exocarp portion of the berry. During juice expression, the pigments are extracted by juice liberated from the crushed berries [71]. In commercial juice production, cranberries are subjected to a freeze-thaw treatment before pressing. This disrupts the cellular structure in the berries and results in increased juice yield and greater pigment extraction.

Cranberries were frozen and stored at −18°C. Samples were thawed at 3°C overnight prior to chopping and pressing. Juice yield, soluble solid content, and total anthocyanin content were compared with those obtained from unfrozen cranberries stored at 3°C. Morphological studies were carried out by cutting thin slices perpendicular to the surface of both the fresh and freeze-thaw–treated berries. The slices were examined by light microscopy to localize pigment-containing cells.

The freeze-thaw treatment of cranberries prior to pressing increased juice yield by nearly 50% for one variety. Further, anthocyanin content of juice increased by as much as 15-fold. Also a more rapid liberation of juice during pressing of freeze-thaw–treated cranberries was found. Migration of anthocyanins from the exocarp into the mesocarp and endocarp was observed, resulting in improved anthocyanin extraction during pressing. The observed maximum 60% pigment extraction suggests that some pigment-containing cells still retained their anthocyanin content after the freeze-thaw treatment. This theory has been confirmed by morphological studies of the berry tissue. In fresh tissue (Fig. 4) two layers of

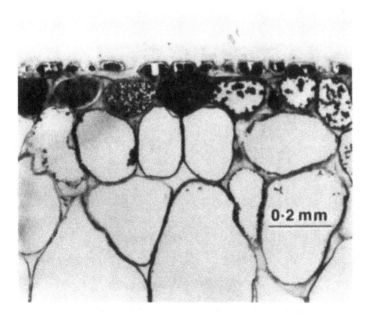

**Figure 4**  Epoxy section through the skin of fresh cranberry (80x). Two pigment-bearing cell layers beneath the cuticle are visible. (From Ref. 71.)

pigment-bearing cells were seen beneath the cuticle. The outer layer consisted of small, elongated epidermal cells, densely packed with pigments, while the inner layer contained larger cells also packed with pigments, either dispersed in the cell or precipitated along the cell walls as a result of fixation and dehydration. The authors reported that the cells of the epidermal layer were very dark red under the light microscope, while those of the inner layers were lighter red, indicating less concentrated pigments.

After the freeze-thaw treatment (Fig. 5), the outer cell layer retained a high pigment content, while the inner layer was almost devoid of pigmented cells. Light microscopy confirmed the loss of red color from the inner cell layer.

The interior cells of the cranberries were severely disrupted after thawing, according to the authors. The outer three to four cell layers, however, were less damaged, and the epidermal cells were nearly unchanged in shape after freezing and thawing. The cells of the second pigment layer were only slightly flattened. Thus, pigment liberation from the cell layer just below the epidermis is enhanced by the freeze-thaw process, while pigments are still retained in the epidermal cells. Anthocyanins in the epidermal layer are not extracted during pressing of cranberry juice. Homogenization of unfrozen berry tissue prior to juice extraction resulted in pigment recoveries similar to those obtained by freezing and thawing the berries.

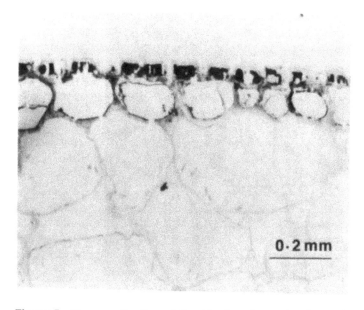

**Figure 5**   Epoxy section through the skin of cranberry given freeze-thaw treatment (80x). Pigments are retained by epidermal cells but are lost from the second layer of cells. (From Ref. 71.)

## 5. Raspberries

Bushway et al. [72] evaluated five raspberry cultivars, examining their suitability for frozen storage. The fruits were harvested at the red-ripe stage and frozen in a blast freezer at –30°C before being packed into polyethylene freezer bags and being stored at –20°C for 1, 6, or 9 months. Analyses were performed with fresh and thawed berries. The samples were rated by instrumental color analysis and by a sensory panel for color preference. The effect of freezing on instrumental hue and sensory color was not significant for any of the five raspberry varieties investigated. Three out of five varieties displayed decreased color intensity (chroma) in the thawed fruits after extended frozen storage.

## 6. Passion Fruit

Lee et al. [32] reported no significant change in color hue (a/b) of pasteurized and fresh passion fruit juice stored at –20°C for a period of 12 months.

## B. Chlorophylls

Chlorophyll is present in unripe fruits but tends to be broken down during ripening [55]. In fruits like avocados, pears, grapes, muskmelons, and kiwis, chlorophyll is also retained in the ripe fruits. Chlorophyll forms the ground color in fruits like apples and pears, while in others it predominates and the fruits appear totally light green. Chlorophylls are found in high quantities in the deep green peel of the ripe lime and in the flesh of other fruits like kiwi. Chlorophyll imparts a green color, which does not change visibly during ripening. During ripening of kiwi, total chlorophyll decreases by 30% from 28.3 to 20 µg/g [55].

## 1. Kiwi

Growth in utilization and international trade of kiwi and kiwi products has increased interest in chlorophyll compounds and their behavior during processing and frozen storage [29,33,59]. Robertson [22] emphasized the importance of controlling the color of frozen commercial products to avoid disputes between processors and buyers with regard to the color of frozen fruit products.

Robertson [22] studied the stability of chlorophyll and pheophytin in kiwi puree during frozen storage. Kiwi puree was prepared by hand-peeling and homogenizing fruits in a blender for 3 minutes. The puree was packed into plastic pots, sealed, and frozen immediately at –18°C. After frozen storage for a period of up to 68 days at –18°C, the puree was thawed overnight at room temperature.

Pheophytins are conversion products of chlorophylls. The rate of reaction depends upon factors such as temperature and pH. Therefore it is important to control pH during chlorphyll analysis to prevent undesired conversion of chlorophylls to pheophytins. Enforced chlorophyll degradation may also occur during blending of kiwi.

The rate of chlorophyll degradation at freezer temperatures has been investigated by measuring the chlorophylls and pheophytins in frozen pureed kiwi that had been stored for up to 68 days at −18°C. Results are presented in Table 1 and demonstrate that extensive chlorophyll degradation occurred even at a storage temperature of −18°C. Within 36 days, the chlorophyll concentrations were reduced to less than one third of their initial concentrations.

The effect of temperature on chlorophyll and color of kiwi pulp during frozen storage has been studied by Venning et al. [33]. Kiwi, including seeds, was pulped, packed, and frozen in foil laminate or polyethylene bags of aerated or nonaerated pulp. The bags were stored at −9, −18, or −25°C for periods of up to one year. Samples were thawed by immersing the frozen bags in water at 15°C for 15 minutes. During this time, the pulp reached temperatures of −1.5 to −2.5°C.

The total chlorophyll content of fresh pulp containing seeds ranged from 14.0 to 15.5 mg/kg. The changes in percent total chlorophyll for the three storage temperatures during the 52 weeks of storage are shown in Fig. 6. No significant change in total chlorophyll content of the kiwi pulp was observed at −18 or −25°C. In contrast, a significant decrease in chlorophyll content was found during storage at −9°C.

Changes in Hunterlab −a/b values of the pulp at the three storage temperatures were similar. At −25°C no change occurred, while there was a slight decrease in −a/b at −18°C. Storage at −9°C produced a pronounced decrease in this color valve during storage. Decreased −a/b values correspond to samples becoming less green and more yellow, until it switches (−a/b = 0) to a brownish-red color. Visual comparison with Munsell colors confirmed this change in color from moderate yellow-green (5.0 GY 5/6) to a more brown and yellow khaki color (7.5 Y 4/4).

The authors [33] discussed their results in relation to those obtained by Robertson [22]. Substantial degradation of chlorophyll occurred during frozen storage of kiwi pulp at −18°C. Observed discrepancies were suggested to be due to differences in thawing methods and chlorophyll degradation greatly exceeding that arising from many weeks of storage at −18°C, when pulp was held at 0°C or higher after thawing.

Cano and Marín [29,59] studied qualitative pigment differences between fresh and frozen kiwi slices to determine pigment degradation due to freezing. Kiwi were

**Table 1**  Changes in the Chlorophylls and Pheophytins During Storage of Frozen Kiwi Puree at −18°C[a].

| Times of storage (days) | Chlorophyll a (mg/kg) | Clorophyll b (mg/kg) | Pheophytin a (mg/kg) | Pheophytin b (mg/kg) |
|---|---|---|---|---|
| Fresh | 3.1 | 3.7 | 2.2 | 0.5 |
| 1 | 2.2 | 1.7 | 4.1 | 5.0 |
| 36 | 1.0 | 0.9 | 5.6 | 6.0 |
| 68 | 0.6 | 0.3 | 5.9 | 6.2 |

[a]Each result is the average of three determinations.
*Source*: Ref. 22.

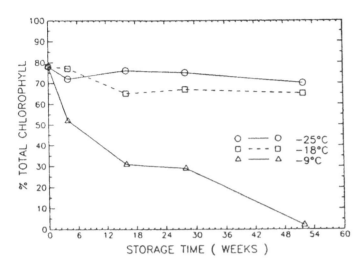

**Figure 6** Changes in percent total chlorophyll of kiwi pulp with seeds during frozen storage at three different temperatures. (From Ref. 33.)

peeled, sliced (6–8 mm), and frozen (–40°C; 5.5 m/s air flow) in an air-blast freezer until the temperature of the product reached –20°C. Frozen slices were packed in polyethylene bags, vacuum sealed, and stored at –18°C for 6 months. Samples were thawed in a refrigerator prior to analysis.

HPLC chromatograms of kiwi extracts showed the presence of xanthophylls, chlorophylls, and hydrocarbon carotenoids. Individual compounds were identified by spectroscopy, chromatography, and chemical methods. Stored frozen kiwi displayed a similar pattern to fresh kiwi. Chlorophyll $a$ and $b$ were accompanied by their most common derivatives, pheophytin $a$ and pheophytin $b$. Pheophytin $b$ was not observed in chromatograms of fresh kiwi extracts.

For color analysis, kiwi slices were pureed and placed in a 5-cm-diameter plastic dish to a depth of 2 cm or more. The sample dish was placed on the light port of a Hunter Lab Model D25-9 colorimeter, standardized on a white plate, and the sample dish was covered to avoid stray light.

After 6 months of storage, frozen kiwi fruit slices did not differ significantly in lightness (L = 42.53) from fresh slices (L = 41.94). Yellowness (b) differed slightly between frozen (b = 22.24) and fresh (b = 21.73), while a more pronounced difference was found between the greenness (a) of frozen fruits (a = –9.02) and fresh fruit (a = –12.26), indicating that the color of the frozen and thawed kiwi slices was less green and slightly more yellow than that of fresh slices.

Cano et al. [29] also examined freezing performance of four kiwi cultivars with regard to pigment stability and color changes. Fruit slices were frozen at –40°C and stored at −18°C for periods of up to one year. Freezing produced a decrease in total chlorophyll content in samples from all cultivars (Fig. 7), and the total chlorophyll decrease due to the freezing process was very similar in all four kiwi cultivars studied (7–10%). The decreases in chlorophyll *a* and *b* during frozen storage were also very similar. As the chlorophyll content gradually decreased, the color changed towards a more yellowish hue.

## 2. Citrons

Cancel and Hernández [73] evaluated the effect of freezing on color stability of citrons to be used in the production of candied citron. Freezing of the citrons prior to fermentation was desired because it would reduce storage space and shipping costs, since transportation of brine is avoided. The green color of candied fruits was assessed by absorbance measurements of methanol extracts at 650 nm. Samples were frozen unblanched and after blanching with and without 0.5% $CaCl_2$ for periods of up to 3 minutes and stored for 10 or 22 months at −23.3°C prior to curing. Products with an intense green color were obtained from all treatments except 3-minute blanching and samples cured without blanching.

## C. Carotenoids

The yellow, orange, and red colors of many fruits are due to the presence of carotenoids. Carotenoids are widespread compounds, which together with chloro-

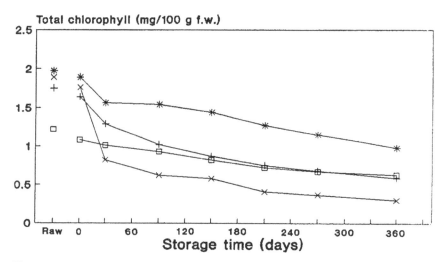

**Figure 7**  Changes in total chlorophyll content of slices of four kiwi varieties during frozen storage at −18°C. Monty —X—; Bruno —'—; Abbot —*—; Hayward —#—. (From Ref. 29.)

phyll are found in all organisms capable of photosynthesis [55]. The amount of carotenoids found in various types of fruits varies from low levels, e.g., in strawberries (0.4 µg/g) and cranberries (5.8 µg/g), to substantially higher levels, e.g., in tomatoes (922–1700 µg/g) and mangoes (89.2–125.0 µg/g). Most fruits contain a variety of different carotenoids, and composition tends to vary among cultivars. Content generally increases as fruits ripen. However, carotenoids are easily oxidized. Therefore, the content is prone to decrease during freezing, frozen storage, and thawing.

## 1. Mangoes

Ramana et al. [56] compared carotenoid retention of frozen mango pulp with that of canned pulp. The pulp was packed in polyethylene bags and frozen as slabs in a plate freezer at –40°C. The slabs were stored at –18°C for periods of up to 14 months. Ascorbic acid was added to a portion of the pulp samples prior to freezing. Pulp sealed in cans and heated for 30 minutes in boiling water was stored at 4°C and served as a control. Prior to analysis, samples were thawed for one day in a refrigerator.

After 10 months of storage, total carotenoid retention was greatest in the canned pulp (93.2%), followed by frozen ascorbic acid–fortified pulp (85%) and natural frozen pulp (64%). Losses increased during further storage to 14 months. Color (percent chroma retained) also decreased during storage, and the trend was similar to the carotenoid losses. After 6, 10, and 14 months of storage, color of frozen mango pulp without added ascorbic acid was significantly inferior to the color of canned pulp when assessed by sensory evaluation, while frozen pulp with added ascorbic acid did not deviate from canned pulp in color. For nectars produced from these pulps, corresponding differences in color were only observed after 10 and 14 months of storage.

During frozen storage (–18°C) of mango slices, β-carotene content decreased, with the extent varying between the four varieties investigated [15]. The highest pigment retention was 4.27 mg β-carotene/100 g fresh weight in the raw fruit, 3.75 mg β-carotene/100 g in fruits just after freezing, and 3.50 mg β-carotene/100 g in fruits after 120 days of frozen storage. Another variety with lower pigment stability had β-carotene contents of 3.20, 3.11, and 0.23 mg/100 g, respectively. Since β-carotene is fat soluble, it is not likely to be subject to leaching losses, but reactions like *cis-trans* isomerization or epoxidation during food processing and any oxidative changes may decrease pigment levels.

## 2. Tomatoes

Tomato is a fruit rich in carotenoids. The performance of partially ripe and fully ripe fruits of three tomato cultivars was evaluated during frozen storage for periods of up to one year [17]. Fully ripe tomatoes were diced, packed in polyethylene bags, and frozen at –30°C. Samples were stored at –20°C for one year. Frozen tomato cubes were defrosted in a water bath and then pulped in a homogenizer. The pulp was deaerated under vacuum prior to analysis of CIELAB color and carotenoid content and composition.

During the one-year storage period, red (a*) diminished in all three cultivars, while the yellow (b*) character increased, maintaining a similar total saturation. During storage, all samples became lighter (increased L*). In one cultivar, total color difference ($\Delta$ E*) was 13.49, while this difference was 7.33 and 6.47 in the two other varieties.

Significant correlations between most color tristimulus characteristics and storage time at $-20°C$ were obtained. The change in color during frozen storage was also dependent on tomato variety and stage of maturity, according to the authors.

Total carotenoid content also decreased with storage time. Retention was highest in samples with the highest degree of maturation. Significant correlations were found between total carotenoid content and duration of frozen storage at all stages of maturity. Total carotene content decreased by 40% from 2725.1 to 1647.3 $\mu g/g$ during 175 days of frozen storage of ripe tomatoes. In less ripe tomatoes, about 55% of the initial carotenoid content was lost during this period. Significant correlations were found between specific carotenoids ($\gamma$-carotene, neurosporone, and lycopene) and storage time. These were the most prevalent carotenoids in the samples.

## D.  Browning Reactions

Color changes in frozen fruits are occasionally due to the loss of the true color with formation of an off-color or the loss of true color to the surrounding medium and the development of off-color from originally noncolored constituents. However, color changes in frozen fruits are predominantly caused by oxidative reactions [24,37].

Enzymatic browning is due to polyphenoloxidase reacting directly with oxygen from the air and occurs in most stone and many seed fruits. Peaches, apricots, plums, prunes, cherries, apples, and pears contain polyphenolic substrates, which in the presence of enzymes and oxygen are first changed into quinones and later into brown pigments. At the quinone stage, the reaction can be reversed by a reducing agent like ascorbic acid, while at the polymer stage reducing agents have no effect. In vegetables and some fruits, the problem of enzymatic browning is overcome by blanching the product until the enzymes are partially or fully inactivated.

### 1.  Mangoes

Color stability of mango pulp during frozen storage has been investigated [74]. With one cultivar (Dashehari), gradual browning was observed by sensory evaluation within a few months. However, this change was not observed with pulp from the Langra cultivar. The problem was not overcome by deaeration, but addition of citric acid to the pulp suppressed browning during frozen storage. Pulp frozen and stored at −29°C retained its color during 4 months of storage, while pulp frozen at −20°C and stored at −18°C did not. The addition of 0.3% citric acid partially prevented the formation of brown color. A chelating effect of the added citric acid preventing the browning reactions was suggested by the authors. The mango cultivar Langra contained citric acid concentrations similar to those found in the citric acid enriched pulp of the Dashehari cultivar.

## 2. Citrus Fruits

Orange juice in concentrated form (42–65°Brix) has been reported to remain constant in color when stored at −20°C for 6 months [75]. Color values were evaluated by the CIE parameters (X, Y, Z).

Askar et al. [76] investigated the storage suitability of lime juice concentrates produced by the serum-pulp method. Serum and pulp of lime juice is separated by centrifugation, and only the serum fraction is concentrated in this method. The two fractions are then recombined into the concentrate. The concentrated juice was stored at −12 and 4.5°C for 330 days. Samples were analyzed initially and after 30 days, and then periodically every 60 days thereafter. Color index at 420 nm was determined colorimetrically in the supernatant after addition of acetone and centrifugation. During storage a reduction in color index occurred. After 330 days of storage, the reduction was lower (14%) in concentrates stored at −12°C than in those stored at 4.5°C (21%). These reductions in color index during storage were attributed to the Maillard reaction between free amino acids and reducing sugars, as well as to the destruction of ascorbic acid. Various theories have been proposed to explain the mechanism of browning in citrus products, but ascorbic acid degradation is thought to be the governing one. Ascorbic acid is oxidized in several steps into active substances, which react with nitrogenous materials to give brown pigments and off-flavor. This browning in citrus juices does not occur until at least 10–15% of the ascorbic acid has disappeared. This level was reached after about 270 days of storage at −12°C and after 150–200 days in concentrates stored at 4.5°C [76]. Samples in which the serum fraction had been freeze-concentrated during processing displayed the least ascorbic acid degradation.

## 3. Grapes

The effect of freezing on color quality of grape must and wines has been investigated [77]. Grape bunches were frozen and stored at −10°C for one month. The grapes were then thawed, crushed, and made into wine, which was compared with wine made from unfrozen grapes. Wines made from frozen grapes had a lower tannin content than wines made from fresh grapes. Among the three white grape varieties tested, the wines from two had better color (lower abs 420 nm) than wines made from control grapes. With regard to the red varieties, the color of the control wine from one variety was better than the wine from frozen grapes. However, the contrary was found with the other variety.

## 4.  Peaches

Philippon et al. [78] investigated browning in unpeeled clingstone peaches to be used for canning during frozen storage at −8, −12, −16, and −20°C. Sensory analysis was conducted and extent of enzymatic browning was determined on thawed fruits and canned product. Production of a canned product with similar quality to the best products produced from fresh peaches was possible, if the following guidelines were observed: storage temperature was −20°C, thawing was completed within a few minutes (e.g., by microwave thawing), and the time interval between chemical peeling and pasteurization was kept to a minimum. To prevent enzymatic browning, thawed peaches must be lye-peeled by brief (45 s) immersion in 7% NaOH, rather than by immersion for a longer period (90 s) in 1% NaOH.

## IV.  EFFECTS OF FREEZING, FROZEN STORAGE, AND THAWING ON PALATABILITY ATTRIBUTES

The main components of the overall sensation of flavor are taste and aroma [50]. Soluble and volatile constitutents in food reach receptors on the tongue and in the upper nasal cavities and trigger responses perceived as flavor. Aroma generally makes the major contribution to total flavor. The range of compounds capable of stimulating these senses is very wide. Since the receptors are extremely sensitive, very small changes in the amounts of the various substances may influence the overall flavor perceived. Fruit flavors are assessed by sensory analysis in which the actual sensitivities of humans are utilized. Analytical techniques, like gas chromatography and mass spectrometry, based on chemical identification and quantification of the compounds present in the fruits are also utilized. Effects of processing, freezing, frozen storage, and thawing are determined chemically by the changes produced in these compounds.

Texture may be defined to include both the properties perceived by the sense of touch in the mouth, and the appearance of fruits. In a more physically oriented definition, textural properties like firmness and toughness as measured by instrumental techniques may also be included in the term. In the following discussion, palatability is used to describe both sensory and chemically assessed flavor and textured properties.

## A.  Sensory Assessment of Flavor

El-Ashwah et al. [79] used sensory evaluation to determine changes occurring during frozen storage (−12°C) of lime juice. Generally, unpasteurized juice retained more desirable flavor for a longer period of time than pasteurized juice. The content of volatile oils decreased in both pasteurized and unpasteurized lime juice during 10 months of storage. Lime juice could be successfully stored at –12°C without either pasteurization or chemical treatment for about 10 months. Somewhat in contrast to these results, Lee et al. [32] reported flavor deterioration of passion fruit juice during frozen storage. Juice heated to 75°C for 80 seconds prior to freezing was compared with unheated juice. Both types of juice were stored at –20°C for 12 months. After 12 months the heated juice had reduced but acceptable flavor, while unheated passion fruit juice had unacceptable flavor. Hernandez and Villegas [80] stored pulps of some tropical fruits at –15°C for 2 and 6 months. Sensory evaluation after 6 months revealed moderate decreases in quality of custard apple, pineapple, guava, and mamey colorado pulps and large decreases in sensory quality of sapodilla and papaya pulps.

Sensory changes in mango nectar made from quick-frozen pulp at –40°C and stored at –18°C for periods up to 14 months have also been studied [56]. Pulp was frozen either with or without added ascorbic acid (50 mg/100 g). During sensory evaluation, frozen pulps were compared with canned pulp stored at 4°C for the same period of time. At the end of 6 and 10 months of frozen storage, the sensory characteristics of the frozen pulp without any ascorbic acid added were significantly inferior to the canned pulp with respect to color, aroma, flavor, and overall quality. However, ascorbic acid–fortified pulp compared favorably with the canned pulp in these sensory variables. After 14 months of storage, frozen pulp with added ascorbic acid also was inferior to the canned pulp with regard to aroma, flavor, and overall quality. When nectars (20% pulp, 15°Brix, 0.3 % acidity) from these pulps were evaluated, similar results were obtained. Nectar from ascorbic acid–fortified pulp had higher retention of color, aroma, flavor, and overall quality than nectar from pulp with no ascorbic acid addition. When nectar from frozen pulps was subjected to heating (85°C, 5 min) prior to sensory evaluation, no differences in sensory characteristics between nectars from ascorbic acid–fortified and canned pulp were observed even after 14 months of storage. Thereby, it was concluded that mango pulp frozen without pasteurization developed off-flavors during storage, but the intensity of these off-flavors was reduced if the pulp was fortified with ascorbic acid. The off-flavor was also found to be heat labile. Similarly, Abufom and Olaeta [40] investigated the stability of chirimoya fruit pulp during frozen storage. The variables studied included pasteurization (75°C, 10 min) and addition of ascorbic acid (0.15%), citric acid (0.2%) and EDTA (0.02%). The fruit pulp was frozen at –38°C and stored at –18°C for periods of up to 120 days. Pasteurization degraded sensory quality, while the additive mixture gave good control of enzymic browning and stable sensory quality, when compared with the control pulp without pasteurization or any additives.

Gradziel [23] reported sensory ratings of texture, flavor, appearance, and overall quality characteristics of tomatoes held at –40 or –10 °C for periods up to 330 days.

Off-flavor ratings were consistently higher with fruits held at –10°C compared with fruits held at –40°C. Flavor and overall quality ratings were also higher for fruits held at –40°C vs. –10°C when analyzed after 150 and 330 days of freezing. No single objective measurement (ascorbic acid, soluble solids, pH, titratable acidity, or percent moisture) was found to correlate consistently with sensory measurements.

Venning et al. [33] studied sensory changes in kiwi pulp stored under various conditions for periods up to 12 months. Pulps treated under "ideal" conditions (deaerated, packed in foil laminate, storage temperature –35°C) were compared with pulps treated under more "commercial" conditions (nondeaerated, packed in polyethylene, storage temperature –18°C). Sensory evaluation revealed no difference in total odor intensity or total flavor intensity at any of the individual sampling intervals between samples of pulp receiving the two treatments. Similarly, based on pooled data from the experiment, no differences were found between fresh pulps and those stored for 12 months. The frequency of use of specific descriptor terms for odor and flavor did not change significantly during storage or differ between samples from the two storage treatments. It was, therefore, concluded that the sensory properties of the kiwi pulps were relatively stable during prolonged storage under appropriate commercial conditions and that no advantage was evident from the use of deaeration, improved barrier packaging, or lower storage temperatures.

Stanley [81–84] studied freezing performance of soft fruit cultivars using sensory assessment of color, flavor, and texture characteristics. Fruits from strawberries, blackberries, gooseberries, black currants, and blueberries were frozen at –34°C and stored at –18°C for 3 months. Significant differences between cultivars were found, and cultivars best suited for freezing according to their sensory performance after thawing were reported.

## B.  Chemical Assessment of Flavor

### 1.  Enzymic Studies

Burnette [86] reviewed the importance of the influence of peroxidase content to fruit quality. Enzymes not destroyed during freezing continue to produce off-odors, off-flavors, and color changes during frozen storage. Since peroxidase is one of the most heat-stable enzymes in fruits, it often serves as a prime index when blanching fruits. If peroxidase is inactivated, all other enzymes in the product are usually inactivated also.

Marín et al. [15] studied chemical and biochemical changes during freezing and frozen storage of mango slices from four varieties. The mango slices were frozen in a blast freezer at –40°C without any pretreatment and stored in plastic bags at –18°C for 120 days. Fruits were thawed at 20°C for 3 hours prior to analysis. Peroxidase (POD, EC 1.11.1.7) and polyphenoloxidase (PPO, EC 1.14.18.1) activities were reduced during freezing, and reactivation was not observed during frozen storage. After 4 months of storage at –18°C, up to 40% of the initial POD activity and 20% of the initial PPO activity was detectable in the thawed fruits. The extent of inactivation varied between the four mango varieties tested. Since POD activity is a major cause of deterioration in the objective and subjective quality of frozen food, the importance of choosing a mango cultivar with low POD activity was emphasized to obtain good product quality. PPO, an enzyme generally regarded as having low thermostability, is regarded as being more heat stable than POD in fruits. This was confirmed by demonstrating that the inactivation rate of PPO during freezing was lower than that of POD. However, for three out of four mango cultivars, only 20% of the initial PPO activity was detectable after the 120-day frozen storage period. Consequently, it was concluded that deteriorative changes in nutritional and sensory attributes from PPO activity were not important in frozen mango fruits.

Calderón et al. [87] described a technique to localize the enzyme peroxidase during freezing/thawing of fruit tissue. Fruit tissue was cut in the frozen state, and, after thawing, sections were blotted on a nitrocellulose membrane. A histochemical peroxidase-staining technique (4-methoxy-α-naphthol, 4-MN) was then applied on the blotted tissue, resulting in a visualized location of the enzyme in the tissue. In grapes the peroxidase was principally associated with the skin and to a lesser extent with the pericarp, where discrete areas of reaction products were located in the vascular bundles. In immature fruits, peroxidase activity in the skin was low and similar to that found in the pericarp. The reliability of this technique in the histochemical localization of peroxidase in grapes was confirmed by fractionation and determination of peroxidase activity in the various tissue. Peroxidase in conjunction with β-glucosidase may be responsible for anthocyanin turnover and degradation. This observation is supported by the finding of co-localization of peroxidase and anthocyanins in grapevine vacuoles [88]. In many fruits anthocyanins are located in the epidermal and subepidermal layers that constitute the berry skin.

## 2. Bitterness in Citrus

Maotani et al. [89] investigated the mechanism of increased bitterness in citrus fruits (*Citrus natsudaidai*) caused by freezing. This bitterness is produced by the water-soluble glycoside naringin. The increase in bitterness of frozen pulp segments depended on an increase in naringin content of the exuded juice. Large quantities of naringin were exuded from the juice sac membranes and from the stalk tissue when frozen fruits were chewed. Some years later, Matsumoto et al. [90] published a study of changes in naringin content in the juice of *Citrus natsudaidai* fruit determined by HPLC. The flavanone glycosides detected were naringenin 7β-rutinoside (narirutin), naringenin 7β-neohesperidoside (naringin), and hesperitin 7β-neohesperidoside (neohesperidin). When detached fruits were treated at

–6.5°C for 11 hours, the content of naringin increased from about 18 mg% after the treatment to about 30 mg% after about 4 weeks of storage at 4°C. Upon further storage the content of naringin remained at this level. When the fruit was frozen while still on the tree, the level of naringin increased similarly. After about 4 weeks on the tree after the freezing injury, the level of naringin again decreased to about 24 mg%.

## 3.  Flavor Compounds

Strawberries have been the subject of several studies dealing with the effect of freezing, frozen storage, and thawing on palatability. This fruit is widely grown and undergoes many changes during the freezing process. Freezing strawberries is probably the best way to preserve them [10]. Aroma substances of freshly harvested and deep-frozen strawberries have been examined in several studies using gas chromatography and mass spectroscopy (GC-MS). Methyl and ethyl butanoate, methyl and ethyl hexanoate, *trans*-2-hexenyl acetate, *trans*-2-hexenal, *trans*-2-hexen-1 ol, as well as 2,5-dimethyl-4-methoxy-3(2H)-furanone have been identified as the main volatile components [91]. Deep-freezing generally resulted in reduced concentration of most of the aromatic substances, but higher amounts of 2,5-dimethyl-4-methoxy-3(2H)-furanone were found in deep-frozen berries compared with fresh ones. The content of readily volatile components dropped during frozen storage of strawberries, while there was a rise in less volatile substances [92]. In this study analyses were carried out immediately after freezing and then after 3, 6, and 9 months of storage at –12°C. Packing material was shown to have a significant effect on the retention of aroma in the frozen stored strawberries.

Ueda and Iwata [93] studied the aroma deterioration of strawberries from two cultivars after freezing and during frozen storage. The berries were frozen at –20°C. Off-odor was detected by a taste panel after only one day of frozen storage. For one cultivar this off-odor reached a maximum after storage for one month, while in the other cultivar this off-odor reached its maximum after only one week. Longer storage reduced off-odor and reduced natural fragrance. Volatiles from fresh and frozen strawberries were trapped and analyzed. Results revealed that esters that dominate among volatiles in fresh strawberries were barely detectable in the frozen/thawed fruits. Carbonyl compounds were maintained at levels similar to those found in fresh berries. No changes in the fatty acid composition of neutral or polar lipids extracted from the fresh and frozen berries were detectable.

Hirvi [94] studied aroma of fresh and deep-frozen strawberries from eight varieties by mass fragmentography and sensory evaluation. The concentrations of 17 major aroma components were determined as well as intensity of odor, character of odor, overall impression of odor, sweetness, overall impression of taste, sourness, off-odors, and off-tastes. In general, the concentration of volatile acids increased during freezing and thawing. In particular, hexanoic acid content of the frozen/thawed strawberry fruits increased. This acid and other C6 compounds are regarded as "secondary" aromatic substances, formed by enzymatic-oxidative cleavage of linoleic and linolenic acids in the presence of oxygen after rupture of tissues [91], which does not occur in intact berries. Ethyl and methyl butanoates were also found in higher concentrations in deep-frozen strawberries than in fresh berries.

Sensory analysis results for thawed strawberries were generally quite different from those for fresh berries [94]. Polygons drawn so that the attractive properties are shown on the righthand side in Fig. 8 reveal that strawberries of all varieties lost some intensity in the positive characteristics (intensity, character, and overall impression of odor, sweetness, and overall impression of taste), while the less positive character sourness increased during freezing and thawing. Significant increases in off-odor and off-flavor due to freezing and thawing were reported only for the Ostara variety.

Further, statistical analyses revealed significant correlations between titratable acids and sourness as well as between soluble solids and sweetness of the deep-frozen/thawed strawberries. Significant correlations between sensory parameters of odor and total amounts of ethyl hexanoate, ethyl butanoate, *trans*-2-hexenal, 2,5-dimethyl-4-methoxy-3(2H)furanone, and linalool were also detected. Correlations were higher for the fresh than for the frozen/thawed fruits.

A few years later, Douillard and Guichard [95] conducted similar experiments with six strawberry cultivars using GC-MS for identification and quantification of aromatic components. Determination of volatiles in stored frozen fruits showed only a slight influence on mesifurane (2,5-dimethyl-4-methoxy-2,3-dihydrofuran-3-one) and furaneol (2,5-dimethyl-4-hydroxy-2,3-dihydrofuran-3-one), while the concentration of nerolidol increased and the concentration of esters decreased when compared with unfrozen fruits. No increase in acids due to the freezing/thawing process was observed. For this study, Douillard and Guichard [95] used strawberries thawed to 0°C. No increase in content of methyl and ethyl butanoates was recorded, in contrast with the results of Hirvi [94], who reported an increase in these butanoates. This disparity in results may be attributed to differences in experimental methods, as the berries in the earlier study were thawed at 25°C for 5 hours and then pressed in a hydraulic press prior to analysis.

Studies similar to those performed on strawberries have also been conducted using other fruits. A total of 126 compounds were identified by combined GC-MS in raspberries, but no significant modifications of chromatographic profiles between fresh and frozen raspberries were detected [96]. Aska et al. [76] studied the effect of storage at −12°C on the chemical composition of concentrated lime juice. The juice had been concentrated using four different methods. Oxygenated terpenes, limonene content, carbonyl, and ester compounds were monitored during storage. Decreases in ester and limonene content and increasess in terpenes and carbonyl compounds were detected. The reduction in limonene during storage was attributed to the conversion of limonene to some oxygenated terpenes like α-ter-

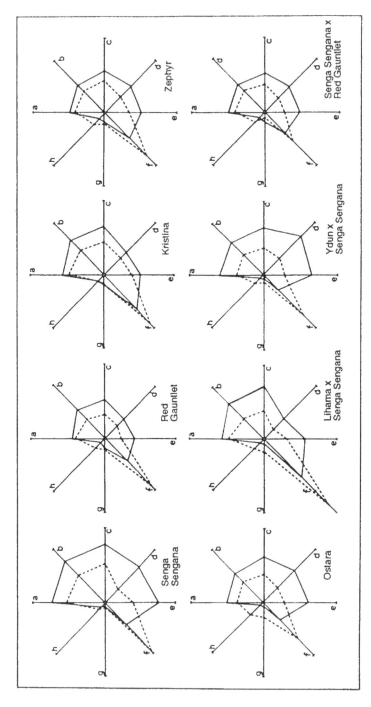

**Figure 8**  Polygons illustrating eight sensory properties evaluated in eight strawberry varieties (solid line = fresh berries; broken line = deep-frozen berries). Properties: a = intensity in odor; b = character of odor; c = overall impression of odor; d = sweetness; e = overall impression of taste; f = sourness; g = off-odor; h = off-taste. (From Ref. 94.)

pineol. The authors refer to previous works where storage of orange juice and orange juice concentrate resulted in reduction of esters, limonene, and linalool and formation of α-terpene and some aldehydes.

Etievant et al. [97] studied flavor components of fresh and frozen/thawed mirabelle plums using GC and GC-MS to characterize headspace composition before and after deep-freezing. Fruits were analyzed fresh and after 2 weeks of storage at –30°C. Thawing was performed by leaving the fruits at 4°C for 36 hours. After thawing, a quick and deep browning was observed. A sensory test was performed to find out whether the browning reaction could be linked to a parallel modification of the aroma. Odor intensity, fresh flavor, oxidized flavor, and cooked flavor were evaluated in the fresh and frozen/thawed mirabelle fruits. Thawed fruits were significantly less intense in fresh mirabelle flavor and higher in intensity of oxidized and cooked flavor than fresh fruits. No difference in odor intensity between the two types of fruits was observed. As a consequence of these observations, investigations were carried out to explain whether variation in the concentration of any volatile constituents could explain the changes in odor. A headspace-collection technique was used to quantify the major constituents of mirabelle aroma. The analysis showed that esters accounted for about 88% of total volatiles. The major esters were hexyl ester (40%, with 36% of hexyl butanoate alone), butyl ester (32%, with 22% for *n*-butyl butanoate alone), and ethyl esters (16%).

To visualize the differences between samples and to interpret more easily the influence of deep-freezing in terms of chemical attributes, a principal component analysis (PCA) was performed. The projection of the two first principal axes are shown in Fig. 9. The discrimination between the extracts obtained from fresh and thawed fruits can be seen from the first axis, explaining 67% of the variance. This analysis demonstrated an increase in C6 alcohols and aldehydes (hexanol, *cis*-3-hexenol, *trans*-2-hexenol, hexanal, *trans*-2-hexenal), which arose from enzymic oxidation of linolenic and linoleic acids during thawing. An intense oxidation was also demonstrated by the increase in benzaldehyde. *Trans*-2-hexenyl acetate, propanoate, and butyrate, as well as *cis*-3-hexenyl propanoate, were also found to be significantly more abundant in the headspace of thawed fruits, presumably as a result of the formation of the corresponding alcohols. Some other esters were reported to be more abundant in the headspace of fresh fruits. A possible explanation is that fresh fruits had synthesized these esters during the analytical aroma collection period, while the dead cells of the thawed fruits had not. In addition, deep-freezing and thawing may favor the activity of esterases or, more probably, the denaturation or inhibition of enzymes involved in the biosynthesis of esters. Significant differences in hydrocarbons were also demonstrated in the headspace of fresh and thawed fruits. Alkanes are more abundant in fresh fruits, while α- and β-pinene were only detected in thawed fruits. This difference suggests that terpene precursors occur in mirabelles but are not, or only

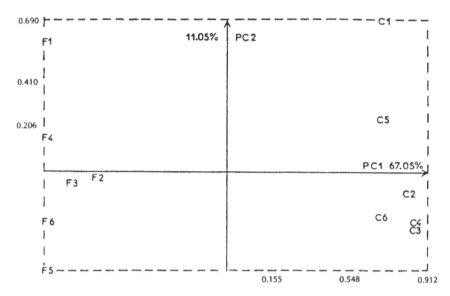

**Figure 9**   Projection of the sample space of the two first principal axes for fresh and defrosted mirabella plums. F1–6: Fresh plums; C1–6: defrosted plums. (From Ref. 97.)

to a small extent, converted to α- and β-pinene in the fresh fruits. As for C6 alcohols and aldehydes, these compounds are produced almost wholly during thawing, indicating abnormal enzymic activities favored by cell structure disruption.

Yen et al. [34] examined changes in volatile flavor components of guava puree during processing and frozen storage. The changes were evaluated by Linkens-Nickerson distillation and capillary GC-MS after processing (85–88°C, 24 s) and subsequent frozen storage of the puree for up to 4 months at 0, –10, and –20°C. Results were compared with those of fresh, unpasteurized guava puree. Pasteurized guava puree is known to undergo deterioration during frozen storage, resulting in development of off-flavors and decreased sensory quality of guava puree to be used for further processing into various juice blends. In the unpasteurized fruits, 27 compounds, including 2 acids, 5 alcohols, 6 aldehydes, 7 esters, and 7 hydrocarbons, were identified from the gas chromatogram. During pasteurization, there was a decrease in esters and increases in aldehydes, terpenes, and acids, but almost no change in alcohols was observed (Fig. 10). During storage, esters, alcohols, and acids remained unchanged or increased during frozen storage, while alcohol and terpene content decreased slightly during storage at –10 and –20°C.

By making use of the high gas permeability of fruits, experiments resulting in improved flavor of frozen strawberries have been conducted [98]. Enhancement of

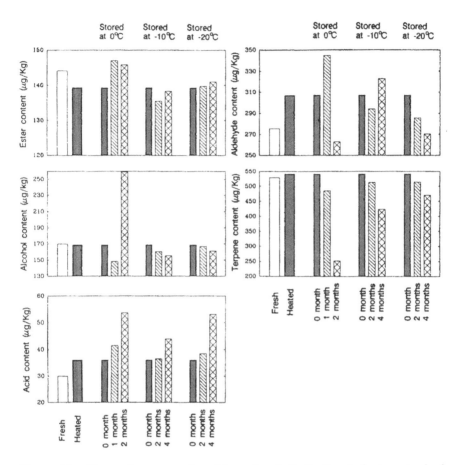

**Figure 10** Changes in concentration of the main volatile fractions in fresh, pasteurized and stored guava puree. (From Ref. 34.)

aroma was obtained by blocking rapid catabolism of strawberries in a $CO_2$ atmosphere during freezing. During storage of strawberries in normal atmosphere, catabolism produced extensive losses in important volatile components within a few hours. In a $CO_2$ atmosphere, the catabolic reactions were blocked and the genuine strawberry aroma was preserved.

Experiments in which fruits are treated with an atmosphere containing various aroma precursors (PA storage) have also been carried out, although not in direct connection with freezing. Strawberries stored for 24 hours in an atmosphere containing 8 mmol ethanol/kg fruit appeared to increase considerably in various ethyl esters, which suggest potential future combinations of biochemical knowledge and practical food technology.

## C.  Assessment of Texture

Fruit texture is greatly influenced by freezing, frozen storage, and thawing, and extensive investigations have been performed within the area of evaluating and improving texture performance during freezing [4,10,21,24]. The general principles of ice formation and the influence of structural deformation on individual cells and tissues are discussed in Chapter 1. Therefore, only conditions specific for fruit and fruit products are included in the present discussion.

Brown [99], in a review, pointed out the important works of Woodroof [100] on the effects of freezing at various temperatures on fruits. Woodroof, in his microscopic observations, revealed the extent of tissue damage and the location and size of ice masses in the fruit tissue. Hoser et al. [101] used microscopy to obtain microphotographs of frozen fruit tissue. Frozen strawberries were cut in a cryostat and stained with gentian violet prior to microscopy, and the histological structures of tissues were studied after 3 and 6 months of storage at –20°C. A new, noninvasive approach to texture evaluation of fresh and frozen tissue has been presented by Duce et al. [102]. Using nuclear magnetic resonance (NMR) imaging, they showed changes in the appearance of the structure due to freezing and thawing of courgettes. The image of the fresh courgettes demonstrated a clearly delineated internal structure with distinct regions corresponding to skin, vascular tissue, cortex, seed, and seed beds. After freezing and thawing, the overall intensity of the image was greater and the contrasts between different types of tissue were less distinct. In the future, two- or three-dimensional NMR imaging may prove to be an interesting way to study physical and chemical changes in fruit tissues during frozen preservation.

Turgor, the pressure of the cell protoplasts against the cell walls, is important for crispness and succulence of fruits [103]. It varies with osmotic movements of water through the cell walls and cannot be restored when cells are destroyed. Rupture and fragmentation of cell membranes and cell walls are therefore important mechanisms in decreased liquid retention, i.e., increase in drip losses, during freezing and thawing of fruits [24].

Frozen storage may cause dehydration of a product due to fluctuations in storage temperature below the freezing point [25]. As the temperature rises, water evaporates from the fruit. As the temperature drops again, the water vapor condenses on the surface of the packaging material because it cools more quickly than the fruit. It is important to prevent thawing during storage because the freezing that will subsequently occur under these conditions will be slow. Large ice crystals will then form and damage the fruit tissue. Melting of ice is an endothermic reaction, i.e., heat is absorbed during thawing. Heat is usually taken from the surrounding air or thawing medium, but it may also be extracted from within the tissue. A slow, partial refreezing may therefore occur during thawing, resulting in further damage to cell walls and membranes. The amount of drip loss during freezing and thawing is a frequently used measure for evaluating freeze damage to fruit tissue.

## 1. Cell Structure

Szczesniak and Smith [104], in a detailed study of frozen strawberries, demonstrated a relationship between structural properties and the extent of intracellular damage. Fibrous tissues, such as vascular cells and other thick-walled tissues, are naturally resistant to freezing damage [103].

Fruit cell walls, and in particular the middle lamella between cells, are rich in pectic substances. During ripening, deesterification of pectins occurs and softens the fruit tissue [10]. Wolford et al. [105] reported that addition of $Ca^{2+}$ ions prior to freezing increased firmness of fruits after thawing. These ions form intermolecular linkages in the pectin and thus strengthen cell walls. This contribution to texture, unlike the turgor effect, remains after the initial breakage of cells. Crivelli et al. [106] reported reduced drip loss in one strawberry variety when plants were sprayed with calcium nitrate during the blooming season. Another variety, however, did not display improved texture from this $Ca^{2+}$ treatment.

Reid et al. [107] investigated softening of plant tissues during freezing and frozen storage in relation to changes in the pectic fraction of the cell walls. Ripe strawberries and peaches were frozen with an air-blast temperature of $-70°C$ or in still air in a cold room. Texture was measured as back-extrusion with an Instron Universal Testing Machine, and pectin materials were extracted and characterized by various chemical procedures. Cell wall material was found to contribute to texture, and pectin was an important contributor to the texture of tissues that had been frozen. In strawberries, the water-soluble pectic fraction decreased in parallel with the reduction in firmness. Compositional studies indicated that the changes were associated with the pectin rhamno-galacturonic backbone.

Sapers et al. [54,66,108] used fruits from various cultivars of thornless blackberry to study causes of variation in drip loss during thawing after 7–9 months of frozen storage at $-13°C$. In one season, drip losses for various cultivars ranged from 1 to 30%. Drip losses were greater in riper fruits but were not dependent on the size of the fruits. A negative correlation between drip loss and the amount of insoluble pectin was found, but results were not consistent during the 2 years studied. Blackberry cultivars varied considerably in soluble, insoluble, and total Ca content, but no indication of a relationship between endogenous soluble or insoluble Ca and drip loss was observed. The insoluble pectin fraction was consistently greater during the growing season with lower drip losses (13.0%) compared with a year with higher drip losses (23.8%), the insoluble pectin content was 0.19% and 0.13%, respectively. The extent of droplet collapse in blackberries was not related to pectin content. Microscopic examination revealed an inverse relationship between epidermic cell layer thickness and the tendency of the thawed fruits to drip. However, no consistent relationship was detected between cell density, i.e., cells per running mm of cuticle, and the amount of drip loss. In fresh but not frozen blackberry fruits, depth of epidermic and subepidermic cell layers appeared to be reduced in a cultivar that leaked more. In addition, some shrinkage of epidermal tissues during freezing and/or thawing was observed.

## 2. Freezing Rate

In general, higher freezing rates produce smaller ice crystals, less migration of water, and less breakage of cell walls in tissue. Consequently, less texture damage occurs under these conditions [99,109]. Holdsworth [57] referred to studies where strawberries frozen in liquid nitrogen or Freon resulted in better texture and lower drip loss than berries frozen by air blast freezing. The positive effect of the high freezing rate was maintained during 6–12 months of storage at temperatures between –20 and –30°C. However, freezing rates higher than 1.5 cm/h did not improve quality retention significantly [4]. In addition, the positive effect of rapid freezing will disappear if the conditions during frozen storage are poor, for example, when large fluctuations in storage temperatures occur [24].

Ramamurthy and Bongirvar [27] froze mango slices under various conditions prior to freeze-drying. Freezing was accomplished by placement in a deep freezer (–20°C), in a contact plate freezer (–40°C), or by immersion in liquid nitrogen (–196°C). After reconstitution of the dried mango slices, rehydration properties, water-holding capacity, and texture (shear press) were evaluated. By decreasing freezing temperatures, an increase in water uptake and a decrease in water loss were found. Texture also clearly improved with decreasing temperature, since the shear press of the reconstituted mango slices from the liquid nitrogen treatment was higher than the shear press of slices frozen at –40 and –20°C. Although the liquid nitrogen–frozen mango slices retained the most desirable textural properties, they never regained the texture of fresh mango slices.

Phan and Mimault [110] studied the effect of freezing and thawing by evaluating the exudate and texture parameters of apples and peaches in relation to fruit quality. Fruits were rapidly frozen either in a freezing tunnel, which decreased the temperature to –20°C in one hour, or in a conventional freezer, where 13 hours were needed to reach the same temperature. After frozen storage at –20°C for 6 months, fruits were thawed in air at 4 and 20°C and by microwaves. The thawing processes were stopped when the fruits reached –4°C. Rate of freezing played a major role in maintenance of texture and retention of liquid in both thawed apples and peaches. Thawing method proved to have a smaller effect on quality, especially for the peaches. However, rapid freezing in combination with rapid thawing using microwaves appeared to be the most appropriate combination of treatments to provide the most firm texture and the lowest amount of exudate from the fruits.

Martí and Aguilera [111] evaluated three freezing methods for blueberries and wild blackberries (static freezing at –23 ± 2°C, plate freezing at –50°C and immersion in liquid nitrogen). Relative freezing rates between the three methods were 1:5:15 for static, plate, and liquid nitrogen, respectively. No effect of freezing rate was found on drip loss from blackberries. However, higher texture values were obtained in berries frozen using the two fastest freezing methods compared with the slowest method. Freezing in liquid nitrogen best retained texture and minimized drip loss from the thawed blackberries. Histological studies showed plate freezing at –50°C, and immersion in liquid nitrogen produced almost no damage to cell walls. In contrast, slow freezing produced significant damage to cell walls and membranes. Cell wall and membrane damage correlated well with changes in texture and drip losses (Fig. 11).

Prolonged immersion in a cryogenic solution like liquid nitrogen, after completion of freezing, produces a very rapid drop in temperature starting at the surface of the fruit. The contraction of this cold exterior while the interior is still near the freezing point leads to cracking and shattering of fruits frozen individually [105, 112]. Freezing techniques that avoid direct immersion of the fruit in the freezing medium are, therefore, preferable.

Studies by Chuma et al. [113] demonstrated that strawberries could be efficiently cooled to cryogenic temperatures by spraying with liquefied nitrogen gas. Visual observation of the berries revealed no damage in the rapidly frozen fruit. A critical temperature between –40 and –60°C was observed, below which the

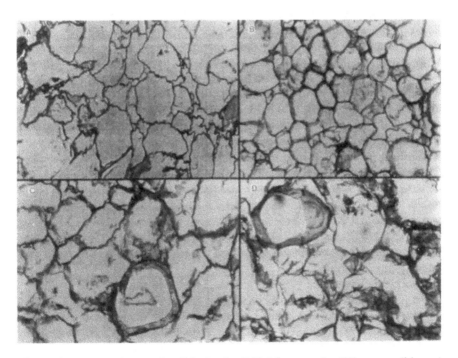

**Figure 11**   Photomicrographs of blueberries (150x) frozen under different conditions. A = fresh; B = liquid nitrogen; C = plate freezer; D = static freezer. (From Ref. 111.)

material could be crushed to smaller pieces. Temperatures between –100 and –150°C are recommended if a ground product is desired. In agreement with this, Watanabe et al. [114] found that juice sacs of citrus fruit could not be separated individually by crushing at temperatures above –30°C. The degree of separation also increased with decreasing temperture to –120°C. At even lower temperatures, no improvement in separation ability was observed.

## 3. Variety and Maturity Level

Factors such as variety, maturity, growing area, and seasonal variations influence frozen storage performance of fruits to an extent that may override the positive effect of a high freezing rate.

Drip loss and accompanying droplet collapse in thawing blackberries have been shown to be more extensive in fully ripe than in slightly underripe fruits [54,66, 108]. However, fruit size had no effect on the extent of drip losses. Therefore, the authors suggested that differences in fruit condition may be more important than cultivar in determining drip losses. Crivelli and Rosati [115], however, reported increased drip loss when fruit weight of strawberries increased.

Lenartowicz et al. [116,117] studied the suitability of red and black currants for freezing. Significant differences in drip loss between varieties were reported. Drip losses of red currants ranged from 9.8 to 26.7%, while those of black currants ranged from 0.0 to 7.17%. Conclusions from previous studies with strawberries, raspberries, sour cherries, and blueberries were that the best quality raw fruit with regard to cultivar and ripeness exhibited the least extent of quality degradation during freezing [119]. In studies with black currants, Plockarski et al. [120] reported drip losses ranging from 1.0 to 8.2% after thawing of frozen fruits, while Rosati and Crivelli [121] found drip losses between 4.7 and 46.7% after 6 months frozen storage of fruits from 37 strawberry varieties.

Bushway et al. [72] evaluated five raspberry cultivars harvested at the red-ripe stage at harvest and after 1, 6, and 9 months of frozen storage at –20°C. On the day of harvest, fruits were frozen in an air-blast freezer at –30°C or analyzed to obtain results for raw fruits. After frozen storage at –20°C, textural evaluation was performed with a Kramer shear press connected to an Instron Universal Testing Instrument. Results revealed a significant decrease in shear values for all raspberry varieties during the 9-month storage period, corresponding to less firm fruits after freezing and frozen storage.

Cancel and Hernández [73] showed that freezing of citron (*Citrus medica* L.) cubes prior to the production of candied citron produced a softer product than when the citrons were produced directly from fresh fruits. When citron cubes were blanched before freezing and fermentation, products similar to those obtained by fermentation of fresh fruits were obtained. The texture of breadfruits that were blanched and frozen prior to cooking proved to be inferior to fresh-cooked breadfruits when assessed by sensory analysis [26]. Gradziel [23] studied firmness of thawed tomatoes after freezing and frozen storage at –10 and at –40°C. Puncture tests after 5 days of frozen storage revealed an 80% loss in firmness. Additional losses were recorded with longer storage periods.

## V.  EFFECTS OF FREEZING, FROZEN STORAGE, AND THAWING ON NUTRITIONAL VALUE

Nutritional value of fruits is related to the content of minerals, sugar, protein, fat, and vitamins. Among these components, vitamins may be regarded as the most important from a dietary point of view. The vitamins are also the compounds most susceptible to degradation and may, therefore, be rendered physiologically ineffective before they are consumed.

During the process of freezing, storage, and thawing, vitamins and minerals may dissolve in liquid exuding from the fruits. If the drip is included when the fruits are consumed or further processed, no loss in nutritional value during freezing and thawing will occur, however. This same phenomenon applies for all chemically stable components dissolved in the drip of thawed fruits. Losses of water-soluble compounds will also occur to some extent when fruits are blanched prior to freezing.

A comprehensive summary of fruit sources for the various nutrients is given by Young and How [122].

## A.  Vitamin C

Ascorbic acid was one of the first compounds to be studied in connection with the quality of frozen fruits [57,123]. This was because of the importance of ascorbic acid as a vitamin and also because of the availability of an analytical method. As ascorbic acid is a reactive compound, it also serves as an indicator substance for chemical reactions taking place in the product [20,124]. The oxidation of ascorbic acid may be enzymic or nonenzymic and proceeds in the presence of oxygen [125]. Early work showed ascorbic acid to be oxidized to dehydroascorbic acid and then to 2,3-diketogulonic acid [10]. The latter have no vitamin C activity [125]. The low pH values of fruit and berries have a positive effect on ascorbic acid stability [126].

### 1.  Berries

In a review, Herrmann [124] refers to ascorbic acid losses in deep-frozen strawberries, peaches, and raspberries as ranging from 15 to 25% after 12 months of storage at –18°C. Thawing was reported to produce a loss of 2–10% in cherries, red currants, blackberries, and raspberries.

Changes in vitamin C stability during fast freezing (12–15 cm/h), frozen storage, and thawing of strawberries and bilberries were studied by Malinowska et al. [127]. Dehydroascorbic acid was determined separately, and the vitamin C content was calculated as the sum of ascorbic acid and dehydroascorbic acid. The initial total vitamin C content was 70–72 mg% and 32 mg% for strawberries and bilberries, respectively. Ascorbic acid content was correspondingly 63–70 mg% and 22.5–28 mg%. No significant change in vitamin C content due to freezing was observed. During 12 months of frozen storage at temperatures ranging from –19 to –24°C, vitamin C content of strawberries decreased to 31.7% of the level in raw fruits. With bilberries, the decrease was 50%. After 9 months of frozen storage, thawing of fruits in the dark at room temperature caused a small decrease in vitamin C content for strawberries (1%) and bilberries (5%). Thawing in daylight gave higher losses—12.7% for strawberries and 20.9% for bilberries—compared with the vitamin C content of the frozen, stored fruits. Similar results were obtained when fruits were defrosted after 12 months of storage.

Fraczak and Zalewska-Korona [36] studied the effect of freezing methods on ascorbic acid retention in fruits of various strawberry and raspberry varieties. The freezing methods were –30°C in a freezing tunnel for several hours, fluidization for 1–3 minutes, or freezing in sugar for 75–90 minutes. Fruits were also frozen in liquid nitrogen at –180°C or in liquid carbon dioxide at –78°C for 15 seconds. Fruits frozen in the tunnel were packed in paper bags and stored for 6 months at –25°C. Samples frozen in sugars were stored in cans at –18°C, and the remaining samples were stored in polyethylene bags at –20 to –22°C. Results revealed that the vitamin C content of fruits was significantly influenced by the freezing procedure. The greatest losses were found when fruits were frozen in the freezing tunnel (3.6% after freezing and 38.4% after storage for strawberries, and 27.6% after freezing and 61.3% after storage for raspberries). Highest vitamin C retention in strawberries occurred when berries were frozen in sugars, with a loss of only 2.5% after freezing and 25.0% after storage. The highest vitamin C retention for raspberries was obtained when fruits were frozen in liquid nitrogen, with a loss of only 1.3% after freezing and 8.6% after 6 months of storage.

## 2.  Citrus Fruits

El-Ashwah et al. [79] studied ascorbic acid retention in frozen lime juice. During 12 months of frozen storage, the ascorbic acid content of lime juice pasteurized at 76°C for 1 minute prior to freezing decreased from 37.5 to 25.4 mg/100 ml juice. In frozen unpasteurized lime juice, ascorbic acid content decreased similarly from 37.8 to 21.2 mg/100 ml. Higher ascorbic acid stability was obtained when the lime juice was treated with sodium benzoate and potassium metabisulfite. In this instance, the vitamin C levels decreased from 37.8 to 29.6 and 26.5 mg/100 ml of juice in the pasteurized and unpasteurized juice, respectively. This effect was attributed to the antioxidative effect of $SO_2$.

Bisset and Berry [47] studied the retention of ascorbic acid in concentrated orange juice (45°Brix) as related to type of container. The most oxygen-permeable containers were cans of fiber/polyethylene, while the least permeable were foil cartons. Orange juice concentrates were frozen at –20.5°C and transferred to storage temperatures of –20.5, –6.7, and 1.1°C. Only minor differences in ascorbic acid retention were observed after one year when juices were stored at the lowest temperature (–20.5°C). Retention was 91.5 and 93.5% for the fiber/polyethylene and foil container, respectively (Fig. 12). At higher storage temperatures, however, the foil barrier gave superior retention. After 8 months at –6.7°C, retention of ascorbic acid in juice subjected to the highest oxygen permeability was 43%, while the other was 79% of the initial level. At 1.1°C the retentions were 44 and 89% after only 3 months of storage.

El-Baki et al. [128] evaluated changes in ascorbic acid content during freeze-concentration of lime juice. After being pasteurized at 85°C for 1 minute, the juice was frozen in containers at –12°C for 24 hours. The ice crust formed during the freezing process was removed, crushed, and centrifuged to remove the juice. This

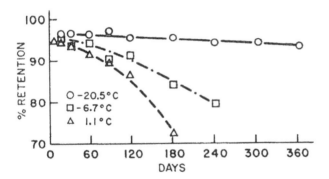

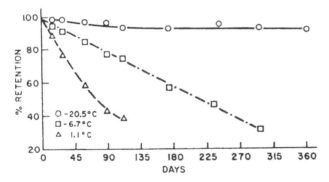

**Figure 12** Ascorbic acid retention in frozen concentrated orange juice stored in foil-lined fiberboard cartons (top) and in polyethylene lined fiberboard cans (bottom). (From Ref. 47.)

process was repeated twice, and the concentrated lime juice (41°Brix) was finally stored at −12°C for a period of up to 8 months. Ascorbic acid concentration decreased by about 35% during the process of freeze-concentration. This loss likely occurred partly as a result of oxidation during the centrifugation step and partly as a loss in the ice phase. Storage of the concentrate at −12°C for 8 months revealed a further loss of 20% in the ascorbic acid content of the juice. During the entire process, a reduction from 33.20 to 13.63 mg/100 ml in reconstituted lime juice was reported.

Askar et al. [76] also studied ascorbic acid degradation in frozen concentrated lime juice. A more pronounced degradation was observed when the juice was concentrated by a method in which serum and pulp of the original juice was separated during processing (serum-pulp method), compared with concentration without separation of serum and pulp. More atmospheric oxygen was probably included during the serum-pulp concentration method, thus producing a more comprehensive ascorbic acid oxidation in the lime juice.

The flavedo of oranges are rich in ascorbic acid. Valadon and Mummery [129] reported 127 and 322 mg/100 g in Spanish and Turkish oranges, respectively. After 24 months of frozen storage in polyethylene bags at −20°C, 81% of the ascorbic acid was retained in the Spanish variety and 94% was retained in the Turkish variety. After thawing, ascorbic acid was more liable to degradation. After 4 days of storage of frozen-then-thawed flavedo at room temperature, the ascorbic acid content of the Spanish variety was reduced to about 40% of the level in the fresh flavedo, and the reduction in the Turkish variety was about 75%.

## 3. Tomatoes

In an interlaboratory study, Buret et al. [130] questioned the practice of storing frozen material to be used for compositional analysis at a later date. Tomato fruits were stored whole or as puree in triple polythene (thickness $3 \times 62.5$ μm) bags at −20 to −25°C. Samples were thawed by placing the plastic bags in hot water (80°C). After thawing, the whole fruits were blended into a puree for 2–3 minutes and submitted to ascorbic acid analysis. The ascorbic acid content of both whole fruits and puree decreased significantly during an 11-week storage period. On average, ascorbic acid retention was significantly higher when the tomatoes were frozen as whole fruits (12.96 mg/100 g) compared with puree (10.28 mg/100 g). During this study a much higher content of ascorbic acid was detected when samples were thawed directly in the extraction medium for the ascorbic acid assay. A second trial was, therefore, conducted to more closely examine the thawing process. A spectrofluorometric method was used, allowing the detection of both ascorbic acid and dehydroascorbic acid, in contrast to the initial procedure with 2,6-dichlorophenol-indophenol, where only ascorbic acid was detected. While tomato fruits were crushed under liquid nitrogen and aliquots were left at 40°C for 0–35 minutes before extraction. The content of ascorbic acid fell during the thawing period (Fig. 13).

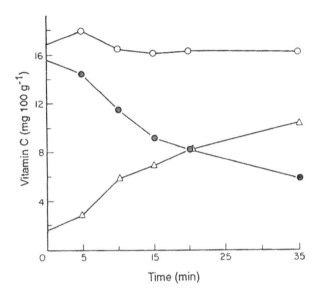

**Figure 13**  Changes in vitamin C content of pureed tomato fruit during thawing at 40°C. Ascorbic acid ●—●; ascorbic acid + dehydroascorbic acid O—O; dehydroascorbic acid △—△. (From Ref. 130.)

Simultaneously, the content of dehydroascorbic acid increased. Thus, the sum of ascorbic acid and dehydroascorbic acid was constant during the 35-minute thawing period. Since the thawing method influenced the ascorbic acid content, the practice of holding tomatoes and other fruits in frozen storage for delayed analysis of ascorbic acid and other more unstable constituents like vitamins must be questioned. Analysis of fresh fruits should be performed when possible. Storage in 80% ethanol may be an alternative to frozen storage.

Gradziel [23] studied the effect of frozen storage time, storage temperature, and packaging on ascorbic acid retention in two tomato cultivars. Higher levels of reduced ascorbic acid were maintained in all fruits stored at –40°C compared with fruits stored at –10°C when analyses were performed between 120 and 330 days of storage. The thickness of the plastic containers (0.8, 2.7, and 20 mm) did not significantly influence ascorbic acid content of the frozen tomatoes.

## 4. Kiwi

The production of kiwi has grown considerably during recent years, and consequently efforts have been made to examine the changes taking place during freezing, storage, and thawing of kiwi and kiwi products. Venning et al. [33] conducted a study wherein changes in ascorbic acid and dehydroascorbic acid in kiwi pulp were followed during frozen storage. At storage temperatures of –18 and

–25°C, ascorbic acid content decreased from an initial 89–104 mg/100 ml juice to about 75 mg/100 ml during 55 weeks of storage (Fig. 1). No changes in the content of dehydroascorbic acid (about 25 mg/100 ml) were observed. Pulp packed in polyethylene and in foil laminate material behaved similarly at these storage temperatures. When stored at –9°C, however, packaging type had a major effect on ascorbic acid stability. While pulp packed in foil laminate retained about 75% of its ascorbic acid content, pulp packed in polyethylene bags lost 88% of its ascorbic acid content during 55 weeks of storage. A corresponding increase in dehydroascorbic acid was observed (Fig. 2). Only minor effects of deaeration of the kiwi pulp were observed. However, comprehensive deaeration of pulp is difficult to achieve [131]. Therefore, oxygen might have been present despite the deaeration process. Roy and Singh [132] reported a decrease in ascorbic acid from 48.0 to 43.0 mg/100 g during 6 months of frozen (–15°C) storage of pulp from bael fruit. This decrease in ascorbic acid was associated with oxidation by oxygen trapped in the container during freezing of the pulp.

Cano et al. [29,59] studied changes during freezing and storage of kiwi slices. Ascorbic acid content of fruits from four cultivars varied from 95 to 170 mg/100 g fresh weight. Prior to analyses, samples were stored for 3 hours at 20°C. The freezing process produced a decrease of 10–25% in all varieties except one, where no detrimental effect on ascorbic acid content was found. After 12 months of storage at –18°C, losses in ascorbic acid ranged from 22 to 37%. The variety with the highest initial ascorbic acid content also retained the highest level after frozen storage.

## 5.  Mangoes

Freezing performance of mango cultivars has been studied by Marín et al. [15]. Fruit slices were frozen without previous treatment in an air-blast freezer operating at –40°C. As the inner part of the slices reached –24°C, the slices were vacuum-packed in plastic bags and stored at –18°C for 120 days. Thawing was performed under controlled conditions (3 h, 20°C). The ascorbic acid content of fresh mango ranged from 18.08 to 38.69 g/100 g. Losses in ascorbic acid content during freezing ranged from 25 to 50% for fruits from three cultivars. A fourth variety showed no significant drop in ascorbic acid content due to the freezing process. During frozen storage, ascorbic acid suffered a reduction between 57 and 73%.

## B.  Other Vitamins

Vitamins other than ascorbic acid are also susceptible to chemical degradation. Therefore, steps must be taken to protect these valuable components from deterioration during processing. As pointed out earlier, ascorbic acid is often considered to be an indicator substance when evaluating chemical changes in fruits. Consequently, limited recent literature concerning other vitamins is available.

Spiess [20] reported early investigations concerning the retention of the B vitamins and carotene during frozen storage. The stability of folic acid, niacin, pantothenic acid, riboflavin, thiamine, and pyridoxine in orange juice and sliced strawberries during frozen storage at –18°C are reported. After 12 months of storage only thiamine and pyridoxine were reported to diminish. All the others were stable. Therefore, it was concluded that frozen storage of fruits at –18°C for periods up to 1 year produced no detectable decrease in the B vitamins.

Minasyan and Astabatsyan [33] studied B-group vitamins in various frozen fruits. The fruits were frozen (–40°C) and stored (–18 to –20°C) for 8 months in polyethylene bags. B-group vitamins (thiamine, pantothenic acid, pyridoxine, nicotinic acid, and inositol) were determined prior to freezing, after freezing, and after frozen storage. The fruits studied were reported to contain 729.1–2654 mg/g DM of B-group vitamins. After frozen storage the vitamin content decreased only slightly (94.9–99.3% of inositol and nicotinic acid persisted). Pyridoxine was retained at 78.3–97.6%, and 85.1 and 88.2% retention was measured for pantothenic acid and thiamine, respectively.

Extensive work has been undertaken regarding the persistence of provitamin A after freezing and frozen storage, as it also functions as a pigment in fruits. β-Carotene and other closely related carotenoid pigments, such as α- and γ-carotene and cryptoxanthin, are referred to as provitamin A [50]. The content of β-carotene has been shown to decrease during frozen storage of tomatoes [17] and mango [15]. Further discussions on the effect of freezing, frozen storage, and thawing on carotenoid stability are given in the test dealing with appearance.

## VI. EFFECTS OF FREEZNG, FROZEN STORAGE, AND THAWING ON INTRINSIC CHEMICAL REACTIONS

Fresh fruits consist of living cells. Consequently, an abundance of chemical and biochemical reactions take place in fruit tissue. During the process of freezing, several reactions connected to the metabolic activity of the cells are terminated or drastically reduced. However, various chemical reactions like oxidations, hydrolysis, and polymerizations, as well as a number of enzymatically catalyzed reactions occur, even in dead cells. These changes, known to be connected with defined changes in appearance, palatability, and nutritional quality of fruits, have been discussed in other parts of this chapter. In the following section, only reactions relating to the content of dry matter, acids, fiber, amino acids sugars, and pH are included.

Goratti Netto et al. [28] reported no significant changes in pH, °Brix, and total or reducing sugars when mango slices were stored frozen in syrup. Various syrups, sucrose (25–40°Brix), glucose, and a combination of both sucrose and glucose were evaluated. Storage conditions were –20°C for periods of up to 4 months. Total acidity decreased slightly during storage. Only minor changes in pH, amino nitrogen, and total reducing sugars were found during storage of pasteurized and unpasteurized passion fruit juice [32]. No significant changes in total soluble solids and acidity were reported in mango pulp stored at –18°C for periods of up to 14 months [56]. Total soluble solids and total acidity were also constant in lime juice during 10 months of frozen (–12°C) storage [79]. Content of amino nitrogen varied inconsistently, while content of volatile oils decreased in both pasteurized and unpasteurized lime juice during storage. Buret et al. [130], in an interlaboratory study, tested changes in soluble solids, dry matter content, electrical conductivity, titratable acidity, potassium, pH, and total N during freezing and frozen storage of whole tomatoes. The levels remained constant during frozen storage for most constituents and were close to those found in the fruits prior to freezing.

Ross et al. [134] investigated the dietary fiber constituents of fresh and frozen strawberries. No differences in neutral detergent fiber (0.94 g/100 g wet weight), cellulose (0.50–63 g/100 g wet weight), hemicellulose (0.21–0.10 g/100 g wet weight), lignin (0.23–0.21 g/100 g wet weight), or pectin (0.36–0.42 g/100 g wet weight) occurred during freezing and thawing.

Changes recurring in grapes during freezing, frozen storage, and thawing have been investigated in relation to off-season wine production. Suresh et al. [77] found that frozen storage of grapes at –10°C for 1 month did not alter sugar content of the must. However, a slight decrease in acid content was observed. This decrease in acid caused a simultaneous increase in pH in wines made from the grapes. Variations in pH were marginal in aged wines. Spayd et al. [135] also reported decreased acid content in musts and wine and increased pH levels when grapes were frozen prior ot processing. No difference in °Brix was observed, but the potassium concentration of the wine increased.

Gallander and Hill [136] investigated the freezing performance of 14 strawberry varieties and selections. Fruits were sliced in half, sugared, and packed in moisture-vapor–proof containers prior to freezing. Due to the addition of sugar to the berries for freezing, direct comparisons of all chemical parameters from fresh and frozen fruits could not be made. After 6 months of storage, all frozen/thawed fruits had higher pH and lower acid content than fresh fruits.

Bushway et al. [72] studied physical, chemical, and sensory properties of five raspberry cultivars at harvest and after 1, 6, and 9 months of frozen storage. Fruits were frozen in a blast freezer at –30°C, packed in polyethylene bags, and stored at –20°C. Significant differences in pH, soluble solids, and titratable acidity were observed between varieties, but the effect of freezing on these parameters was not significant over time.

Sugar content has been shown to greatly depend upon thawing conditions in strawberries, since sucrose hydrolysis with the liberation of glucose and fructose readily occurs [137]. Fruits were thawed at 4°C until the inner part of the fruits reached the temperature of the surroundings. They were then transferred to 20°C. A substantial amount of sucrose was lost during thawing and the after-thaw period (Fig. 14). Along with the decrease in sucrose, there was an increase in fructose and glucose. No significant change in the ratio between glucose and fructose during the entire thawing process was observed, and the metabolization of glucose and fructose in frozen/thawed fruits was stopped by the freeze damage imposed on the tissue. All remaining sucrose was found in the juice exuded from the fruits. No sucrose was detectable in drained fruits following 3 hours of after-thaw storage.

As an alternative to thawing at 4°C, strawberries were thawed by microwaves, in a current of air (30°C), or in water (35°C). Subsequently, the fruits were given an after-thaw period of 5 hours at 20°C. The sucrose content of the drained fruits just after thawing was reduced to 57–67% of the amount found in the frozen fruits by all methods. Five hours at 20°C reduced the sucrose content to 10–15% of the initial amount. Thawing in warm air, microwaves, or water produced a less comprehensive sucrose inversion than thawing at 4°C. Although the time required

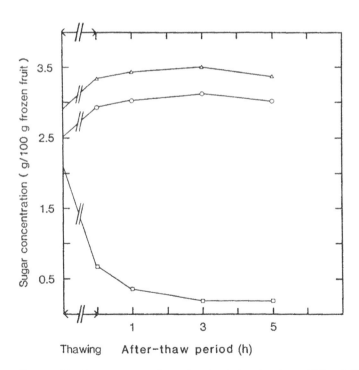

**Figure 14**   Sugar content of strawberries during thawing at 4°C and after-thaw storage at 20°C. Sucrose ▢—▢; glucose O—O; fructose △—△. (From Ref. 137.)

for these methods was shorter, invertase activity was retained during the thawing period in all methods.

Bushway et al. [72] also reported significant decreases in sucrose content and corresponding increases in glucose and fructose content of raspberries after frozen storage.

Thus, for greatest preservation of the initial free sugar composition, processors should blanch fruits or process them directly from the frozen state. If this is not possible, a fast thawing method should be chosen, and the fruits should be processed without delay. In general, inactivation of invertase should be performed prior to analysis, a precaution that is not always implemented.

## VII.  EFFECTS OF FREEZING, FROZEN STORAGE, AND THAWING ON MICROBIOLOGICAL QUALITY AND SAFETY

Most ripe fruit tissues are soft and susceptible to invasion by yeasts, mold, and bacteria [1,6]. However, Marth [138], after a review of the literature on microorganisms in fruits and fruit juices, concluded that food-poisoning microorganisms were not a problem in frozen fruit and fruit juices. *Clostridium botulinum* will not multiply in acid media and does not form toxin at pH levels below 4.5. The low pH of fruits favors growth of acid-tolerant bacteria like lactobacilli if these bacteria are not overgrown by molds.

Death of microbial cells upon freezing has been attributed to the formation of intracellular ice [138]. Slow freezing with the formation of large ice crystals is beneficial from a microbiological standpoint in that it kills more microorganisms. Freezing at −20°C is less detrimental to yeasts and molds than freezing at −10°C [139]. Also, yeast cells were less affected when frozen at −80°C than when frozen at −20°C [140].

Thawing also influences the survival of microorganisms [138]. During slow thawing, recrystallization can occur, causing the period during which the microbial cells are exposed to detrimental concentrations of solutes at high subfreezing temperatures to be substantial. Thus, the conditions best suited for microbial cell damage also produce the most damage to fruit tissues. Thereby, the texture of the thawed fruit is substantially adversely affected. After thawing, fruits are even more susceptible to microbial invasion than the more firm, fresh fruit. Thawed material must, therefore, be processed immediately to prevent the surviving microorganisms from multiplying [139].

Since freezing does not kill all microorganisms, care must be taken during harvesting and handling of the fresh fruits to prevent contamination and growth of bacteria, yeasts, and molds prior to freezing. Rapid cooling after harvesting is important to restrict growth of microoganisms. The primary source of microbial contamination of fruit juices is use of poor-quality raw material and improper sanitary conditions during harvesting [141]. Inadequate processing and handling also contribute to microbial spoilage of the final product.

Dennis and Harris [142] reviewed fungal spoilage of canned and sulfited fruit and reported damage from fungal enzymes formed prior to processing. When the enzymes were more stable than the microorganisms, detrimental effects occurred, even if the microorganisms were destroyed. Pectolytic enzymes are known to cause tissue breakdown in heated products. It is also likely that other microbial enzymes will impose their effects in frozen/thawed fruit tissue.

Limited information on the effect of freezing, frozen storage, and thawing on microoogranisms and safety has been generated during recent years. Gorgatti Netto [28] reported a reduction in microbial count on sliced mangoes in syrup from $10^5$–$10^6$ to $10^2$–$10^4$ microoorganisms/g after 15 days of storage at $-20°C$. Beuchat and Nail [143] reported a dramatic reduction in yeast and mold population after storing fruit purees at $-18°C$ for 6 months. Total viable fungal populations were reduced 1000-fold or more as a result of the cold stress.

Venning et al. [33] examined the effect of freezing and storage on microbiology in kiwi pulp. The tests comprised aerobic plate counts (APC), yeasts, and molds, in addition to fecal coliforms and specific determination of *Staphylococcus aureus*. The freezing process and prolonged frozen storage caused reductions in the viability of all three groups of microorganisms examined. A more extensive reduction in aerobic plate counts were obtained during storage of deaerated pulp stored in foil laminate at $-35°C$ when compared with the more industrial condition of nondeaerated pulp in polyethylene films stored at $-18°C$. No fecal coliforms or coagulase-positive *Staphylococcus aureus* were detected in the kiwi pulp.

## VIII. STABILITY OF FROZEN PRODUCTS

In the 1960s and early 1970s, great effort was made to systematize the knowledge on stability of frozen food products. Time-temperature tolerances (TTT) were established for various products with the purpose of providing a tool for evaluating frozen food quality [144]. To evaluate storage ability, $Q_{10}$ quotients were calculated describing the ratio between the reaction rate of a chemical process at a temperature $(t + 10)$ °C and the rate at temperature $t$ °C. Thus, the $Q_{10}$ temperature quotients indicated how many times the reaction rate increased by a temperature increase of 10°C. For most fruits studied, the $Q_{10}$ values for color were equal to or lower than those for flavor, indicating flavor is influenced by temperature at least to the same extent as color. Munoz-Delgado [24] concluded that the TTT relationships are subjected to wide variability because of variations in type of product, processing, and packaging. In recent years, the effort to establish TTT values for frozen fruit and fruit products has received less attention.

The positive effects of high freezing rates on ascorbic acid retention are also apparent in recent literature. Cryogenic freezing of fruits in liquid nitrogen or carbon dioxide promote higher ascorbic acid retention than slower freezing. The advantage of the high freezing rates can also be traced during subsequent frozen storage.

Storage temperature clearly influences the storage performance of frozen fruits. Odor, flavor, ascorbic acid, and color are all preserved better at storage temperatures in the range of −18 to −25°C, compared with storage at temperatures of about −10°C. Differences in the performance of fruits stored at those temperatures are detectable after 3–4 months of frozen storage. Only minor differences in fruit quality are detectable between the storage temperatures of −18 or −25°C for periods of 12 months or less.

Detrimental reactions in fruits during frozen storage depend to a large extent upon the presence of oxygen. The development of packaging materials with low oxygen permeability, therefore, is of great benefit. Foil-laminated films have especially proven suitable for minimizing quality deterioration of fruits. However, low-permeability packaging material is more important when storage temperatures are not optimal. Polyethylene films may be sufficient at temperatures of about −18°C for storage periods not exceeding 12 months, which most likely represent commercial conditions in the fruit-processing industry today.

Great differences in freezing performance exist between fruit varieties or cultivars based upon both chemically and sensory assessed quality. Differences in cell wall structure, enzyme activity, amounts of pigments, ascorbic acid, and other components are factors contributing to these differences. However, the entire mechanism has not been elucidated. The freezing potential of fruit varieties should be determined in practical trials, with evaluation of the final products after freezing, storage, and thawing of the fruit raw material.

Ripe fruits are generally preferable to unripe fruits as raw material for freezing. Due to the high contents of valuable pigments and vitamin C, relatively higher losses occur during freezing. Nevertheless, final contents are higher than in fruits with lower initial contents. However, for texture, less ripe fruits are firmer, which may be beneficial during freezing, especially if the fruits are subjected to prefreezing processing, like depitting or destoning.

Theoretically, fluctuations in temperature during frozen storage will produce growth of larger ice crystals. As a result, more extensive damage will be imposed on susceptible fruit tissues. Trade in frozen fruits often involves long-distance transportion producing temperature fluctuations sufficient to cause detrimental effects on the final product. Unfortunately, most studies are performed under stable laboratory conditions. Consequently, little emphasis has been placed upon effects of temperature fluctuations during frozen storage. Also, thawing conditions applicable to industrial processing should be considered. Rapid thawing best maintains fruit quality but may be difficult to achieve when large quantities of fruits are handled.

Early studies demonstrated that measurable changes did not occur in fruit products stored for periods of 5 years or longer if the products were held at −29°C or lower [24]. Thus, fruits may be stored frozen for prolonged periods when conditions are optimized. With the current demand for high quality and the development of ever-more-sensitive analytical methods, more detailed knowledge of frozen fruit quality is required. In the future, major emphasis should be devoted not only to minimizing quality deterioration, but also to entirely preserving fresh fruit flavor, appearance, and consistency during freezing, frozen storage, and thawing.

## IX. SUMMARY

Factors influencing the performance of fruits and fruit products during freezing, frozen storage, and thawing have been reviewed. The freezing process causes changes in appearance, palatability, nutrition, and microbiology, which are assessed by sensory and instrumental analysis.

The freezing performance of various types of fruits depend upon variety, maturity level, and the prefreezing treatments. Rapid freezing, low storage temperature, as well as exclusion of oxygen by proper packaging, are beneficial for preservation of fruit quality. Rapid thawing with subsequent direct processing should be included in the process to best maintain the quality of frozen fruits.

## REFERENCES

1. D. K. Tressler and C. F. Evers, Microbiology of frozen foods, *The Freezing Preservation of Foods*, Vol. 1, 3rd ed., AVI Publishing Company Inc., Westport, CT, 1957, p. 1067.
2. O. Fennema and W. D. Powrie, Fundamentals of low-temperature food preservation, *Advances in Food Research*, Vol. 13 (C. O. Chichester, E. M. Mrak, and G. F. Stewart, eds.), Academic Press, New York, 1964, p. 219.
3. M. A. Joslyn, The freezing of fruits and vegetables, *Cryobiology* (H. T. Meryman, ed.), Academic Press Inc., London, 1966, p. 565.
4. J. Gutschmidt, Principles of freezing and low temperature storage, with particular reference to fruit and vegetables, *Low Temperature Biology of Foodstuff, Recent Advances on Food Science*, Vol. 4 (J. Hawthorn and E. J. Rolfe, eds.), Pergamon Press, London, 1968, p. 299.
5. T. N. Morris, The freezing of fruits and vegetables, historic and general, *Low Temperature Biology of Foodstuff, Recent Advances in Food Science*, Vol. 4 (J. Hawthorn and E. J. Rolfe, eds.), Pergamon Press, London, 1968, p. 285.
6. J. D. Ponting, B. Feinberg, and F. P. Boyle, Fruits: Characteristics and the stability of the frozen products, *The Freezing Preservation of Foods*, Vol. 2, 4th ed. (D. K. Tressler, W. B. van Arsdel, and M. J. Copley, eds.), AVI, Westport, CT, 1968, p. 107.
7. D. K. Tressler, Prepared and precooked fruit products, *The Freezing Preservation of*

*Foods*, Vol. 4, 4th ed. (D. K. Tressler, W. B. van Arsdel, and M. J. Copley, eds.), AVI Westport, CT, 1968, p. 182.

8.  K. Herrmann, Pflanzliche Lebensmitteln, *Tiefgefrorene Lebensmitteln* (K. Herrmann, ed.), P. Parey, Berlin, 1970, p. 249.

9.  W. D. Powrie, Characteristics of food phytopsystems and their behaviour during freeze-preservation, *Low-Temperature Preservation of Foods and Living Matter*, Vol. 3 (O. R. Fennema, W. D. Powrie, and E. H. Marth, eds.), Marcel Dekker, New York, 1973, p. 352.

10. M. S. Brown, Frozen fruits and vegetables: Their chemistry, physics, and cryobiology, *Advances in Food Research*, Vol. 25 (C. O. Chichester, E. M. Mrak, and G. F. Stewart, eds.), Academic Press, New York, 1979, p. 181.

11. M. Kalbassi, Changes during processing of frozen vegetables and fruit, *Quality in Stored and Processed Vegetables and Fruit* (P. W. Goodenough and R. K. Atkin, eds.), Academic Press, London, 1981, p. 373.

12. M. Jul, The PPP-factors, *The Quality of Frozen Foods* (M. Jul, ed.), Academic Press, London, 1984, p. 112.

13. D. Arthey, Freezing of vegetables and fruits, *Frozen Food Technology* (C. P. Mallett, ed.), Blackie Academic and Professional, Glasgow, 1993, p. 237.

14. K. K. Singh, Refrigerated storage and freezing of mango fruit in India, *Punjab Hort. J.* 17(1/2):6 (1977).

15. M. A. Marín, P. Cano, and C. Fuster, Freezing preservation of four Spanish mango cultivars (*Mangifera indica* L.): Chemical and biochemical aspects, *Lebensm. Unters. Forsch.* 194(6):566 (1992).

16. K. Polyak-Feher and A. Szabo-Kismarton, Weeping of frozen fruit, *Huetoeipar* 30(1):22 (1984).

17. G. Urbànyi and K. Horti, Colour and carotenoid content of quick-frozen tomato cubes during frozen storage, *Acta Aliment.* 18(3):247 (1989).

18. P. Cano, M. A. Marín, and C. Fuster, Freezing of banana slices. Influence of maturity level and thermal treatment prior to freezing, *J. Food Sci.* 55(4):1070 (1990).

19. W. Plocharski, Strawberries—quality of fruits, their storage life and suitability for processing. Part VI. Quality of fruit frozen immediately after picking or frozen after cold storage under controlled atmosphere conditions, *Fruit Sci. Rep.* 16(3):127 (1989).

20. W. E. L. Spiess, Changes in ingredients durig production and storage of deep-frozen food—a review of the pertinent literature *ZFL* 8:625 (1984).

21. F. P. Boyle and E. R. Wolford, The preparation for freezing and freezing of fruits, *The Freezing Preservation of Foods*, Vol. 3, 4th ed. (D. K. Tressler, W. B. van Arsdel, and M. J. Copley, eds.), AVI, Westport, CT, 1968, p. 70.

22. G. L. Robertson, Changes in the chlorophyll and pheophytin concentrations of kiwifruit during processing and storage, *Food Chem.* 17:25 (1985).

23. P. H. Gradziel, The effects of frozen storage time, temperature and packing on the quality of the tomato cultivars Pink-red and Nova, *Diss. Abstr. Int. B* 48(9):2609 order no DA8725826 (1988).

24. J. A. Munoz-Delgado, Effects of freezing, storage and distribution on quality and nutritive attributes of foods, in particular of fruit and vegetables, *Food Quality and Nutrition* (W. K. Downey, ed), Applied Science Publishers Ltd, London, 1978, p. 353.

25. M. Edwards and M. Hall, Freezing for quality, *Food Manuf.* 63(3):41, 43, 45 (1988).

26. H. C. Passam, D. S. Maharaj, and S. Passam, A note of freezing as a method of storage of breadfruit slices, *Trop. Sci.* 23(1):67 (1981).
27. M. S. Ramamurthy and D. R. Bongirwar, Effect and freezing methods on the quality of freeze dried Alphonso mangoes, *J. Food Sci. Technol. India* 16(6):234 (1979).
28. A. Gorgatti Netto, E. W. Bleinroth, and L. C. Lazzarini, Quality evaluation of frozen sliced mangoes in syrup, *Proceedings of the XIII International Congress Refrigeration*, Vol. 3, AVI, Westport, CT, 1973, pp. 265–270.
29. M. P. Cano, C. Fuster, and M. A. Marín, Freezing preservation of 4 Spanish kiwi fruit cultivars (Actinidia-chinensis, planch)—chemical aspects, *Lebens. Unters. Forsch.* 196(2):142 (1993).
30. J. R. Morris, G. L. Main, and W. A. Sistrunk, Relationship of treatment of fresh strawberries to the quality of frozen fruit and preserves, *J. Food Qual.* 14:467 (1991).
31. H. R. Serratos, C. H. Mannheim, and N. Passy, Sulphur dioxide and carbon monoxide gas treatment of apples for enzyme inhibition prior to freezing, *J. Food Proc. Pres.* 7(2):93 (1983).
32. Y. H. Lee, J. S. Chen, and C. C. Chen, Quality improvement of passion fruit juice, *Food Industry Research and Development Institute, Taiwan*, Research Report E-95, 1983.
33. J. A. Venning, D. J. W. Burns, K. M. Hoskin, T. Nguyen, and M. G. H. Stec, Factors influencing the stability of frozen kiwifruit pulp, *J. Food Sci.* 54(2):396–400, 404 (1989).
34. G. C, Yen, H. T. Lin, and P. Yang, Changes in volatile flavor components of guava puree during processing and frozen storage. *J. Food Sci.* 57(3):679–681, 685 (1992).
35. J. Kulisiewicz and J. Kolasa, Sugar as a factor enhancing the quality of frozen strawberries, as exemplified by the freezing plant of the agricultural centre of the "Samopomoc Chlopska" cooperative in Sochaczew, *Przemysl Spozywezy* 28(10):433 (1974).
36. T. Fraczak and M. Zalewska-Korona, Effects of different freezing procedures applied to berries and of frozen products' storage conditions on the vitamin C content, *Przemysl Fermentacyjny i Owocowo-Warzywny* 34(10):19 (1990).
37. R. E. Wrolstad, G. Skrede, P. Lea, and G. Enersen, Influence of sugar on anthocyanin pigment stability in frozen strawberries, *J. Food Sci.* 55(4):1064–1065, 1072 (1990).
38. J. Kozup, W. A. Sistrunk, and J. R. Morris, Chemical stabilization of color in machine harvested strawberries, *Arkansas Farm Res.* 29(4):5 (1980).
39. C. Fuster and G. Prestamo, Quality of frozen strawberries during refrigerated storage, *Alimentaria.* 114(47):49 (1983).
40. J. Abufom and J. A. Olaeta, Frozen chirimoya (*Annona cherimola, Mill*) I. Pulp, *Alimentos* 11(2):27 (1986).
41. G. Pinnavaia, M. Dalla Rosa, and C. R. Lerici, Dehydrofreezing of fruit using direct osmosis as concentration process, *Acta Aliment. Pol.* 14(1):51 (1988).
42. A. Monzimi and E. Maltini, New trends in processing of horticultural products. Dehydro-freezing and osmotic treatment, *Ind. Conserve* 61(3):265 (1986).
43. D. Torreggiani, E. Forni, G. Crivelli, G. Bertolo, and A. Maestrelli, Researches on dehydrofreezing of fruit, Part I. Influence of dehydration levels on the products quality, *Proceedings XVII Int. Cong. of Refrigeration*, Vienna, Austria, 1987, Vol C.
44. D. Torreggiani, Osmotic dehydration in fruit and vegetable processing, *Food Res. Int.* 26:59 (1993).

45. M. Tomasicchio, R. Andreotti, and A. de Giorgi, Osmotic dehydration of fruits. II. Pineapples, strawberries and plums, *Ind. Conserve* 61(2):108 (1986).
46. J. Philippon, Packaging and maintaining the quality of frozen fruits and vegetables, Literature review, *Rev. Gen. Froid* 71(3):127 (1981).
47. O. W. Bisset and R. E. Berry, Ascorbic acid retention in orange juice as related to container type, *J. Food Sci.* 40(1):178 (1975).
48. Z. Blonski, A. Jedrzejczak, and L, Wasowicz, Natural losses of frozen fruit and vegetables during storage. II. Effect of drying on quality changes, *Chlodnictwo* 22(2):15 (1987).
49. F. J. Francis and F. M. Clydesdale, *Food Colorimetry: Theory and Applications*, AVI, Westport, CT. 1977.
50. D. K. Salunkhe, H. R. Bolin, and N. R. Reddy, Sensory and objective quality evaluation, *Storage, Processing and Nutritional Quality of Fruits and Vegetables*, Vol. 1, 2nd ed. (D. K. Salunkhe, H. R. Bolin, and N. R. Reddy, eds.), CRC Press, Boca Raton, FL, 1991, p. 181.
51. R. S. Hunter and R. W. Harold, *The Measurement of Appearance*, 2nd ed., John Wiley & Sons, New York, 1987.
52. R. E. Wrolstad, Color and pigment analysis in fruit products, Bulletin No. 624, *Oregon Agricultural Experiment Station*, Corvallis, OR, 1976.
53. A. Polesello and C. Bonzini, Observations on pigments of sweet cherries and on pigment stability during frozen storage. I. Anthocyanin composition, *Confructa* 22(5/6):170 (1977).
54. G. M. Sapers, A. M. Burgher, and J. G. Phillips, Composition and color of fruit and juice of thornless blackberry cultivars, *J. Am. Soc. Hort. Sci.* 110(2):243 (1985).
55. J. Gross, *Pigments in Vegetables, Chlorophylls and Carotenoids*, van Nostrand Reinhold, New York, 1991.
56. K. V. R. Ramana, H. S. Ramaswamy, B. Aravinda Prasad, M. V. Patwardhan, and S. Ranganna, Freezing preservation of Totapuri Mango pulp, *J. Food Sci. Technol.* 21:282 (1984).
57. S. D. Holdsworth, Fruit preservation developments reviewed, *Food Manuf.* 45(8):74 (1970).
58. P. Markakis, Stability of anthocyanins in foods, *Anthocyanins as Food Colors*, (P. Markakis, ed), Academic Press Inc., New York, 1982, p. 163.
59. M. P. Cano and M. A. Marín, Pigment composition and color of frozen and canned kiwi fruit slices, *J. Agric. Food Chem.* 40(11):2141 (1992).
60. J. Gross, Pigments in fruits, *Food Science and Technology, A Series of Monographs* (B. S. Schweigert, ed.), Academic Press, Inc., London, 1987.
61. G. Skrede, Evaluation of color quality in black currant fruits grown for industrial juice and syrup production, *Norw. J. Agric. Sci.* 1(1):67 (1987).
62. D. B. Carrick, *Some Cold Storage and Freezing Studies on the Fruit of the Vinifera*, Cornell University Agricultural Experiment Station Memoir 131, Itacha, NY, 1930.
63. F. Pizzocaro, C. Crotti, and A. Polesello, Observations in the pigmentation of cherries and the stability of this pigmentation during frozen storage: Influence of variety, *Bull. Ins. Int. Froid* 59(4):1174 (1979).
64. G. Urbànyi, Changes in colour and anthocyanin content of quick-frozen sour cherries during freeze-drying and subsequent storage, *Huetoeipar* 33(3):118 (1987).

65. G. Urbànyi and K. Horti, Changes of surface color of the fruit and of the anthocyanin content of sour cherries during frozen storage, *Acta Aliment.* 21(3–4):307 (1992).
66. G. M. Sapers, A. M. Burgher, J. G. Phillips, and S. B. Jones, Effects of freezing, thawing, and cooking on the appearance of highbush blueberries, *J. Am Soc. Hort. Sci.* 109(1):112 (1984).
67. C. S. T. Yang and P. P. A. Yang, Effect of pH, certain chemicals and holding time-temperature on the color of lowbush blueberry puree, *J. Food Sci.* 52(2):346–347, 352 (1987).
68. W. Lenartowicz, J. Zbroszczyk, and W. Plocharski, The quality of highbush blueberry fruit. Part I. Fresh fruit quality of six highbush blueberry cultivars and their suitability for freezing, *Fruit Sci. Rep.* XVII(2):77 (1990).
69. D. L. Jennings and E. Carmichael, Colour changes in frozen blackberries, *Hort. Res.* 19(1):15 (1979).
70. A. Polesello, G. Crivelli, S. Geat, E. Senesi, and P. E. Zerbini, Research on the quick freezing of berry fruit. II. Colour changes in frozen blackberries, *Crop. Res. (Hort. Res.)* 26:1 (1986).
71. G. M. Sapers, S. B. Jones, and G. T. Maher, Factors affecting the recovery of juice and anthocyanin from cranberries, *J. Am. Soc. Hort. Sci.* 108(2):246 (1983).
72. A. A. Bushway, R. J. Bushway, R. H. True, T. M. Work, D. Bergeron, D. T. Handley, and L. B. Perkins, Comparison of the physical, chemical and sensory characteristics of five raspberry cultivars evaluated fresh and frozen, *Fruit Var. J.* 46(4):229 (1992).
73. L. E. Cancel and E. R. de Hernandez, Effect of blanching and freezing on the texture and color of candied citron, *J. Agric. Univ. Puerto Rico* 63(3):309 (1979).
74. P. K. Mukerjee and R. B. Srivastava. Control of browning in frozen mango (*Mangifera indica* L.) cv. Deshehari and improvement of its quality, *Sci. Culture* 45(4):166 (1979).
75. A. van der Heijden, H. G. Peer, L. B. P. Brussel, and J. G. Kosmeijer, Colour measurement of orange, peach and apricot fruit drinks and of their raw materials in relation to pasteurization and storage, *Confructa* 24(5/6):195 (1979).
76. A. Askar, S. K. El-Samahy, M. M. Abd El-Baki, S. S. Ibrahim, and M. G. Abd El-Fadeel. Production of lime juice concentrates using the serum-pulp method, *Alimenta* 20(5):121 (1981).
77. E. R. Suresh, S. Ethiraj, and H. Onkarayya. A note on the effect of freezing grape bunches on the composition of musts and wines, *J. Food Sci. Technol. India* 18(3):119 (1981).
78. J. Philippon, M. A. Ronet-Mayer, D. Gallet, and A. Herson, Pre-freezing of whole peaches for canning. Effects of various stages of a new processing technology on the quality of thawed fruits and the canned product, *Refrig. Sci. Technol.* 4:302. (1982).
79. F. A. El-Ashwah, H. K. El-Manawaty, H. N. Habashy, and M. A. El-Shiaty, Effect of storage on fruit juices: Frozen lime juice, *Agric. Res. Rev.* 52(9):79 (1974).
80. T. M. Hernández and M. I. Villegas, Effects of frozen storage on the quality of some tropical fruit pulps. Preliminary study, *Tecnol. Quim.* 7(2):33 (1986).
81. R. Stanley, *The Suitability of Soft Fruit Cultivars for Processing: 1983 Trials*, Technical Memorandum, Campden Food and Drink Res. Ass., No. 368, 1984, p. 41.
82. R. Stanley, *The Suitability of Soft Fruit Cultivars for Processing: 1984 Trials*, Technical Memorandum, Campden Food and Drink Res. Ass., No. 408, 1985, p. 49.

83. R. Stanley, *The suitability of Soft Fruit Cultivars for Processing: 1985 Trials*, Technical Memorandum, Campden Food and Drink Res. Ass., No. 390, 1986, p. 34.

84. R. Stanley, *The suitability of Soft Fruit Varieties for Processing: 1987 Trials*, Technical Memorandum, Campden Food and Drink Res. Ass., No. 494, 1988, p. 32.

85. R. Stanley, *The Suitability of Soft Fruit Varieties for Processing: 1988 Trials*, Technical Memorandum, Campden Food and Drink Res. Ass., No. 526, 1989, p. 20.

86. F. S. Burnette, Peroxidase and its relationship to food flavor and quality: A review, *J. Food Sci.* 42(1):1 (1977).

87. A. A. Calderón, J. M. Zapata, R. Muñoz, and A. Ros Barceló, Localization of peroxidase in grapes using nitrocellulose blotting of freezing/thawing fruits, *Hort. Sci.* 28(1):38 (1993).

88. A. A. Calderón, E. Garcia-Florenciano, M. A. Pedreño, R. Muñoz, and A. Ros Barceló, The vacuolar localization of grapevine peroxidase isoenzymes capable of oxidizing 4-hydroxystilbene, *Z. Natursforsch.* 47c:215 (1992).

89. T. Maotani, Y. Hase, and R. Matsumoto, Studies on bitterness in citrus fruits. I. Method for determination of naringin in natsudaidai (*Citrus natsudaidai Hayata*) fruits as related to palate, *J. Jpn. Soc. Hort. Sci.* 47(4):546 (1978).

90. R. Matsumoto and N. Okudai, Changes in naringin content of natsumikan fruit (*Citrus natsudaidai Hayata*, cv Kawano-natsudaidai) after freezing, *J. Jpn. Soc. Hort. Sci.* 52(1):1 (1983).

91. P. Schreier, Quantitative composition of volatile constituents in cultivated strawberries, *Fragaria ananassa* cv. Senga Sengana, Senga Litessa and Senga Gourmella, *J. Sci. Food Agric.* 41:487 (1980).

92. B. Banev, Effect of cold storage on the aroma constituents of strawberries, *Konservna Promishlenost* 3:19 (1982).

93. Y. Ueda and T. Iwata, Undersirable odour of frozen strawberries, *J. Jpn. Soc. Hort. Sci.* 51(2):219 (1982).

94. T. Hirvi, Mass fragmentographic and sensory analyses in the evaluation of the aroma of some strawberry varieties, *Lebensm. Wiss. Technol.* 16(3):157 (1983).

95. C. Douillard and E. Guichard, The aroma of strawberry (*Fragaria ananassa*): Characterisation of some cultivars and influence of freezing, *J. Sci. Food. Agric.* 50(4):517 (1990).

96. E. Guichard, Identification of the flavoured volatile components of the raspberry cultivar Lloyd George, *Sci. Aliment.* 2(2):173 (1982).

97. P. X. Etievant, E. A. Guichard, and S. N. Issanchou, The flavour components of Mirabelle plums. Examination of the aroma constituents of the fresh fruits: Variation of head-space composition induced by deep-freezing and thawing, *Sci. Aliment.* 6(3):417 (1986).

98. R. G. Berger, Modern storage technologies for the postharvest flavour preservation of fruits, *Ernährung/Nutrition* 16(9):487 (1992).

99. M. S. Brown, Texture of frozen fruits and vegetables, *J. Text. Stud.* 7:391 (1977).

100. J. G. Woodroof, *Microscopic studies of frozen fruits and vegetables*, Georgia Agricultural Experiment Station Bulletin, 1938, p. 201.

101. A. M. Hoser, K. Zelakiewicz, and J. Chocianowisk, Technique of preparing microscopies preparations from frozen fruit, *Przemysl Spozywczy* 33(1):21 (1979).

102. S. L. Duce, T. A. Carpenter, and L. D. Hall, Nuclear-magnetic-resonance imaging of fresh and frozen courgettes, *J. Food Eng.* 16(3):165 (1992).

103. R. M. Reeve, Relationships of histological structure to texture of fresh and processed fruits and vegetables. *J. Text. Stud.* 1:247 (1970).

104. A. S. Szczesniak and B. J. Smith, Observations on strawberry texture. A three-pronged approach. *J. Text. Stud.* 1:65 (1969).

105. E. R. Wolford, R. Jackson, and F. P. Boyle, Quality evaluation of stone fruits and berries frozen in liquid nitrogen and in Freezant 12, *Chem. Eng. Prog. Symp. Series* 67(108):131 (1971).

106. G. Crivelli and P. Rosati, Research on quick freezing of strawberry. VII. Influence of pre-freezing treatment on berry quality, *Ann. Inst. Sper. Valorizzazione Tecnol. Prod. Agric.* 4:73 (1973).

107. D. S. Reid, J. M. Carr, T. Sajjaanantakul, and J. M. Labavitch. Effects of freezing and frozen storage on the characteristics of pectin extracted from cell walls, *Chemistry and Function of Pectins* (M. L. Fishman, J. J. Jeh, eds.), ACS Symposium Series 310, Washington, DC, 1986, p. 200.

108. G. M. Sapers, A. M. Burgher, S. B. Jones, and J. G. Phillips, Factors affecting drip loss from thawing thornless blackberries, *J. Am. Soc. Hort. Sci.* 112(1):104 (1987).

109. J. Kuprianoff, Einige aktuelle Probleme des Gefrierens von Lebensmitteln, *Ernährungswirtschaft* 15:326 (1968).

110. P. A. Phan and J. Mimault. Effects of freezing and thawing on fruit. Evaluation of some texture parameters and exudate. Relation to fruit quality, *Int. J. Refrig.* 3(5):255 (1980).

111. J. Martí and J. M. Aguilera. Effects of freezing rate on the mechanical characteristics and microstructure of blueberries and wild blackberries, *Rev. Agroquim. Tecnol. Aliment.* 31(4):493 (1991).

112. E. R. Wolford, D. W. Ingalsbe, and F. P. Boyle, Freezing of peaches and sweet cherries in liquid nitrogen and in dichlorodifluoromethane and behaviour upon thawing of strawberries and raspberries, *Proceedings of the 12th International Congress of Refrigeration*, Madrid, 1967, pp. 459–469.

113. Y. Chuma, S. Uchida, and K. H. H. Shemsanga, Cryogenic properties of fruits and vegetables, *Trans ASAE* 26(4):1258 (1983).

114. H. Watanabe, Y, Hagura, M. Ishikawa, and Y. Sakai. Cryogenic separation of citrus fruit into individual juice sacs, *J. Food Proc. Eng.* 9(3):221 (1987).

115. G. Crivelli and P. Rosati, Research on quick freezing of strawberry. VIII. Influence of pre-freezing treatment on berry quality, *Ann. Inst. Sper. Valorizzazione Tecnol. Prod. Agric.* 5:93 (1974).

116. W. Lenartowicz, J. Zbroszczyk, and W. Plocharski, Processing quality of currant fruit. I. Basic chemical composition of the fruit of seven varieties and four hybrids of black currant and the quality of stewed and frozen products made from them, *Prace Inst. Sadownictwa i Sadownictwa i Kwiaciarstwa w Skierniewicach. Seria A, Prace Doswiadczalne z Zakresu Sadownictwa* 29:195 (1990).

117. W. Lenartowicz, J. Zbroszczyk, and D. Chlebowska, Processing quality of currant fruit. II. Basic chemical composition of the fruit of red currant and the quality of frozen and stewed fruit obtained from them, *Prace Inst. Sadownictwa i Sadownictwa i Kwiaciarstwa w Skierniewicach. Seria A, Prace Doswiadczalne z Zakresu Sadownictwa* 29:203 (1990).

118. W. Lenartowicz, J. Zbroszczyk, and W. Plocharski, Processing quality of currant fruit. III. Characteristics of the fruit of new black currant varieties and clones and the quality of stewed and frozen fruit obtained from them, *Prace Inst. Sadownictwa i Sadownictwa i Kwiaciarstwa w Skierniewicach. Seria A, Prace Doswiadczalne z Zakresu Sadownictwa* 29:213 (1990).

119. W. Lenartowicz, W. Plocharski, J. Zbroszczyk, and J. Piotrowski, The effect of freezing method on the quality of frozen fruit, *Bull. Inst. Int. Froid* 59(4):1170 (1980).

120. W. Plockarski, J. Banaszcyk, and D. Chlebowska, Quality characteristics of a few black current cultivars and clones and the quality of obtained compotes and frozen fruit, *Fruit Sci. Rep. (Skierniewice)* 19(3):125 (1992).

121. P. Rosati and G. Crivelli, Researches on quick freezing of strawberry. IX. Suitability of varieties, *Ann. Inst. Sper. Valorizzazione Tecnol. Prod. Agric.* 6:67 (1975).

122. C. T. Young and J. S. L. How, Composition and nutritive value of raw and processed fruits, *Commercial Fruits Processing*, 2nd ed. (J. G. Woodroof and B. S. Luh, eds.), AVI Westport, CT, 1986, p. 531.

123. R. Ulrich and N. Delaporte, Ascorbic acid in fruits held in cold storage, either in air or in a controlled atmosphere, *Ann. Nutr. Aliment.* 24(3):B287 (1970).

124. K. Herrmann, Quality maintenance in vegetables and fruit by processing and freezing, *Qual. Plant. Mater. Veg.* 21(1/2):1 (1971).

125. J. Taylor, What a lifetime, *Food Flav. Ingred. Packag. Proc.* 6(11):43 (1984).

126. A. Bognar, Effects of domestic preservation by freezing or thermal sterilization (bottling) on the quality of fruits and vegetables, *Verbraucherdienst* 35(7):143 (1990).

127. I. Malinowska, C. Myslinska, and E. Urbanska, Stability in the quality of fruit frozen by means of the fluidizing method, *Int. Bull. Inf. Refrig.*:417–426 (1966).

128. M. M. Abd El-Baki, S. K. El-Samahi, and A. Askar, Concentration of fruit juice. I. Concentration of lime juice, *Flüssiges Obst* 47(6):234 (1980).

129. L. R. G. Valadon and R. S. Mummery, Effect of canning and storage on carotenoids (vitamin A activity) and vitamin C in Spanish and Turkish oranges., *J. Sci. Food Agric.* 32(7):737 (1981).

130. M. Buret, R. Gormley, and P. Roucoux, Analysis of tomato fruit: Effect of frozen storage on compositional values—an inter-laboratory study, *J. Sci. Food Agric.* 34(7):755 (1983).

131. H. T. Chan Jr. and G. Cavaletto, Effects of deaeration and storage temperature on quality of aseptically packaged guava puree, *J. Food Sci.* 51(1):165 (1986).

132. S. K. Roy and R. N. Singh, Studies on utilization of bael fruit (*Aegle Marmelos Correa*) for procesing: IV. Storage studies of bael fruit products, *Indian Food Pack.* 33(6):3 (1979).

133. S. M. Minasyan and G. A. Astabatsyan, Retention of B-group vitamin in vegetables and melons stored in the frozen state, *Konservnaya i Ovoshchesushil'naya Promyshlennoost'* 7:10 (1978).

134. J. K. Ross, C. English, and M. S. Perlmutter, Dietary fiber constituents of selected fruits and vegetables, *J. Am. Diet. Assoc.* 85(9):1111 (1985).

135. S. E. Spayd, C. W. Nagel, L. D. Hayrynen, and M. Ahmedullah, Effect of freezing fruit on the composition of musts and wines, *Am. J. Enol. Vitic.* 38(3):243 (1987).

136. J. F. Gallander and R. G. Hill Jr., Effect of variety and harvest date on the quality of frozen strawberries, *Ohio Agric. Res. Dev. Center Res. Cir.* 271:43 (1982).

137. G. Skrede, Changes in sucrose, fructose and glucose content of frozen strawberries with thawing, *J. Food Sci.* 48(4):1094 (1983).

138. E. H. Marth, Behaviour of food microorganisms during freeze-preservation, *Low-Temperature Preservation of Foods and Living Matter* (O. R. Fennema, W. D. Powrie, and W. H. Marth, eds.), Marcel Dekker, New York, 1973, p. 386.

139. E. J. Hsu and L. R. Beuchat, Factors affecting microflora in processed fruits, *Commercial Fruits Processing*, 2nd ed. (J. G. Woodroof and B. S. Luh, eds.), AVI, Westport, CT, 1986, p. 129.

140. J. V. Gil, E. Campos, M. Jimenez, T. Huerta, and J. J. Mateo, Effect of freezing on the yeast-like population of grape must, *Rev. Esp. Cienc. Tecnol. Aliment.* 32(2):2113 (1992).

141. S. S. Deshpande, M. Cheryan, S. K. Sathe, and D. K. Salunkhe, Freeze concentration of fruit juices, *CRC Crit. Rev. Food Sci. Nutr.* 20(3):173 (1984).

142. C. Dennis and J. E. Harris, Effect of spoilage fungi on the quality of canned and sulphited fruit, *Quality in Stored and Processed Vegetables and Fruit* (P. W. Goodenough and R. K. Atkin, eds.), Academic Press, London, 1981, p. 275.

143. L. R. Beuchat and B. V. Nail, Evaluation of media for enumerating yeasts and molds in fresh and frozen fruit purees, *J. Food Prot.* 48(4):312 (1985).

144. D. G. Guadagni, Cold storage life of frozen fruits and vegetables as a function of time and temperature, *Low Temperature Biology of Foodstuff, Recent Advances in Food Science*, Vol. 4 (J. Hawthorn and E. J. Rolfe, eds.), Pergamon Press, London, 1968, p. 399.

# 7
# Vegetables

**M. Pilar Cano**

*Instituto del Frío, (C.S.I.C.)*
*Madrid, Spain*

## I. INTRODUCTION

Market studies of the consumption of frozen foods [1] indicate that frozen vegetables and potatoes form a very significant proportion of the world frozen food market. In the past few years, the consumption of these products has exceeded all other frozen food categories (excluding ice cream) in European countries (Denmark, Finland, France, Germany, Italy, Norway, Sweden, Switzerland, the United Kingdom, and Austria) and the United States. In the United Kingdom, the most consumed frozen vegetables are peas, green beans, and Brussels sprouts; in the United States they are corn, peas, and green beans; in Japan, they are corn, pumpkin, and spinach; and in continental Europe, they are spinach and peas [2].

Freezing is one of the most important methods for retaining quality during long-term storage. Frozen vegetables are safe to eat and nutritious as long as high-quality raw materials are used, good manufacturing practices are employed, and products are kept at temperatures that comply with current legislation.

The quality of frozen vegetables delivered to the consumer can be no better than their quality at the time of freezing. Nonetheless, the importance of this basic fact is all too often not recognized. This chapter is designed to create a greater awareness of the importance of prefreezing operations to frozen vegetable quality.

For frozen vegetables marketed though retail channels, frozen storage is the main cause of damage. Damage during freezing and thawing, if conducted in accordance with industry standards, is relatively insignificant when compared to damage typically encountered during frozen storage. Microbiological quality is also an important aspect of frozen vegetables. Although frozen vegetables are not sterile, pathogenic organisms cannot grow at appropriate freezer temperatures. Therefore, frozen vegetables are not a hazard to health, provided they were clean and free from contaminants when frozen. Despite the very high quality of frozen vegetables now available to consumers, there will continue to be some competition between fresh and frozen products. The frozen vegetable industry clearly faces the prospect of losing market share to vegetables preserved by alternative means because of technological advances in competitive preservation methods, especially thermal processing [3].

Freezing is often considered to be the simplest and most natural way to preserve vegetables. The frozen food industry is sophisticated and firmly based in modern science and technology. Therefore, the products that reach the consumer are consistently of high quality. However, to retain and improve market share for frozen vegetables, it might be necessary to apply new technologies and to investigate poorly understood factors that influence the quality of frozen vegetables.

## II.  EFFECTS OF PREPARATION AND PACKAGING

### A.  Preparation of Vegetables for Freezing

The complete scheme of preparation of vegetables for the freezing process has several important steps related to quality retention in the final product (Fig. 1). One of these steps, blanching, has intrinsic relevance for frozen vegetables because of the necessity to inactivate the enzymes in vegetable tissue, which cause the quality changes that occur during prolonged storage, even at subfreezing temperatures. For this reason, blanching has been extensively researched. However, blanching has not been sufficiently defined for all species of vegetables, because research studies have been conducted only on specific varieties of each vegetable, specific blanching methods, and certain blanching parameters. The enormeus amount of available literature on blanching dictates that special attention be given to this process in relation to the preparation and packaging of frozen vegetables.

### 1.  Characteristics of Raw Materials

Raw material characteristics is possibly the most important factor related to final frozen vegetable quality. These characteristics are related to the vegetable cultivar, crop production, crop maturity, harvesting practices, crop storage, transport, and factory reception.

Crop cultivars must be selected based upon their suitability for frozen preservation in terms of factory yield and product quality. For this reason, the vegetable cultivar must have the following qualities:

1.  Exceptional flavor
2.  Exceptional and uniform color
3.  Desirable texture
4.  Simultaneous and uniform maturity
5.  Resistence to fungal attacks and viral and other diseases
6.  High yield
7.  Suitability for mechanical harvesting practices

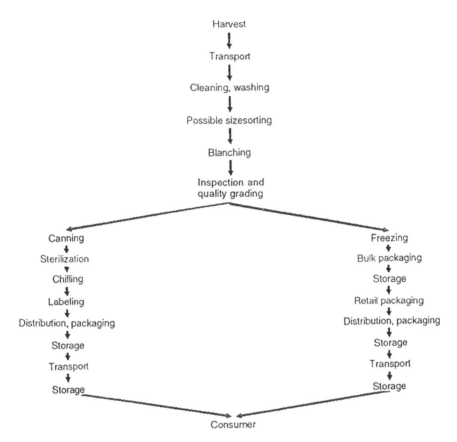

**Figure 1** Flow chart of pea processing: up to the step of quality grading, the lines are identical for the frozen and heat-sterilized products. (From Ref. 28.)

Consequently, the varietal and cultural requirements for optimum product quality depend upon the vegetable to be processed and the end use. For carrots, tenderness and flavor are of major importance to the quality of the retail product. Frozen carrots used for soups should be able to withstand postprocessing heating without losing their shape. Thus, a firmer-textured variety grown to greater maturity is best for this use. For both uses, a bright, uniform orange color is desired. Varieties recommended for freezing are primarily Chantenay and Danvers strains [4].

Pea cultivars for processing should be selected for their suitability for freezing. Maturity, yield, disease resistence, and sieve size requirements are the most important factors to take into account. Pea cultivars preferred for freezing have a small, wrinkled seed with a dark green color. Many cultivars grown for freezing are not suitable for other processing methods, such as canning, because of the tendency for their skin to develop a bronze color during processing or storage. Darskin Perfection and its many related types, such as Venus, are typical full-season cultivars, which, along with Thomas Laxton, are widely used for freezing [5].

The above are some examples of the importance of cultivar selection in obtaining a high-quality end product. Frozen vegetable processors are interested in producing a uniform product throughout the processing season with respect to sieve size, color, flavor, and texture. Uniformity in nutritional composition also must be considered, because of nutrient labeling requirements for processed foods.

## 2. Harvesting and Preprocessing Storage

Other aspects, such as harvest and postharvest handling, must be considered to optimize quality. Some vegetables, such as green peas and sweet corn, have only a brief period during which they are of prime quality. If harvesting is delayed beyond this point, quality deteriorates and the crop may quickly become unacceptable. It is important that a constant supply of raw produce be available at the processing plant each day for efficient plant operation [5]. However, because of the perisable nature of peas, pea maturity is an important consideration to ensure processed product quality. Optimum quality is obtained within a narrow range of maturity, which persists only for a short time (a day or two). Initially, maturity, yield, and quality are positively correlated. However, beyond a given point, quality begins to deteriorate and yields decline [6].

Harvest practices have a profound effect on quality retention of frozen vegetables. Occasionally, vegetables are subjected to bruising and tenderization (peas) during harvesting. Consequently, the viner operation influences tenderization, damage, and size-grade distribution [7]. Slight damage to peas results in marked changes in their metabolism, which suggests that bruising during mechanical shelling is largely responsible for the rapid development of off-flavors in vined peas [8].

Delayed development of off-flavors could also be related to the degree of pea bruising, because hand-shelled peas do not deteriorate in flavor as rapidly as other peas [9]. Bruising has been found to modify the respiration of peas and usually decreases the amounts of carbon dioxide, aldehyde, and alcohol produced.

Postharvest delays in handling vegetables to be processed is known to produce deterioration of flavor, texture, color, and nutrients. Considerable changes, physical as well as chemical, take place when shelled peas are allowed to stand. A number of studies have been conducted on the effects of temperature and holding time intervals on the acceptability of peas. Until the late 1950s the methodology used for harvesting frequently involved intervals of several hours. However, the development of mobile viners and the large-quantity transport of vined peas to the processing plant has greatly shortened the interval between the first step of harvest and the initial step of processing. Presently peas used for canning and freezing are processed as soon as possible after harvesting. However, some delay is inevitable. It is feasible to use refrigeration at the moment of harvest, during transport, or upon receipt at the processing plant [5].

However, low-temperature storage does not always preserve quality. Mitchell et al. [10] observed that peas stored in air for 5 or 6 hours had an off-odor and were not suitable for processing, while those stored in ice or ice water had no off-odor. Linch et al. [11], in contrast, observed that large numbers of peas stored in ice water split during storage and longer delays produced a characteristic off-flavor, which is considered to be a complex composed largely of the more odorous unsaturated carbonyls arising from the enzymatic breakdown of unsaturated pea lipids. Therefore, careful handling and proper storage can minimize off-flavor development.

Changes in vegetable texture due to postharvest handling were investigated in the 1940s by Boggs et al. [12]. They found that toughening, particularly of the skin, occurred during the first few hours after shelling. The rate of toughening was reduced by storage at lower temperatures. The freezing process and subsequent storage in the frozen state appeared to produce toughening. The extent depended upon the pea size and maturity. Cellular processes that take place in the tissues after harvest and during delays before processing are believed to closely parallel the processes of natural cell wall elaboration. Therefore, every precaution should be taken to reduce delays between harvest and processing. Possible methods of maintaining quality of some vegetables during postharvest handling, including the application of controlled atmosphere and vacuum cooling, have been studied [13–15] but have yet to be put into commercial practice.

## 3. Preblanching Operations

Preblanching operations include washing, cleaning, peeling, dicing, slicing, and the associated movement of raw product from one piece of equipment to another, by fluming in water, vibratory tables, elevators, or conveyor belts [2]. Depending on the vegetable, crops are subjected to different types of delivery, which are off-loaded into bulk feed hoppers from which they are conveyed to cleaning equipment. Air cleaners and pod-and-stick removers are employed for pea preparation. Grading for size is another feature that is often very important in certain European countries. Codex Alimentarius specifications recommend certain vegetable pod diameters (green beans and peas) and pea diameters.

The osmotic concentration of vegetables prior to freezing is another pretreatment that can improve end product quality. The vegetable industry requires a great amount of energy to freeze the large quantity of water present in fresh produce. Reduction in moisture content of the produce, therefore, would reduce refrigeration requirements during freezing [16]. Other advantages of partially concentrating vegetables prior to freezing include packaging and distribution savings and maintainence of product quality comparable to conventional products. The products obtained from partial concentration and freezing are termed dehydrofrozen. Pre-freezing concentration of plant materials can be achieved by osmotic means [17,18]. The most commonly used osmotic agent for vegetables is sodium chloride. Osmotic dehydration can be achieved at low temperatures and is less energy intensive than air-drying or freezing. Studies conducted by Biswal et al. [19] demonstrated that green beans can be partially concentrated by osmotic dehydration in NaCl-water solutions to produce osmotically concentrated frozen green beans, which are as acceptable as conventionally frozen green beans.

Another preblanching operation used in the processing of potato products is surface freezing. Potato tubers for freezing are often stored for several months at low temperatures (4–7°C) to extend their shelf life. The accumulation of reducing sugars in tubers results in excessive browning of French fries and potato chips. Toma et al. [20] observed that partial freezing of potato strips before blanching improved the quality of French fries produced from potatoes with high levels of reducing sugars. This treatment was also an effective means of decreasing oil absorption, thereby improving the color of the end product.

## 4. Blanching

*Definition of Blanching Process.*    Blanching is one of the most important unit operations preceding other processing techniques, such as canning, freezing, and dehydration. Enzyme activities in raw vegetables are responsible for the undesirable colors and flavors that may develop during processing and storage. In order to prevent enzymatic reactions during processing, most vegetables must first be blanched.

The term *blanching* was originally used to designate heat-treatment operations in the processing of plant foods for frozen storage, which prevent deteriorative changes that occur in foods not so treated. Blanching (70–105°C) is now associated with the destruction of enzyme activity, just as pasteurization (60–85°C) is associated with the destruction of microorganisms. Hot-water blanching is usually carried out between 75 and 95°C for between 1 and 10 minutes, depending on the size of the individual vegetable pieces. Typical times for water blanching at 95°C are 2–3 minutes for green beans and broccoli, 4–5 minutes for Brussels sprouts, and 1–2 minutes for peas [21].

Blanching of vegetables prior to freezing has several advantages as well as a number of disadvantages. The advantages include (1) stabilization of texture (changes caused by enzymes), color, flavor, and nutritional quality, (2) destruction of microorganisms, and (3) wilting of leafy vegetables, which assists in packaging. The disadvantages include (1) some loss of texture (caused also by freezing and thawing), color, flavor, and nutritional quality due to the heating process, (2) formation of a cooked taste, (3) some loss of soluble solids (specially in water blanching), and (4) adverse environmental impact due to the requirements for large amounts of water and energy. Therefore, there are trade-offs in blanching, and process optimization is critical [22].

*Key Enzyme Indicator of the Blanching Process.* Process optimization involves measuring the rate of enzyme destruction, so the blanch time is just long enough to destroy the indicator enzyme. Since catalase and peroxidase are known to be relatively resistant to heat inactivation, these two enzymes have been widely used to determine whether or not the blanching treatment was adequate. However, a number of studies have indicated that the quality of a blanched, frozen, stored product is improved if some peroxidase activity remains at the end of the blanching process [23,24]. Böttcher [25] reported that the highest-quality products were obtained when the following percentages of peroxidase activity remained (the exact value depends on the variety): peas, 2–6.3%; green beans, 0.7–3.2%; cauliflower, 2.9–8.2%; and Brussels sprouts, 7.5–11.5%. It was concluded that the complete absence of peroxidase activity indicated overblanching.

Another of the problems in complete inactivation of peroxidase is the presence of 1–10% more heat-stable isoenzymes of peroxidase in most vegetables [25], since the enzyme can reactivate [26]. Reactivation of peroxidase per se probably does not contribute to quality deterioration. However, the quality specifications for frozen vegetables do not allow for it. The use of peroxidase as an indicator for adequate blanching has an additional problem in that peroxidases of different vegetables and fruits have different heat stabilities (Table 1).

In general, the peroxidases in low-acid foods, such as vegetables, are more resistant to heat treatment than are those in acid foods (fruits), presumably because peroxidases are less stable to heat at pHs below 4.0 [27]. Variation in heat stability of peroxidases among different vegetables requires the processor to determine the time needed for each vegetable with the blanching equipment and conditions that exist.

**Table 1** Temperature Stability of Peroxidase in Plant Materials

| Plant material | z-value[a] (°F) | F-value[b] (min) |
| --- | --- | --- |
| Asparagus | 55–161 | 2–27 |
| Beans, green | 86–88 | 3.8 |
| Beans, lima | 67 | 30 |
| Corn | 70 | — |
| Peas | 48 | 60 |
| Spinach | 59, 81 | 6–10 |
| Peaches | 30–31 | 0.3–0.9 |
| Pears | 20–32 | 0.5–8.2 |
| Tomato juice | 18 | 0.9 |
| Apples, fresh juice | 11–34 | 5–7 |

[a]Slope of the logarithmic thermal inactivation curve.
[b]Time at 82°C to inactivate peroxidase (<1% activity left).
*Source*: Ref. 22.

The presence of isoenzymes of peroxidase may also produce problems in the blanching of most vegetables. Isoenzymes act upon the same substrates, even though the chemical composition of the enzyme is different. This fact is often reflected in different heat stabilities. The relative amounts of isoenzymes vary from vegetable to vegetable and even in the same vegetable and can vary with variety, maturity, environmental factors, etc. Problems with heat inactivation (blanching) of different enzymes and different isoenzymes of the same enzyme occur, because the inactivation begins at different temperatures and proceeds at different rates [22].

After identifying an appropriate indicator enzyme for monitoring the processing of a specific vegetable, it is important that the method for the enzymic assay be as rapid and as simple as possible. The simplicity of the peroxidase assay method has led to its ready adoption by the food industry. Similar methods need to be developed for other indicator enzymes such as lipoxygenase [22].

*Requirements of the Blanching Process.*    The requirement for blanching has been debated in recent years with regard to energy usage and protection of environment [28]. Based on conventional water and steam blanchers, new blanchers have been developed that are well insulated, compact in size, and often carefully controlled in performance. The new blanchers have also been designed to improve transportation of vegetables during blanching, to improve heat penetration, and to minimize damage due to handling.

An acceptable blanching technique must fulfill the following essential demands:

1.    Uniform heat distribution to the individual product units
2.    A uniform blanching time to all product units
3.    No damage to the product during the entire blanching and cooling process
4.    High product yield and quality
5.    Low consumption of energy and water
6.    Tough, reliable design, giving problem-free operation, even in continuous three-shift operations
7.    Quick, easy, and effective cleaning

The design must also provide a pleasant working environment for the staff through the prevention of unnecessary noise, heat irradiation, and loss of steam and process water with foam and product particles onto the floor. All of these demands must be independent of actual flow rate, whether it be high or low.

*Effects of Blanching and Cooling on Product Quality.*    Leaching of soluble components during blanching can be an important problem, depending on the blanching method employed. For whole peas and parts of peas, temperatures above 50°C and longer times gave rise to continously increasing soluble component losses.

Changes in weight and solids content of vegetables occurring during blanching, cooling in air or water, and freezing have been studied [29,30]. It is important that cooling be carried out shortly after blanching, especially for products to be frozen. Frozen vegetables prepared by steam blanching and air cooling lost significantly fewer solids than similar vegetables cooled in water but had lower frozen yields. A relatively large proportion of the water lost during air cooling was regained by the vegetables when they were cooked. Although the loss of solids during cooking was about the same regardless of the method of cooling, the overall loss of solids was greater for water-cooled vegetables. Carroad et al. [31] made a similar examination of broccoli blanched in steam, water, and recycled water (Table 2). Results demonstrated that not only the cooling method, but also the blanching method determined yield and overall loss. As in the previous study, the highest solid content was obtained by evaporative cooling without water spray, and the highest yield was found for broccoli blanched in recycled water, implying the lowest leaching loss.

Quality differences in processed vegetables can be controlled with high pressure steam blanching (HTST). This blanching method requires significantly less energy to blanch vegetables than water blanching [32]. Table 3 shows the quality parameters of HTST-blanched vegetables compared with conventional blanched vegetables. Differences in quality parameters were highly dependent upon the vegetable being blanched and the pressure and duration of the HTST blanch. In addition, substantial energy savings are achieved with HTST blanching. HTST blanching had an 80% calculated efficiency compared to 60% for a typical water blanching [33]. Possible alternative treatments to conventional blanching have also been developed by Steinbuch [34]. Heat-shock treatments are applied in order to inactivate the chlorophyll-converting enzymes (see next section) in the surface of products, such as green beans, to maintain color and to prevent textural damage. This method consists of the exposure of leguminous vegetables, such as beans and peas, to boiling water or condensing steam for 5–15 seconds, thereby giving satisfactory results notwithstanding the residual activities of peroxidase, catalase,

**Table 2** Comparison of Yield and Overall Loss for Broccoli After Various Blanching/Cooling Combinations

| | Yield/overall loss[a] | | |
|---|---|---|---|
| Cooling method | Water blanching | Recycled water blanching | Steam blanching |
| Water | 106.5/20.3 | 110.1/3.3 | 100.6/16.3 |
| Air with condensation | 104.8/4.9 | 104.9/1.3 | 93.7/11.4 |
| Air with water spray | 105.1/15.1 | 103.7/6.7 | 95.7/12.4 |
| Air without water spray | 93.8/10.5 | 94.1/4.7 | 89.3/10.9 |

[a]Yield of frozen broccoli (100 kg) on basis of raw broccoli. Overall fraction of total solids in raw broccoli lost in blanching, cooling, freezing, and cooking.

**Table 3** Quality Attributes of Selected Frozen Vegetables as Influenced by Water or High-Temperature, Short-Time Steam Blanching

| Vegetable | Blanch treatment | | Drip (%) | H$_2$O (%) | Soluble solids (%) | Shear values (N) | Agtron color | | Ascorbic acid (mg/100 g) |
|---|---|---|---|---|---|---|---|---|---|
| | | | | | | | yellow | green | |
| Snap beans | H$_2$O | | 11.7 NS | 90.0 B$^a$ | 5.8 A | 1088 A | 44 B | 42 B | 13.2 B |
| | 45 psi | 35 s | 11.0 | 90.2 A | 5.6 B | 377 B | 42 B | 43 AB | 14.8 A |
| | 45 psi | 45 s | 11.2 | 90.3 A | 5.5 B | 327 C | 49 A | 46 A | 13.3 B |
| | 45 psi | 55 s | 12.1 | 90.3 A | 5.4 B | 218 D | 45 B | 45 A | 13.3 B |
| Sweet peas | H$_2$O | | 7.5 NS | 80.1 AB | 11.4 A | 514 NS | 34 A | 44 A | 14.1 NS |
| | 15 psi | 35 s | 7.8 | 79.9 B | 10.9 AB | 509 | 34 A | 45 A | 12.0 |
| | 15 psi | 40 s | 6.6 | 80.0 AB | 11.2 AB | 504 | 34 A | 44 A | 14.1 |
| | 15 psi | 45 s | 7.6 | 80.4 AB | 10.9 AB | 502 | 32 B | 43 B | 13.4 |
| | 20 psi | 35 s | 6.8 | 80.6 A | 10.7 B | 507 | 33 AB | 44 A | 13.3 |
| | 20 psi | 40 s | 7.5 | 80.2 AB | 10.8 AB | 506 | 34 A | 44 A | 14.5 |
| Lima beans | H$_2$O | | 4.2 A | 64.0 AB | | 817 B | 72 A | 70 A | 15.4 AB |
| | 20 psi | 20 s | 1.9 C | 62.1 C | | 898 A | 68 B | 69 AB | 16.4 A |
| | 20 psi | 40 s | 2.3 BC | 62.1 C | | 898 A | 67 B | 67 AB | 14.7 ABC |
| | 20 psi | 50 s | 2.4 B | 63.6 B | | 821 B | 66 B | 66 B | 13.0 BC |
| | 20 psi | 60 s | 2.4 B | 64.1 A | | 788 C | 65 B | 66 B | 11.9 C |
| | 40 psi | 20 s | 2.4 B | 63.6 B | | 844 AB | 66 B | 65 B | 14.8 ABC |
| | | | | | | | | Red | |
| Carrots | H$_2$O | | 11.9 A | 84.7 NS | 7.8 B | 537 A | 49 NS | 79 A | 12.6 B |
| | 45 psi | 40 s | 6.4 B | 84.6 | 8.0 AB | 304 B | 49 | 78 AB | 13.8 A |
| | 45 psi | 60 s | 10.4 AB | 84.9 | 8.1 A | 115 C | 50 | 77 B | 13.0 AB |
| | 60 psi | 30 s | 7.4 B | 84.6 | 8.0 AB | 301 B | 50 | 78 AB | 13.8 A |
| | 60 psi | 50 s | 9.9 AB | 84.1 | 8.1 A | 138 C | 51 | 77 B | 13.0 AB |

$^a$Means in a column for each vegetable not followed by the same letter are significantly different.
*Source:* Ref. 32.

and lipoxygenase. Beans treated by this method have a firm texture that differs substantially from the softness of conventionally blanched green and wax beans [35].

Several studies of hot-gas blanching and microwave blanching have been conducted in an effort to reduce leaching. However, until recently, operational costs have outweighed the advantages. This situation is likely to continue for hot-gas blanching. However, microwave blanching appears worthy of further study. Detailed studies of the blanching process have been reported by Poulsen [28] and Adams [36].

*pH Influences on Blanching Effectiveness.* The activities of most enzymes are greatly dependent on the pH of the tissue or the solution in which the enzymatic reaction takes place. Diminution of the pH of the water during blanching can reduce the temperature and time requirements of the blanching process, to achieve the same results. Addition of citric acid has been employed to lower the pH during the blanching of potatoes and artichokes. This action could also be improved by the addition of ethylenediaminetetraacetic acid (EDTA). However, this chemical compound is now forbidden by international law. The acidification of blanching water cannot be generally employed due to chlorophyll transformation into pheophytins, which produces an undesirable modification of the color of green vegetables. Other compounds, such as chlorures (10% NaCl) or carbonates (5% $Na_2CO_3$), produced a decrease in thermal inactivation of enzymes in green beans, which increased the blanching time [37]. In general, the use of additives in the blanching water depends upon a study of the blanching time to obtain the optimal enzyme inactivation for each vegetable.

## B. Packaging of Frozen Vegetables

### 1. Packaging Requirements

A successful package protects its contents from the environment for a period of time and at reasonable cost. Factors that must be taken into consideration when selecting a package for frozen vegetables include protection of the contents from: (1) atmospheric oxygen, (2) loss of moisture, (3) flavor contamination, (4) entry of microorganisms, (5) mechanical damage, and (6) exposure to light. In addition to protective function, packaging materials should have a high heat transfer rate to facilitate rapid freezing. The package should be not only attractive and informative, but also easy to open and easy to close when the contents are only partially used. Thus, a well-designed, well-illustrated, and well-printed package is of major importance. Information is needed by the consumer to describe preparation and storage conditions, to give nutritional values, and to provide essential data, such as weight, contents, and the date by which the food should be consumed. In addition, packaging should be environmentally friendly, thereby placing further constraints on the selection of the materials to be used [38].

## 2.  Material Selection

The selection of the appropriate packaging material for frozen vegetables depends on a large number of factors. The most critical factors concern packaging that comes into contact with the product. The following criteria are the most important:

1.  Mechanical properties—e.g., frictional characteristics, stiffness, crush resistance, sealability, ease of opening, ease of separation, cutting and folding characteristics, and resistance to build-up of static electricity.
2.  Storage and transport performance—sufficient barrier properties to prevent staining and dehydration of the contents and sufficient scuff resistance to avoid deterioration of the surface appearance.
3.  Heating performance—often the packaging is required to withstand reheating during preparation of the vegetables by the purchaser (microwave heating or conventional oven heating).
4.  Product resistance—paper-based packaging, in direct contact with the food, must be resistant to various ingredients (principally water and grease).
5.  Environmental impact—selection of packaging material must take into account environmental implications during manufacture, such as effluents, emissions, and waste products, and the ease with which the products can be recycled.
6.  Safety—regulations specify acceptable raw materials, methods of manufacture and conversion, limits for specific components capable of migrating from the package into the product, and limits on the total quantity for all migrants (overall migration) [39].

## 3.  Packaging Types

A European Economic Community statute lists three basic types of packaging: primary, secondary, and tertiary. Primary packaging comes into direct contact with the product until used by the consumer. Secondary packaging is master packaging that contains primary packaged products. Tertiary packaging is used for bulk transportation of primary and/or secondary packaged products and is disposed of by the store.

The most common use of plastic bags in the frozen-food industry is for the packaging of vegetables. A range of plastic materials is used as primary packaging materials. The selection of the plastic package to be used requires knowledge of the conditions that the frozen vegetables are likely to be exposed to and the properties required. Polyolefins and polyesters are materials used for frozen vegetables where thermal characteristics are the most critical requirements. Polyolefins are polymers used to produce plastic bags. However, impact modifiers are often added to improve flexibility during frozen distribution. Lamilates are used if higher temperature resistance is needed for reheating purposes (boil-in-the-bag packaging).

### 4. Temperature Indicators

In recent years, considerable research has been conducted to develop and design a reliable system to determine the temperature conditions of frozen foods during storage and commercial distribution. Physicochemical, chemical, or biological reactions give an irreversible indication (usually visible) of the history of the product [40]. Initial temperature indicators show only exposure above or below a reference temperature, and initial temperature/time indicators indicate the cumulative time and temperature to which a product is exposed above a reference temperature. These indicators are placed on the outside of the packages and combine the time and temperature conditions to which they, and thus the food, have been exposed [38].

## III. EFFECTS OF FREEZING, FROZEN STORAGE, AND THAWING ON COLOR, APPEARANCE, AND CONSUMER ACCEPTANCE

### A. Color and Appearance

Color is the primary quality attribute by which the consumer assesses natural and processed foods. The color of a processed food is often expected to be as close as possible to the natural product. Frozen vegetables are subjected to color modifications, which take place during the blanching process and/or during frozen storage and the final thawing step, which is often replaced by a cooking process. The most important color modifications are related to three biochemical or physicochemical mechanisms:

1.  Changes in the natural pigments of vegetable tissues (chlorophylls, anthocyanins, carotenoids)
2.  Development of enzymatic browning
3.  Breakdown of the cellular chloroplasts and chromoplasts

### 1. Changes in Natural Pigments of Vegetable Tissue

Chlorophylls are mainly responsible for the color in green vegetables (beans, peas, spinach, etc.). The change from bright green to olive green in unblanched frozen vegetables during prolonged storage is a consequence of chlorophylls *a* and *b* being converted to the correspondant pheophytins. Time-temperature conditions during blanching of green vegetables have been found to influence the rate of chlorophyll deterioration during processing as well as during frozen storage [41–43].

The chlorophyll content of vegetables is affected by blanching time and temperature. As shown in Figure 2, a loss in total chlorophyll occurs during blanching, with the conversion of chlorophylls into pheophytins being proportional to the degree of heat treatment applied during blanching [41]. During the initial steps of blanching, chlorophyll *a* appears to be more sensitive than chlorophyll *b*, but after

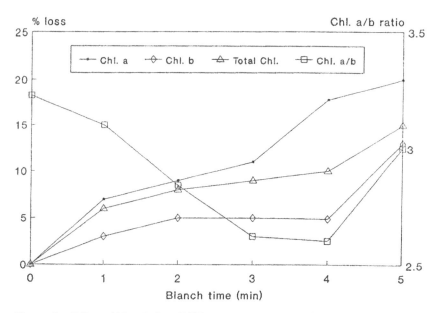

**Figure 2**   Effect of blanch time (98°C) on percent loss of chlorophyll *a*, chlorophyll *b*, total chlorophyll, and chlorophyll *a:b* ratio in green beans. (From Ref. 41.)

blanching for 4 minutes the destruction rate of chlorophyll *b* increases. Thus, the chlorophyll *a*:chlorophyll *b* ratio decreases continuously during the first 4 minutes of blanching and then increases as a result of the change in the rate of destruction of chlorophyll *b*.

Walker [43] studied the effect of blanching time on the changes in chlorophylls in green beans during frozen storage at –10°C (Fig. 3). Unblanched green beans lost more than 45% of their chlorophylls during 20 days of storage, whereas green beans blanched for 2 minutes lost only about 2% during the same period. However, overblanching produced an increase in chlorophyll losses due to the breakdown of chloroplasts. It has been established that these chlorophyll changes continue during prolonged frozen storage and are related to the amount of residual enzyme activity and/or the degree of heat treatment during blanching.

Visual appearance of green frozen vegetables was closely related to the extent of chlorophyll conversion to pheophytin. Initial studies conducted on a series of commercial frozen bean samples confirmed this hypothesis [33]. Mean sensory scores from the color evaluation of frozen green beans blanched for 0–3 minutes and stored for 12 months at –18°C indicated statistically significant differences in chlorophyll conversion between unblanched and blanched samples and between samples blanched for 0–3 minutes [41] (Table 4).

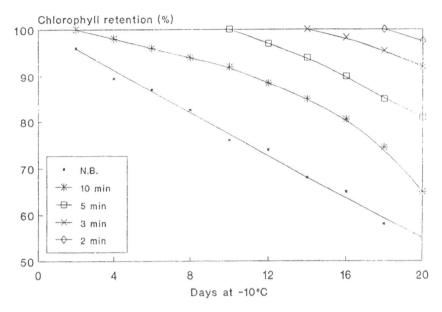

**Figure 3** Changes in chlorophyll retention in green beans blanched at 100°C or unblanched during frozen storage at −10°C. (From Ref. 43.)

The degradation of chlorophylls was progressive during frozen storage. However, chlorophyll conversion to pheophytin was not the only reaction affecting color, since after 12 months of storage chlorophylls and pheophytins were progressively destroyed [42]. Other mechanisms may take place in frozen vegetable tissues, which are related to the increase in fat peroxides.

Moreover, the blanching method is related to the extent of chlorophyll degradation. High-temperature and short-time water blanching produces a significantly smaller loss of chlorophylls than if water blanching occurs at low temperature over a long period of time to produce the same enzymic inactivation [44]. Water blanching of Brussels sprouts produced lower losses of total chlorophylls than steam blanching due to the prolonged time of exposure of the product to the steam necessary to obtain the same inactivation of peroxidase enzyme [45].

Various inorganic salts (NaCl, KCl, $K_2SO_4$, $Na_2SO_4$) have been employed to reduce the pheophytinization of chlorophylls in Brussels sprouts and spinach. These salts, added to the water during the blanching process, did not modify the pH of the medium but did directly affect color modification. This effect was fairly pronounced when lima beans were blanched [46]. Other salts, such as sodium bicarbonate or ammonium bicarbonate, have been used at different concentrations during the blanching of green vegetables with varying degrees of success [47,48].

**Table 4** Mean Sensory Scores for Color Evaluation[a] of Frozen Green Beans

| Storage time (months) | Blanch time (s) | | | | | | | F-test | LSD ($p = 0.05$) |
|---|---|---|---|---|---|---|---|---|---|
|  | 0 | 30 | 60 | 90 | 120 | 150 | 180 |  |  |
| 0 | 3.22 | 6.25 | 4.77 | 5.45 | 4.48 | 4.28 | 3.65 | ** | 0.671 |
| 3 | 3.63 | 6.13 | 4.94 | 4.91 | 4.66 | 3.50 | 3.30 | ** | 0.676 |
| 6 | 3.56 | 5.76 | 5.13 | 5.20 | 4.93 | 3.50 | 2.93 | ** | 0.788 |
| 12 | 2..86 | 4.63 | 4.76 | 4.50 | 3.83 | 3.73 | 2.90 | ** | 0.783 |

[a]Green color intensity and uniformity was scored on a 7-point scale.
*Source*: Ref. 41.

In addition, the blanching process results in the breakdown of chloroplasts, which increases the chlorophyll's lability thus, the color deterioration occurs faster at temperatures above 0°C [49]. In this way, prolonged blanching processes may cause an increase in pigment degradation rate in frozen tissues during storage [43,45]. Table 5 shows the time of frozen storage required to observe a relevant decrease in chlorophylls [45]. Pigment-degradation rate is related to storage temperature, the kind of processed vegetable, and the necessary preparation steps of the crop prior to the blanching and freezing processes.

Green beans and chopped spinach showed the highest rates of chlorophyll degradation due to special operations during the prefreezing steps. Peas appear to be maintain their chlorophyll concentrations under the most adverse storage conditions.

Other natural pigments of vegetable tissues, namely, anthocyanins and carotenoids, are responsible for the color and appearance of raw and processed products. Anthocyanin degradation is not crucial during the freezing process due to the relatively few vegetables in which these pigments are important. However, some vegetable crops exhibit high amounts of carotenoid pigments (xanthophylls, hy-

**Table 5** Months[a] of Storage Required for a 10% Decrease in Chlorophyll

| Product | Temperature | | |
|---|---|---|---|
|  | –18°C | –12°C | –7°C |
| Green beans | 10 | 3 | 0.7 |
| Spincah, leaf | 30 | 6 | 1.6 |
| Spinach, chopped | 14 | 3 | 0.7 |
| Peas | 43 | 12 | 2.5 |

[a]Regression values of two lots each for green beans and chopped spinach and four lots each for leaf spinach and peas.
*Source*: Ref. 45.

drocarbon carotenoids, and carotenol esthers of fatty acids), which give the natural orange or yellow color to the tissue. These pigments are lipid soluble; thus blanching does not affect their loss through a leaching process, with the exception of carrot slices, where some loss was noted during water blanching [50]. During the blanching-freezing process it is very important to preserve the carotenoid compounds, due to their provitamin A activity and their protective effect on chlorophylls against oxidation induced by radical mechanisms [51].

## 2. Development of Enzymatic Browning

Enzymatic browning occurs when some crops (e.g., potatoes, mushrooms) are cut or peeled, due to oxidation mechanisms. These products when they are frozen and stored at temperatures above $-18°C$ without any previous treatment undergo color changes due to enzymatic browning. However, potato varieties vary distinctly in their pigmentation properties during processing, with high- and low-browning potato varieties being available [52]. The amount of browning produced is also influenced by season, cultural practices, and growing and storage conditions. Consequently, the extent of pigmentation produced under similar processing conditions varies. French fries comprise about 85% of all frozen potato products. A large proportion of frozen french fries are now packed for restaurants and other institutional customers. They are prepared for serving by finish-frying in deep fat. For home consumers, they are finished-fried by the processor and are oven-heated in the home. The latter product is usually of lower quality than those finish-fried in deep fat after being frozen [53]. Specific operations to prepare potatoes for frying are washing, peeling, trimming, sorting, cutting, blanching, frying, defatting, cooling, freezing, and packaging. In these processes, there is no development of enzymatic browning. However, after cooking, blackening may occur [54]. The addition of chemical compounds, which chelate free cations, prevents browning reactions when they are added to the blanching water. Disodium dihydrogen pyrophosphate has been used in blanching operations for potato products with great success [55].

The blanching process is the most effective step to prevent enzymatic browning due to the high sensitivity of the polyphenoloxidase enzyme to heat. This sensitivity depends on the pH of the blanching solution and on the intrinsic pH (acidity) of the vegetable tissue. Changes in mushroom color that take place during blanching can be partially avoided by the addition of citric acid (1.5%) to the blanching water [56]. The color of other vegetables may also be improved through addition of metabisulfite salts to the cooling water after blanching. Frozen cauliflower florets were more desirable in appearance when a solution of sodium metabisulfite (2 g/L) was added to the cooling medium before freezing [57].

## 3. Breakdown of Cellular Chloroplasts and Chromoplasts

Green vegetables develop a bright green color after the blanching process. This color differs from the characteristic green color of the raw vegetable. This change may be attributed to breakdown of chloroplasts due to the thermal treatment, which produces migration and diffusion of the pigments within the cellular medium [58]. Another possibility is the leaching of hydrosoluble substances, which results in the more green vegetable tissues. Changes in color and appearance are evident in blanched and frozen carrot slices, since the color changes from orange to intense yellow. Histological studies have shown that the blanching process results in the breakdown of chromoplasts and the release of carotenoid compounds, which are solubilized in the lipids of the cellular medium. Consequently, dark globules develop. Destruction of the cell wall, particularly the pectocellulosic fraction, favors the migration of these globules toward the cellular medium. Decreases in carotenoid concentration in carrot tissues may be verified by the increase in the intensity of the orange color in the blanching water [50]

## B.  Consumer Acceptance

Sensory quality and consumer acceptance are important to the success of frozen vegetables in the market. Consumers rated flavor and texture as the most important sensory attributes with respect to purchase and use of peas in one study [59], while color and appearance played a less important role. Sensory and chemical-physical quality criteria of frozen peas found to be relevant for describing total internal quality variation and consumer acceptance included mealiness, hardness, fruity flavor, and sweetness [61].

The durability of high-quality frozen vegetables during storage and distribution has been questioned by several authors [61]. Dietrich et al. [45] observed no important textural changes in green beans during storage, although appreciable color and flavor changes occurred. This extensive study included constant and fluctuating temperatures and temperature sequences simulating conditions encountered during transportation and distribution. In other studies there were no evaluations of the texture of vegetables exposed to fluctuating conditions during this time-temperature series.

The effect of heat-shock treatments of vegetables to be frozen on consumer acceptance has been extensively studied by Steinbuch [35]. The results were rather surprising, since color maintenance increased without textural damage. Table 6 indicates the required heating time (10 seconds) to obtain high-quality frozen green beans stored for 12 months. The application of heat-shock treatments resulted in residual enzymatic activities. Evaluations of the quality retention and consumer acceptance of processed vegetables indicate considerable difference in the behavior of flavor properties among vegetables.

Complete peroxidase inactivation appears to offer a guarantee for flavor retention in frozen Brussels sprouts. However, studies of the effects of these treatments on various other vegetables used for freezing showed that comprehensive differ-

**Table 6** Relationship Between Heat-Shock Treatment and Quality of Frozen Green Beans

| Heat treatments in water (98°C) | 3 months | | | 7 months | | | 12 months | | |
|---|---|---|---|---|---|---|---|---|---|
| | Color | Flavor | Texture | Color | Flavor | Texture | Color | Flavor | Texture |
| None | Discolored | Strong Off-flavor | Good | Discolored | Strong Off-flavor | Good | Discolored | Strong Off-flavor | Good |
| 2.5 s | Good | Slight Off-flavor | Good | Good | Off-flavor | Good | Discolored | Off-flavor | Good |
| 5 s | Good | Good | Good | Good | Good | Good | Discolored | Off-flavor | Good |
| 10 s | Good | Good | Good | Good | Good | Good | Good | Good | Good |
| 3 min | Good | Good | Good | Good | Good | Good | Good | Good | Good |

*Source:* Ref. 35.

ences in "enzyme inactivation–quality retention" relationships could be ascertained [35] (Table 7).

## IV. EFFECTS OF FREEZING, FROZEN STORAGE, AND THAWING ON PALATABILITY ATTRIBUTES

Frozen vegetables are usually consumed after being prepared using the traditional cooking methods of various countries. Both desirable and undesirable effects of heat processing may result simultaneously. Favorable alteration of color and texture can lead to increased palatability. Time-temperature parameters that produce desirable effects must be used in order to maximize both the sensory and nutritional qualities of the vegetable. However, during prefreezing operations, freezing, and frozen storage, some changes in palatability attributes (mainly flavor and texture) take place, which affect the consumer acceptability of the cooked product.

### A. Texture

Vegetable tissues are living materials. Properties that reflect freshness and turgidity depend largely upon the structural arrangement and chemical composition of the cell wall and the intercellular spaces, where pectic substances are the primary constitutents. The heat applied to vegetable tissues during blanching results in the killing of cells through irreversible damage in cellular structure, breakdown of pectic substances and, as a consequence of these events, the modification of physical characteristics (Fig. 4).

Disruption of the cytoplasmatic membranes increases their permeability and permits water to enter the cells while the intracellular spaces expel gases and other volatile products. Chemical constituents of vegetable tissues are also affected by the blanching process. Denaturation of proteins is the most important and desirable effect of the heat treatment.

Many researchers have studied the effect of blanching and freezing on the textural

**Table 7**  Relationship Between Maximum Residual Enzyme Activity and Minimum High Quality of Frozen Vegetables

|          | Residual enzyme activity (%) | | |
|----------|--------------|----------|------------|
| Product  | Lipoxygenase | Catalase | Peroxidase |
| Peas     | 10           | 10       | 40         |
| Beans    | 20           | 20       | 50         |
| Sprouts  | 0            | 0        | Slight     |

*Source*: Ref. 35.

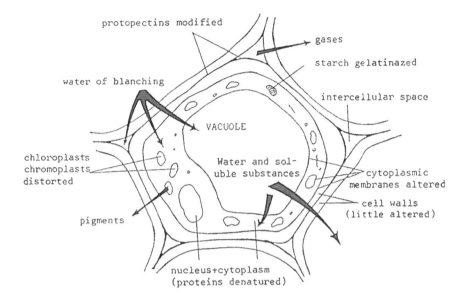

protopectins modified

gases

starch gelatinazed

water of blanching

intercellular space

VACUOLE

chloroplasts
chromoplasts
distorted

Water and sol-
uble substances

cytoplasmic
membranes altered

cell walls
(little altered)

pigments

nucleus+cytoplasm
(proteins denatured)

**Figure 4** Diagram showing the main effects of blanching on a generalized plant cell. (From Ref. 65.)

characteristics of frozen vegetables [61–63]. Destruction of cell walls produced a loss of turgidity due to a decrease in the capacity for water retention in the heat-treated tissues [64]. Heat treatments at temperatures above 65°C produce starch gelatinization, which during the freezing process and frozen storage is retrograded, modifying its hydration capacity [58,65]. Water enters the cells and fills the intracellular spaces, particularly when the vegetables have been bruised or fragmented during preblanching operations (sorting, grading, etc.). This occluded water is crystallized during the freezing process, producing important additional damage to the cell walls, especially if the freezing rate is slow. Solubilization of pectic sustances results in the detachment of cell wall polymers, which may result in an increase in mechanical damage. This deterioration is especially important in certain vegetables, especially green beans. Brown [63] reported that in unblanched green beans, slow freezing produced extensive breaking of cell walls. Thus, if quality is to be maintained, the entire product must be frozen rapidly. A number of other studies have confirmed the influence of freezing rate on tissue integrity, texture, and fluid retention [66].

The blanching process can directly increase the cell wall modifications produced during freezing [67,68]. In some vegetables, such as Brussels sprouts, the blanching time must be quite long to produce the desirable enzymatic inactivation. As a result, cooking of the area near the surface of the sprouts occurs. In these cases, a previous mild heat (52°C) treatment of the product can reduce the necessary blanching time by 20%. Consequently, textural changes in the product are reduced [69]. Earlier research has shown that excessive softness and sloughing in canned or frozen green beans could be overcome by using low-temperature blanching [44,70]. This firming effect, also observed in carrots [63], has been attributed to activation of pectin methylesterase (PME) in the raw vegetable at temperatures of 60–70°C. Therefore, a large number of free carboxyl groups are available on the pectin molecules, and they can be cross-linked through salt bridges with calcium ions present in the tissue (Fig. 5).

The application of heat-shock treatments instead of normal blanching procedures may be feasible with leguminous vegetables, such as beans and peas, to retain quality during frozen storage. Beans treated by this method have a natural, firm texture, which differs substantially from the softness of conventionally blanched green and wax beans [71]. During storage and distribution of frozen vegetables, high temperatures are occasionally encountered where thawing and refreezing occur. Since preservation of cellular structure, and thereby texture, is a fuction of freezing rate, a product thawed and refrozen in storage will be of poorer quality than one maintained in the frozen state [67]. Changes in ice crystal size during temperature fluctuations can influence the loss of fluid during and after thawing and the modification of textural properties. Although the thawing procedure does not appear to influence the

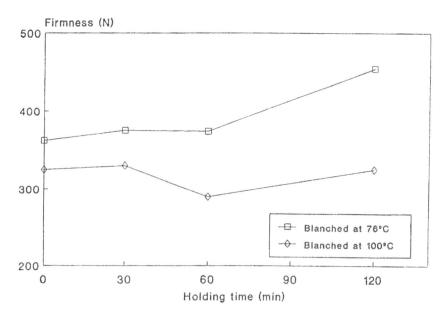

**Figure 5**   Effect of holding time after blanching on canned carrots. (From Ref. 58.)

texture of frozen vegetables, it is usually recommended that they be placed directly into boiling water. Uniform cooking of vegetables frozen in a package will be facilitated by separating them as they thaw. On the other hand, the value of very rapid freezing has been questioned, since it was thought that softening during cooking might obscure textural improvement. However, Brown [72] froze carrots, green beans, and zucchini squash at various rates to produce different degrees of freezing damage. Each of these vegetables was then cooked for three different time periods and served to a sensory panel for texture evaluation. In almost all instances, differences visible in photomicrographs were recognized by the panel. These results document that the softening that takes place during cooking is different from the textural degradation that results from slow freezing.

## B. Flavor

During frozen storage, nonblanched vegetables can have a considerable accumulation of ethanol and other volatile compounds [26,73–75]. The most likely chemical mechanisms for off-flavor development will be more fully described in the section on intrinsic chemical reactions. However, this development has been ascribed to catalyzed oxidative reactions in vegetable tissues. Thus, the blanching process is employed to inactivate the enzymes responsible for this quality modification. A strong blanching, however, could produce a strange flavor and taste in the cooked product. Therefore, vegetables such as green beans, asparagus, cauliflower, and occasionally carrots, when frozen with previous cooking (boiled frozen products) and stored at $-18°C$, developed strong overcooked flavors [76]. In the same way, boiled frozen green peas develop this overcooked flavor faster during frozen storage than do slightly blanched, frozen green peas during the same period [77]. Sensory evaluation indicated a gradual loss of quality due to flavor deterioration, with long-time/low-temperature blanching of cooked frozen peas [78] (Table 8). The stability of most previously cooked frozen vegetables is less than that for blanched products. Only if the cooked vegetables are vacuum-packed or protected by a sauce is it possible to obtain a high-quality frozen product [79].

Water-blanching of aromatic herbs, such as parsley, produces important losses of essential oils but, at the same time, limits the loss of some volatile compounds during frozen storage [80]. The addition of chemicals such as sucrose (1.2%) to the blanching water prevents the loss of compounds responsible for the flavor in Brussels sprouts and onions [47].

## V. EFFECTS OF FREEZING, FROZEN STORAGE, AND THAWING ON NUTRITIONAL VALUE

## A. Vitamins

Blanching is the most important unit operation prior to canning, freezing, or dehydration of vegetables. However, it produces losses of some soluble constitu-

**Table 8**  Average Panel Response to Texture and Flavor (Palatability Attributes) of Cooked Frozen Green Peas[a]

| Months in storage at -23°C | Blanch treatment[b] | | | | | | | |
|---|---|---|---|---|---|---|---|---|
| | 82°C/60 s | | 76°C/60 s | | 71°C/3 min | | 60°C/6 min | |
| | T | F | T | F | T | F | T | F |
| 0 | +0.5 | 0 | +0.4 | -0.3 | -0.1 | -0.6 | -0.4 | -1.0 |
| 3 | -0.1 | -0.4 | 0 | 0 | +0.1 | -0.2 | -0.6 | -0.7 |
| 6 | 0[ab] | -0.4 | -0.2[a] | +0.2 | -0.1[ab] | -0.8 | -0.7[b] | -0.6 |
| 9 | +0.3 | +0.2[a] | 0 | 0 | 0 | +0.2[a] | -0.2 | -1.2[b] |

[a]Means with a different superscript within a row are significantly different at the 95% level.
[b]-4.0 to 0.0, poorer than reference; 0, same as reference; 0.0 to +4.0, better than reference. Reference commercial practice of blanching for peas 96°C/70 s.
T: Texture: F: flavor.
*Source*: Ref. 78.

ents of the vegetable tissues by leaching and through oxidation of thermolabile nutrients, such as vitamins. Vitamin C (ascorbic acid) and its oxidation product, dehydroascorbic acid, are very unstable compounds. Their losses are very important during frozen storage and the blanching process. Unblanched frozen green beans lose more than 75% of their original vitamin C after one year of storage at −20°C [81].

There is a considerable amount of information on the effect of blanching on vitamins. The losses are mainly due to the water solubility of vitamin C and the B vitamins. When water or steam is used for heating, leaching of vitamins, flavors, colors, carbohydrates, and other water-soluble components occurs. If products are going to be frozen after blanching, a chilling step will generally take place before transporting the product into the freezer. If this cooling is done with cold water, additional leaching takes place, resulting in additional losses in nutritional value. Also, losses in nutritional value during meal preparation must be taken into account. Ulrich [65] obtained the relative contents of certain vitamins in peas at various stages from production to consumption (Table 9). Blanched, frozen, and then cooked peas sustained the greatest losses in vitamin C and $B_1$, mainly during blanching and cooking.

Muller and Tobin [82] indicated that the loss of vitamin C (the most heat-labile and most water-soluble vitamin) can be as high as 45% during blanching. However, the deterioration of this vitamin slowed down considerably during frozen storage if the vegetables were previously blanched. Other authors [83] have examined the mechanisms important for leaching during blanching. Temperatures above 50°C and prolonged blanching times gave rise to continuously increasing losses from peas and parts of peas. Another study [84] dealt with the change in concentration of ascorbic acid and dehydroascorbic acid in peas during water blanching. Damage to peas led to increased leaching and oxidation of ascorbic acid. Drake et al. [85] compared the content of vitamin C in asparagus, beans, peas, and corn after blanching at the same temperatures for the same amount of time by water, steam,

**Table 9** Relative Content of Some Vitamins in Peas at Various Steps from Harvest to Consumption

| Step in process | Percent original amount remaining | | | | |
|---|---|---|---|---|---|
| | Vit. C | Vit. $B_2$ | Vit. $B_2$ | Niacin | Carotene |
| Newly harvested | 100 | 100 | 100 | 100 | 100 |
| Blanched | 67 | 95 | 81 | 90 | 102 |
| Blanched, frozen | 55 | 94 | 78 | 76 | 102 |
| Blanched, frozen, heated | 38 | 63 | 72 | 79 | 103 |

*Source*: Ref. 65.

and microwave. Water blanching resulted in the greatest vitamin retention in the three products (Table 10).

Blanched vegetables destined to be frozen should be cooled shortly after blanching. The least leaching and thereby optimal retention of water-soluble vitamins was obtained with evaporative cooling without a water spray [31,84] (Table 11).

Other vitamins, like thiamine (vitamin $B_1$), riboflavin (vitamin $B_2$), and niacin (vitamin PP), are easily destroyed by thermal treatments. Moreover, other hydro-soluble vitamins, such as folic acid, panthotenic acid, and vitamin $B_{12}$, appeared to be stable during frozen storage if the vegetable was previously blanched. Vitamins $B_1$, $B_2$, and PP are mainly degraded during blanching, especially when this process is carried out with water [87]. Losses of 33% ascorbic acid, 20% riboflavin, 10% niacin, and 5% thiamine were obtained from water-blanched peas after 3 minutes [46]. However, these vitamins were stable during frozen storage if the temperature was lower than $-18°C$.

Short-time, high-temperature, blanched products have greater nutritional value retention during frozen storage than products blanched for long times at low temperatures. Long-time thermal treatments increase the loss of soluble compounds [88]. Figure 6 shows the loss of vitamin C content in frozen green beans blanched for different periods and stored at $-18°C$ for 5 and 10 months.

## B.  Other Soluble Compounds

Loss of sugars and soluble proteins may also be important during the blanching and cooling processes. Leafy vegetables (e.g., spinach, Brussels sprouts) or vegetables with young tissues with large amounts of soluble constituents (mushrooms, corn, peas, green beans) are the most affected in nutritional value. Slicing and cutting operations increase the leaching of soluble compounds. Thus, mechanical damage to vegetables that can take place during harvest affects the loss of nitrogen compounds. Peas have shown losses of 25–48% depending on the method employed (mechanical or hand) [89].

**Table 10**   Effect of Three Different Blanching Methods on Vitamin C Content of Four Products

| Blanching method | Vitamin content (mg/100 g) | | | |
|---|---|---|---|---|
| | Asparagus | Beans | Peas | Corn |
| Water | 35.7 | 22.5 | 15.6 | 15.8 |
| Steam | 35.3 | 23.3 | 11.0 | 13.6 |
| Microwave | 18.9 | 13.1 | 9.3 | 12.9 |

*Source*: Ref. 85.

**Table 11** Relationship Between Retention of Vitamin C and Blanching/Cooling Method

| Blanching principle/length of blanch(s) | Cooling method/length of cooling step (min) | Vitamin C level (mg/100 g wet weight) | Vitamin C |
|---|---|---|---|
| Steam/35 | Air/2 | 13.88 | 90.2 |
| Steam/35 | Water/3 | 9.93 | 79.8 |
| Water/90 | Air/2 | 11.9 | 80.4 |
| Water/90 | Water/3 | 9.80 | 69.1 |

*Source*: Ref. 86.

Mineral losses during the blanching-cooling operations may be important depending upon the thermal treatment employed. Odland and Eheart [90] observed that water-blanched broccoli had lower phosphorus and potassium contents than steam-blanched products, with or without the addition of ammonium salts. However, retention of all other minerals analyzed (i.e., sodium, manganese, calcium, magnesium, and copper) was not affected by the blanching method (Table 12). Steam-blanched broccoli was higher in Na, Mn, Ca, Mg, and Cu than water-blanched broccoli, but the differences were not statistically significant. The fact that Na was not significantly higher in the steam-blanched vegetables was evidently

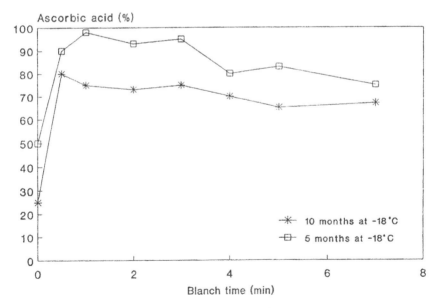

**Figure 6** Evolution of residual ascorbic acid content in green beans during frozen storage at −18°C related to the blanching time. (From Ref. 88.)

**Table 12**  Total Solids, Ash, and Mineral Contents of Blanched Broccoli from Three
Methods of Blanching[a]

| Blanching method[b] | Total solids | Total ash | P | K | Na | Mn | Ca | Mg | Cu |
|---|---|---|---|---|---|---|---|---|---|
| Raw[a] | 9.24 | 0.95 | 61.52 | 330.0 | 10.69 | 0.33 | 56.91 | 28.87 | 23.55 |
| Water | 7.48[y] | 0.54[y] | 47.62[y] | 174.0[y] | 5.17 | 0.22 | 53.02 | 22.27 | 11.91 |
| Steam | 8.08[z] | 0.68[z] | 51.50[z] | 232.0[z] | 5.57 | 0.28 | 57.71 | 25.11 | 16.61 |
| Steam-NH3 | 9.02[z] | 0.67[z] | 53.29[z] | 230.0[z] | 6.31 | 0.28 | 53.07 | 25.01 | 12.95 |

[a]Different superscripts indicate significance at the 15 level.
[b]Mean of five replication.
[c]Means for the raw vegetable (eight samples) are reported for comparison purposes.
[d]Values in 100 g broccoli, wet basis, reported in g for total solids and total ash, in μg for Cu, and in mg for
all other minerals.
*Source*: Ref. 90.

due to the higher loss from the steam-blanched vegetables to the cooling water.
Thus, the loss of minerals during blanching and cooling of broccoli varied with the
mineral analyzed.

During the freezing process there were fewer changes in mineral concentration.
Frozen green beans showed a significant decrease in iron, phosphorus, and potassium due to freezing, but calcium, sodium, and zinc levels increased significantly.
No change in chloride, copper, magnesium, manganese, and silicone were observed
due to freezing [91].

## VI.  EFFECTS OF FREEZING, FROZEN STORAGE, AND THAWING ON INTRINSIC CHEMICAL REACTIONS

### A.  Off-Flavor Development

The development of off-flavors in unblanched vegetables during frozen storage has
been attributed to rancidity development in the lipid matter catalyzed by lipoxidase.
However, little direct evidence is available to support this hypothesis. During
frozen storage, ethanol and other volatile compounds can accumulate in unblanched or insufficiently blanched vegetable tissues. This accumulation coincides
with sensory changes in aroma and taste (development of off-flavors) [92]. Such
off-flavors have been observed to remain in cooked vegetables after prolonged
boiling.

Several studies conducted on unblanched peas [93,94] have established a
relationship between the development of off-flavors and an increase in peroxide
value. Therefore, these off-flavors arise from the oxidation of polyunsaturated fatty

acids by lipoxygenase [95]. Figure 7 shows the amounts of hexenal and ethanol in frozen peas during storage at –5°C. Frozen peas with some harvest damage showed greater hexanal amounts.

However, lipid oxidation alone could not be the mechanism for off-flavor development. Rhee and Watts [96] demonstrated that rancidity was not the main cause of flavor deterioration in frozen peas and possibly other frozen vegetables. The amount of lipid oxidation that occurred in frozen raw peas was insufficient to produce raucid odors. Lipoxidase was rapidly inactivated by the short blanching time, and no regeneration of the enzyme occurred during frozen storage. Moreover, the unbalanced frozen peas were packed in nitrogen atmosphere, where oxidative reactions cannot occur, and developed a taste similar to that observed in peas packed in air [95]. Thus, other compounds, like ethanol and acetaldehyde, appear to contribute to the development of off-flavors [93,97,98]. These compounds are generally products of fermentation reactions and include some alcohols or other volatile organic compounds [97]. In another study, a significant correlation was observed between the content of volatile reducing compounds and the development of off-flavors in frozen vegetables [99]. However, the true relationship between

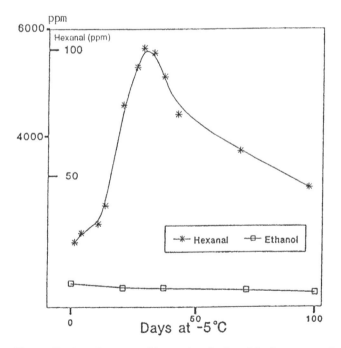

**Figure 7** Development of hexenal and ethanol in frozen peas during storage at –5°C. (From Ref. 146.)

these two factors has been difficult to document, because in some cases the enzymes that produce the acetaldehyde and ethanol can be inactivated at lower blanching temperatures than are required to inactivate the enzymes that catalyze off-flavor development [100]. Fuleki and David [105] observed that, at a storage temperature of −12°C, the reactions that produce acetaldehyde and ethanol cannot take place. However, the reactions implicated in the development of off-flavors are only retarded at this storage temperature. Further basic research is needed to obtain sufficient knowledge of the biochemistry related to off-flavor development in frozen vegetables. Figure 8 shows a simple scheme of proposed off-flavor development mechanisms in vegetable tissues and their relationship to other intrinsic chemical reactions producing quality deterioration.

## B. Color Changes

### 1. Natural Pigment Degradation

*Chlorophylls.* The loss of the natural color during storage of frozen green vegetables has attracted much attention. Campbell [101] showed that color deterioration in frozen peas stored above −18°C was due to conversion of chlorophylls *a* and *b* to the corresponding pheophytins. Other authors [102,103] later proposed that the ratio of chlorophyll to pheophytin be used to measure color changes in green vegetables. In acid solutions the magnesium in chlorophylls is replaced by hydrogen to give the corresponding pheophytins. The reaction rate is first order with respect to acid concentration [102]. Wagenknecht et al. [104] subsequently suggested that the accumulation of free fatty acids in unblanched frozen peas was responsible for their color loss during storage, while Dietrich et al. [103] suggested that the different acidities of peas and beans accounted for the different rates of chlorophyll conversion.

Other forms of chlorophyll degradation are also known. Chlorophylls are bleached during fat peroxidation [105] and during oxidation of glycolic acid by the

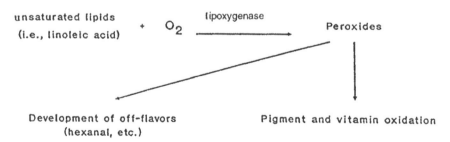

**Figure 8**   Scheme for the proposed mechanism of off-flavor development in vegetable tissues.

enzyme O-hydroxyacid dehydrogenase [106]. In addition, the enzyme chlorophyl-lase hydrolyzes the phytyl ester group of chlorophylls and pheophytins. Figure 9 shows possible mechanisms for chlorophyll degradation in frozen vegetable tis-sues.

Walker [42] demonstrated that the pH of beans remained relatively constant throughout frozen storage and the content of free fatty acids showed no trend with time, but peroxides from fat increased. This increase in peroxides from fat coin-cided with the start of chlorophyll and pheophytin destruction. Several studies have reported chlorophyll solutions to be rapidly bleached in systems containing lipox-ygenase [42,43]. Bleaching results from the breakdown of fatty acid hydroperox-ides, which is catalyzed by a bleaching factor with properties similar to the lipohydroperoxide breakdown factor found initially in soy extracts [107] and also in lower concentrations in pea extracts [108]. Buckle and Edwards [109] observed the storage of hand-podded, unblanched frozen peas for 20 months at –9.4°C and detected considerable conversion of chlorophylls to pheophytins. The formation of phytol-free derivatives results in a decrease in total pigment and the formation of peroxides and TBA-reactive materials. Also, greater storage stability was observed in samples blanched prior to storage and, to a lesser extent, in unblanched peas stored in nitrogen at –23.3°C. Figures 10 and 11 show chlorophyll degradation and pheophytin formation, respectively, during storage of unblanched peas stored under different conditions of temperature and packaging.

Peroxide values of unblanched peas increased during 8 months of storage and then decreased, except in peas stored in nitrogen at –23.3°C (Fig. 12). The decrease in total chlorophyll content is in agreement with a scheme involving peroxide formation by lipoxygenase. Subsequent peroxide destruction and chlorophyll

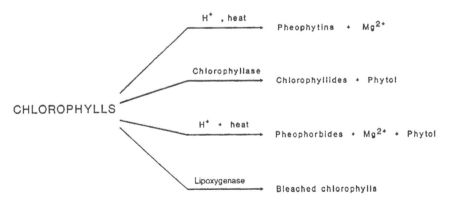

**Figure 9**   Scheme of chlorophyll degradation mechanisms in vegetable tissues.

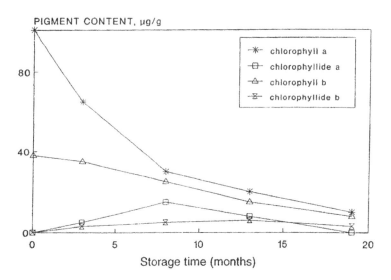

**Figure 10**   Chlorophyll and chlorophyllide contents in frozen unblanched peas stored in nitrogen for 20 months at –9.4°C. (From Ref. 109.)

breakdown involve factors similar to the lipohydroperoxide breakdown factor observed in pea and soy bean extracts, although it is possible that other mechanisms may also be involved [109].

*Anthocyanins.*    Anthocyanins are a group of more than 100 water-soluble plant pigments, which are usually dissolved in the cell sap rather than in the lipoidal bodies [110]. They differ greatly from fat-soluble pigments in their stability to degradation and time of biosynthesis. These pigments usually occur as glycosides (anthocyanin) of glucose, rhamnose, galactose, xylose, and arabinose. They are most stable at low pH and are destroyed in the presence of oxygen at higher pH. Lukton et al. [111] demonstrated that anthocyanin breakdown is dependent on pH in the presence of oxygen. It is also directly related to the level of pseudobase and inversely related to the level of cation. Although other factors contribute to the unacceptable quality of anthocyanin-containing vegetables, pH is the main factor influencing color stability [110]. The effect of heat on the stability of anthocyanin pigments during processing is well known [112,113]. Based upon studies of the thermal degradation of anthocyanins, two possible mechanisms have been proposed: (1) hydrolysis of the 3-glycosidic linkage produces aglucone, which is more labile, and (2) hydrolytic opening of the pyrillium ring forms a substituted chalcone, which degrades to a brown insoluble compound of polyphenolic nature.

Glucosidase and anthocyanase have been shown to destroy the stabilizing sugar. Consequently, the spontaneous decomposition of aglucone occurs. Poly-

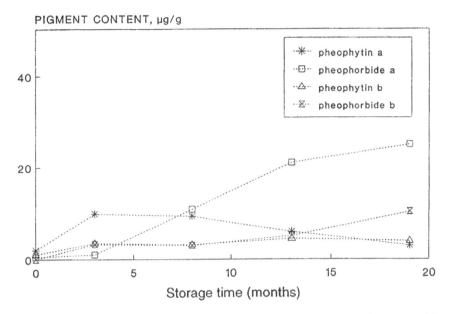

**Figure 11** Pheophytin and pheophorbide contents of frozen unblanched peas stored in nitrogen for 20 months at –9.4°C. (From Ref. 109.)

phenoloxidase, which requires cathecol for activation, has also been described by several authors [110], who have proposed the coupled mechanism shown in Figure 13.

Preservation of the color of anthocyanin-containing vegetables is very complex. Loss of bright red color may be due to loss of the color per se or to browning reactions, which tend to dull the color. Wolstrad [114] described the complexity of maintaining anthocyanin color in processed plant foods by explaining that the intrinsic chemical reactions involved are related to several factors, such as other tissue components (enzymes, acids, metal ions, oxygen, etc.) and the qualitative anthocyanin composition of the vegetable.

Despite the above-mentioned complex mechanisms for color changes in anthocyanin-containing vegetables, the freezing process per se does not affect the stability of these pigments. Preparation operations, mainly blanching, however, produce some anthocyanin degradation due to heat treatment and some loss of anthocyanins through leaching due to the hydrosoluble characteristic of these pigments when water blanching is employed. However, all aspects related to the effects of freezing, frozen storage, and thawing on anthocyanin pigments are more relevant in red fruits than in vegetables, where chorophylls and carotenoids are the most abundant natural pigments.

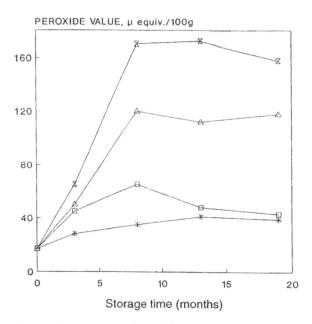

PEROXIDE VALUE, μ equiv./100g

**Figure 12**   Peroxide values of frozen unblanched peas during storage. (*, $N_2$ at –23.3°C; ⊟ , $O_2$ at –23.3°C; Δ, $N_2$ at –9.4°C; ⊠ , $O_2$ at –9.4°C.) (From Ref. 109.)

*Carotenoids.*   Carotenoid pigments (xanthophylls and hydrocarbon carotenoids) are widespread in fruits and vegetables. The loss of color of carotenoid-containing vegetables produces a loss of quality and a potential loss of provitamin A, since a few carotenoids can be split to form vitamin A [110]. The carotenoids are altered or partially degraded by acids. They are usually but not always stable in bases, cleaved by some enzymes, and sensitive to exposure to light. Since these pigments are fat soluble, they are not subject to leaching losses during the blanching and cooling prefreezing operations. However, a slight bleaching of the characteristic orange color has been observed in carrots slices during water-blanching [50]. Carotenoids are generally stable to the heat treatments involved in blanching but are rapidly lost during dehydration due to oxidation. Oxidation producing carotenoid degradation may occur during storage of frozen products due to dehydration during freezing. The conjugated doubled-bond system makes carotenoids especially susceptible to oxidative changes, which usually lead to discoloration or bleaching. Such reactions may be chemical or biochemical (enzymatic). Catalase, peroxidase, or lipoxygenase are enzymes that are inactive with respect to the carotenoids in fresh vegetable.

Various operations involved in processing often activate the enzyme/substrate linkage and carotene loss. Chemically activated oxidative changes are evident in

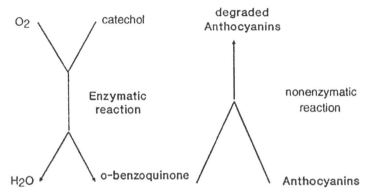

**Figure 13** Scheme of coupled mechanisms for anthocyanin degradation in plant tissues.

frozen vegetables [110]. Carotenoid oxidation is generally associated with unsaturated fatty acids and is usually autocatalytic [115]. According to the lipid peroxide theory, lipids, including carotenoids, undergo an initiation step to produce free radicals, which can be propagated and terminated, resulting in the bleaching of the vegetable tissues Although these chemical reactions are restricted in frozen vegetables, they must be controlled due to the protective role of chlorophylls against carotenoid degradation [116].

## 2. Enzymatic Browning

Some vegetables, such as potatoes and mushrooms, may develop color changes during frozen storage if they are frozen without previous treatment (e.g., blanching, some additives). These color changes result in nonreversible browning or darkening of the tissues. The enzyme related to this chemical reactions is polyphenoloxidase, which can be found in most plant tissues but in especially high amounts in mushrooms and potato tubers. Enzyme activity can vary markedly among varieties of the same plant, different stages of maturity, cultivation conditions, etc. [117]. Polyphenoloxidase is of particular importance in food science, not only for its role in enzymatic color changes but also in loss of nutritional quality and development of undesirable taste. Because of the broad substrate specificity of this enzyme, it has been called tyrosinase, polyphenolase, phenolase, catechol oxidase, catecholase, and cresolase at various times [117].

Some polyphenoloxidases catalyze two quite different types of reactions: the hydroxylation of monophenols to form o-dihydroxyphenols (cresolase-type activity) and the further oxidation of o-dihydroxyphenols to benzoquinone (cathecolase-type activity). The mechanism of polyphenoloxidase is poorly understood. Figure 14 shows a possible scheme for related chemical reactions.

Polyphenoloxidase activity in vegetables can be controlled by the addition of sulfite, ascorbic acid, and thiol compounds, partially due to reduction of o-

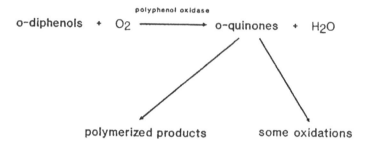

**Figure 14**   Scheme of enzymatic browning mechanism in vegetable tissues.

benzoquinone to o-diphenol, preventing color formation, and partially due to a direct effect of these compounds on the enzyme [118]. However, blanching is the best method to avoid the development of enzymatic browning due to low thermal sensitivity of this enzymatic system [119], which is dependent upon the acidity of the vegetable tissue and the acidity of the blanch-water solution.

## C.  Textural Changes

The chemistry of compounds that affect the texture of frozen vegetables is concerned largely with the cell wall and middle lamella components (pectins, hemicelluloses, and celluloses). The most important chemical modifications of these constitutents are related to the thermal process of blanching or to the ultimate cooking of these products. During frozen storage the texture modifications that can take place are mainly due to the physical processes of ice recrystallization. Thus, in this section an explanation of the principal chemical reactions affecting the texture of frozen vegetable products will concentrate on the effects of thermal treatments and addition of certain salts on the chemistry of vegetable texture.

Vegetables soften when they are heat-treated, due partially to the loss of turgor (particularly with leafy vegetables), but also due to a variety of chemical changes in the polysaccharides of the cell wall matrix. These modifications can be influenced by a number of factors—principally pH and the amounts and types of salts that are present [120]. Enhancement of softening at low pH has been attributed to hydrolytic cleavage of glycosylic bonds of the neutral sugar components of the cell wall. Enhanced softening at neutral pH has also been associated with polymer cleavage, in this case through the β-elimination reaction involving polyuronides, which has been demonstrated in potatoes [121].

The hydration of cell wall components that accompanies blanching reduces the cohesiveness of the matrix, softens the cell wall, and decreases intercellular adhesion. Matrix composition can play a role in the degree of hydration. It has been suggested that a high degree of methoxylation of cell wall pectin could increase water uptake [122]. The influence of calcium ions on texture has been investigated in a variety of products and process conditions. Many cases have been documented in which the addition of calcium either increased the firmness of tissues or prevented the loss of firmness. The most frequent observation is that vegetable firmness is improved by calcium addition after a mild blanch treatment. Such firming has been observed for frozen snap beans, cauliflower, potatoes, tomatoes [120], and carrots [58]. It is possible that the blanching process at 50–80°C activates pectinesterase activity, and pectin demethylation occurs while calcium ions cross-link the demethylated pectin, which increases tissue firmness [123,124]. However, other authors [125] have concluded that the firming effect of the precooking heat treatment of potatoes was more related to the increased availability of calcium ion released from the starch than to changes in pectin methylation produced by pectin methylesterase. Van Buren [126] found a 12% increase in calcium binding by green bean components when the vegetable was blanched at 71°C instead of 93°C prior to retorting. This finding was attributed to an increase in the number of calcium-binding sites after partial pectin demethylation. Huang and Bourne [127] demonstrated that the rate of thermal softening of vegetable tissues was consistent with two simultaneous first-order kinetic mechanisms acting on two substrates with different apparent rate constants. However, further work will be required to elucidate the biochemical changes responsible for the complex kinetics of these processes in vegetable tissues.

## D. Vitamin Degradation

Evaluation of the chemical effects of blanching, freezing, storage, and thawing on vitamins is a difficult task because of the diverse nature of the various vitamins as well as the chemical heterogeneity within each class of compound. Many factors affect retention of vitamins during processing and must be considered: (1) the time and temperature of processing and storage, (2) the concentration dependence and temperature dependence of the degradation reaction, (3) environmental variables such as the pH and the concentration of oxygen, metal ions ands various reducing or oxidizing agents, (4) the rate of competition or sequential reactions, (5) the relative stability of the various forms of vitamin present, (6) the chemical nature of the other food components, (7) the mechanism(s) of chemical loss of vitamin activity in the food, and (8) the water activity of the food system [128].

In order to facilitate the explanation of chemical reactions involved in vitamin losses in frozen vegetables, only the vitamins most abundant in these food products will be taken into consideration. Vitamin A and carotenes can undergo conversion of all-*trans* forms to *cis* isomers, which causes a reduction in the net vitamin A activity of the product. Losses of 15–35% of the vitamin A activity of vegetables have been reported during typical thermal processing and home-cooking procedures as a result of carotene stereoisomerization [129]. Therefore, other chemical reactions related to fatty acid enzymatic oxidation can degrade provitamin A carotenoids in vegetable tissues (see Sec. VI.B.1).

Ascorbic acid (vitamin C) is the most important vitamin in vegetable products. This vitamin was employed as a nutritional quality indicator of time-temperature-tolerance (TTT) studies of frozen vegetables [104]. Unblanched frozen vegetables suffer considerable loss of vitamin C during frozen storage, and such loss is clearly dependent on the storage temperature. Figure 15 shows the retention of vitamin C in unblanched peas during frozen storage at different temperatures [130].

The stability properties of ascorbic acid vary markedly as a function of environmental conditions such as pH and the concentration of trace metal ions and oxygen. Tannenbaum [131] reviewed the chemical mechanisms of ascorbic acid degradation in foods and explained the differences between aerobic and anaerobic conditions. In addition to the effects of pH, oxygen, and metal ions, the stability of ascorbic acid has been found to be affected by photochemical degradation and by the presence of other food components, such as sugar or correspondent sugar alcohols [132].

The kind of material used for packaging frozen vegetables significantly affects the stability of ascorbic acid. Factors to take into account include the effectiveness of the material as a barrier to moisture and oxygen and the chemical nature of the surface exposed on the frozen vegetable.

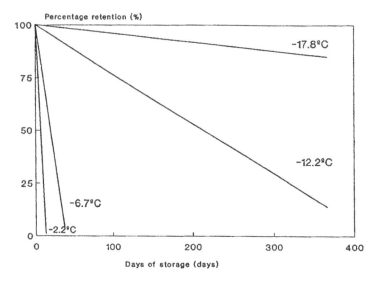

**Figure 15**   Ascorbic acid retention in frozen peas stored at different temperatures. (From Ref. 130.)

## E. pH changes

During the freezing process organic and inorganic salts tend to concentrate in the aqueous phase of the tissue. These salts can precipitate during cooling and change the pH of the unfrozen tissue water. Figure 16 shows the pH changes in minced cauliflower during frozen storage at $-10$ and $-18°C$. After the freezing process, pH is initially reduced but then undergoes a significant increase, which is greater in frozen products stored at higher temperatures. Similar trends have been obtained for frozen peas and green beans [133]. The pH changes during frozen storage of vegetable products may be important because they may directly affect the rate of some (bio)chemical reactions involved in quality changes of frozen products during storage and distribution.

## VII. EFFECTS OF FREEZING, FROZEN STORAGE, AND THAWING ON MICROBIOLOGICAL QUALITY AND SAFETY

## A. Microbiological Quality of Frozen Vegetables

Processing of vegetables by freezing does not cause any important microbiological and safety problems. The blanching process is the most relevant preparation step prior to freezing and destroys most microorganisms, except for bacterial spores.

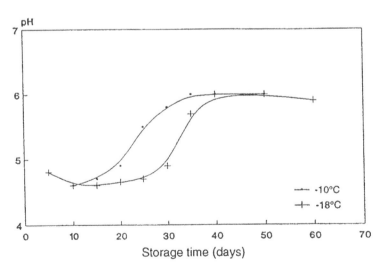

**Figure 16** pH changes in frozen cauliflower during storage at $-10$ and $-18°C$. (From Ref. 133.)

Consequently, blanching favors later development of the most thermoresistant stocks. Water blanching involves a greater microbiological risk than steam blanching, mainly due to the difficulty encountered in cleaning some parts of the blanching equipment [134]. Heat treatment produces cell death. Consequently, vegetable tissue could easily be recontaminated due to increased sensibility. Therefore, cooling after blanching must be accomplished as soon as possible to reduce the temperature of the blanched product to 15°C, and the transit time between blancher-cooler and freezer must be shortened. Integrated blancher-coolers perform well when compared with other equipment. Frozen vegetable products processed by this new equipment displayed mean values for total bacteria of 86,000/g and a mean value for coliforms of 320 (only 41 samples included) [135].

Proliferation of microorganisms in the lower areas of the blanching-cooling equipment may be an important factor in the microbiological quality of the finished product. Some product residues may stick to the most difficult-to-clean parts, producing an enormous increase in microbiological counts, mainly in vegetable products with high amounts of sugar and high moisture contents (i.e., corn). Such residues may come off during the flow of new blanched products, thereby contaminating them [136]. Recontamination could also occur due to the wearing away of belts. Total microorganism counts may increase more than 100- or 200-fold for lima beans and sweet corn. A continuous clean procedure using water pulverization containing 1–6 ppm of free chlorine is essential for safety purposes [136,137]. Cooling water must also be subjected to strict controls, since it can produce substantial contamination. Water recycling is an important step in the blanching-cooling of vegetable products to be frozen, not only for energy savings but also to reduce effluent volume and improve the microbiological quality of the frozen product. Bacterial contamination checked after various stages in the production line of quick-frozen peas is displayed in Table 13 [138]. These values give a simple view of the microbiological reductions during the blanching-cooling-freezing process.

**Table 13**  Bacterial Contamination at Various Stages of Production for Quick-Frozen Peas

| Stage of production | Thousands of germs/peas |
| --- | --- |
| Platform | 11,346 |
| After washing | 1090 |
| After blanching | 10 |
| At the end of water cooling flow | 239 |
| At the end of inspection belt | 410 |
| Freezer intake | 736 |
| After freezing | 560 |

*Source*: Ref. 138.

It is possible to process vegetables with less than 100,000 bacteria/g [139]. However, greater total counts of microorganisms are frequently detected under commercial conditions, particularly in frozen peas, green beans, and sweet corn [137]. New equipment designed to take into account heat regeneration, reduction of effluent water, and leaching of water-soluble components, as well as the microbiological aspects of finished products have been constructed. The integrated blancher-cooler is the most logical and the most advanced [28].

Microbiological studies of frozen peas processed by an integrated blancher-cooler have been carried out. Table 14 shows the mean total counts for processed peas and the distribution of coliforms and yeast/molds [28]. Differences in mean total counts were found between peas of different sizes, mainly in yeast/mold counts. Other microbiological studies related to frozen vegetables have examined the possible incidence of *Listeria monocytogenes* in market samples [140]. Isolates from pea pods and green peas were both catalase-negative, while those from green beans and spinach were gram-positive cocci. There are a variety of organisms present on fresh and frozen vegetables. However, it appears that the microflora of both types of vegetables is diverse. *L. monocytogenes* was not detectable in any of these samples.

## B. Safety Related to Chemical Contaminants

The use of pesticides has been an important factor for increasing the yield of crops throughout the world. Unfortunately, residues of pesticides sometimes remain on the products and such persistence has received considerable attention. Research has been directed toward ascertaining the effects of processing on the pesticide levels of vegetables. Agrochemicals have been developed to a very sophisticated level. Selective herbicides can be used in vegetable crops and leave the crop unaffected while effectively eradicating all undesirable weeds. Similarly, pesticides and fungicides have been developed to control insects and fungal diseases, respectively.

The use of pesticides does have other implications for crop quality. Some

**Table 14** Mean Value of Total Count, Coliforms, and Yeast/Mold for a Large Number of Examinations

| Product | Mean total bacteria/g | Mean coliform count/g | Mean count of yeast/g | Mean count of mold/g |
|---|---|---|---|---|
| Very fine peas | 30,684 | 45 | 234 | 262 |
| Fine peas | 27,752 | 31 | 81 | 154 |
| Medium fine peas | 30,116 | 18 | 67 | 119 |
| Rather coarse peas | 34,787 | 23 | 83 | 138 |

*Source*: Ref. 28.

pesticides will remain in the product after harvest, and these so-called residues can occasionally lead to quality problems. The level of residue may be unacceptable, or the presence of the residue may lead to the development of undesirable taints or off-flavors in the processed products. However, such instances are rare in countries where appropriate husbandry practices are employed [141]. The absence of taints from the use of pesticides in processed foods in recent years is perhaps an indication of the rigorous testing that pesticides now undergo. Many processors that contract crops specify the pesticides that may be used, most of which have been demonstrated not to cause taint problems. A wide range of possible options exists to minimize contamination problems in raw materials: (1) contract production, as mentioned above, (2) removal through processing—pesticides can be removed from vegetables by preparation steps prior to freezing (trimming, washing, blanching, etc.), (3) specifications, and (4) good manufacturing practices [142].

## VIII.  STABILITY OF FROZEN PRODUCTS

During frozen storage there is a gradual cumulative and irreversible loss of quality with time. The loss of quality arises from the individual or combined effects of physical, physicochemical, chemical, or biochemical changes (see Sec. VI). However, there is usually one limiting factor that influences the storage life of frozen foods [143]. Storage temperatures and temperature fluctuations and storage time are the principal factors affecting the quality of frozen vegetables, referred to as TTT (time-temperature-tolerance) factors. Assessment of quality is also dependent upon the requirements of consumers, which differ notably between countries.

The storage life of nearly all frozen foods, even vegetables, increases with colder storage temperatures to at least between –25 and –40°C (–13 and –40°F). The storage life/temperature relationship for a given vegetable product is illustrated in the time/temperature diagram (TTT curve) shown in Figure 17. A storage temperature of –18°C (0°F) or colder is specified for deep-frozen foods and for frozen foods, in general, in the legislation of many countries and in commercial practice [143]. Storage temperatures should be reasonably uniform in bulk storage rooms where frozen products are held between freezing and consumption. Increasing and fluctuating air temperatures may occur during transport and retail display. The most critical period for packaged frozen vegetables is during retail display and in the home freezer. In both, temperatures may be relatively warm and fluctuate.

TTT data for a given product may be determined by experiments where identical samples of the product are frozen and subsequently stored at a number of different temperatures (minimum four or five) normally in the range of –10 to –40°C (14 to –22°F). At suitable intervals, samples are examined for quality using sensory assessment, occasionally combined with appropriate objective tests. In the case of frozen vegetables, different parameters have been used, but the main one has been the content of vitamin C, considered to be one of the most labile constituents [144]. Chemi-

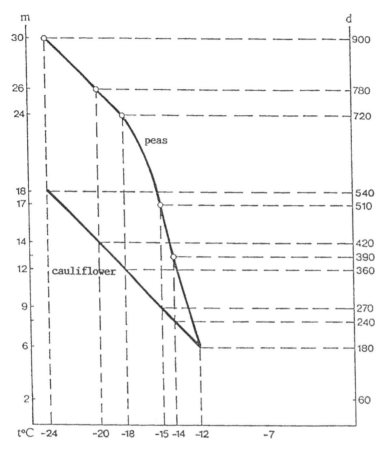

**Figure 17** Practical storage life (PSL) of frozen peas and frozen cauliflower based on TTT data. (t°C, temperature; m, months of storage; d, days of storage.) (From Ref. 143.)

cal tests for other variables may be employed to correlate with sensory observations, for example, reflectance measurement of the color and the chlorophyll content.

From the tests for quality at each storage temperature, the corresponding high quality life (HQL) or the practical storage life (PSL) can be established. Under ordinary conditions, time/temperature influences are cumulative over the entire storage life of the product.

Studies conducted on the changes occurring in vitamin C content in frozen vegetables, such as cauliflower, stored under different conditions indicate if a constant temperature of −22°C is maintained during storage, the frozen product loses only 25% of its vitamin C content after 13 months. Acceptable sensory quality was maintained for 10 months, but thereafter a decrease in the characteristic flavor occurred. Storage of the frozen cauliflower in a display freezer did not affect vitamin C, but temperature fluctuations during storage were very detrimental, especially to sensory characteristics [145]. For other frozen vegetables, the limiting quality factor for PSL may change with the product.

A temperature of −18°C (0°F) is commonly accepted as the upper limit for storing most vegetables from one season to another, allowing for a reasonable overlap. In the case of most vegetables, the frozen storage life can substantially exceed one year, providing the packaging material used affords the product sufficient protection against moisture migration and temperature fluctuation in the store is minimized [143].

The most fragile vegetables, such as mushrooms and white asparagus, have a product life of less than one year at −18°C (0°F). Storage at −25°C (−13°F) or colder is required to give such vegetables a product life of one year. In most vegetables, if longer storage periods are required, a blanching process may be performed. As shown in Figure 18, blanching treatment at 98°C for 30 seconds effectively prevents off-flavor development in green beans throughout 12 months of frozen storage at −18°C. However, this treatment leaves approximately 7% of the initial peroxidase activity of the fresh product, which indicates in green beans (as in most vegetable products) that the sensory quality is achieved prior to complete inactivation of this enzyme [41].

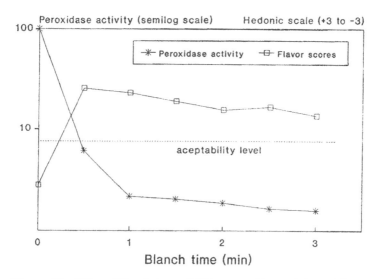

**Figure 18**    Effect of blanch time at 98°C on peroxidase activity and flavor score of green beans stored for 12 months at −18°C. (From Ref. 41.)

## IX. SUMMARY

The quality of frozen vegetables may be affected during each of a series of operations, including harvest, transport, cleaning, trimming, sorting, blanching, freezing, and packaging. The physical changes during freezing and subsequent thawing and cooking constitute a great hazard to the retention of freshness. Nevertheless, frozen vegetables are fresher than any other preserved product. At low enough temperatures (below −10°F, −23°C), vegetables can retain freshness for very long periods of time. The technical knowledge exists to successfully freeze almost all vegetables to high-quality standards, but this is not always achieved in commercial operations. In the last few years, a great deal of effort has been devoted to the development of new blanching equipment, as the first important operation in industrial freezing of vegetable products. Integrated blancher-cooler equipment may solve many of the problems of traditional blanching processes by taking into account heat regeneration, water consumption, production of effluent water, and leaching of water-soluble components. Storage life of frozen vegetables is longer at lower temperatures. Consequently, temperatures of −20°C or below are recommended for long-term storage. Temperature fluctuations can reduce storage life by accelerating the physical processes related to migration of water in the packaged frozen product.

Freezing is often preferred to other preservation methods for vegetables, such as canning, because quality changes are often less extensive due to the slower rates of intrinsic chemical reactions at subfreezing temperatures.

## ACKNOWLEDGMENTS

I acknowledge Dr. Fco. José Alguacil for guidance and helpful assistance in the composition and drawing of tables and figures and Mrs. M. Carmen Rodriguez for manuscript review and typing.

## REFERENCES

1. C. P. Mallet, *Frozen Food Technology* (C. P. Mallet, ed.), Blakie Academic and Professional, London, 1993, p. 339.
2. D. Arthey, Freezing of vegetables and fruits, *Frozen Food Technology* (C. P. Mallet, ed.), Blakie Academic and Professional, London, 1993, p. 237.
3. O. Fennema, The frozen food industry: Key to competitiveness, *Technical Innovations in Freezing and Refrigeration of Fruits Vegetables* (D. S. Reid, ed.) International Institute of Refrigeration, Paris, 1989, p. 1.
4. G. Mazza, Carrots, *Quality and Preservation of Vegetables* (N. A. Michael Eskin, ed.), CRC Press, Boca Raton, FL, 1989, p. 75.
5. C. Y. Lee, Green peas, *Quality and Preservation of Vegetables*, (N. A. Michael Eskin, ed.), CRC Press, Boca Raton, FL, 1989, p. 159.
6. V. R. Boswell, Factors influencing yield and quality of peas—biophysical and biochemical studies, *Md. State Agric. Exp. Stp. Bull.* 306:54 (1929).

7.  D. J. Casimir, R. S. Mitchell, L. J. Lynch, and J. C. Moyer, Vining procedures and their influence on yield and quality of peas, *Food Technol.* 21:427 (1967).

8.  C. Erickson and E. von Sydow, Postharvest metabolism of green peas with special reference to glutamic acid and related compounds, *J. Food Sci.* 29:18 (1964).

9.  R. V. Makower and A. Ward, Role of bruising and delay in the development of off-flavors in peas, *Food Technol.* 4:46 (1950).

10. R. S. Mitchell, P. J. Rutledge, and P. W. Board, Cooling of vined peas in Australia, *Aust. C.S.I.R.O. Food Preserv. Q.* 38:56 (1978).

11. L. J. Lynch, R. S. Mitchell, and D. J. Casimir, Preservation of green peas, *Advances in Food Research*, Vol. 9 (C. O. Chichester, E. M. Mrak, and G. F. Stewart, eds.), Academic Press, New York, 1959, p. 61.

12. M. M. Boggs, H. Campbell, and C. D. Schwartze, Factors influencing the texture of peas preserved by freezing. I., *Food Res.* 7:272 (1942).

13. J. K. Steward and W. R. Barger, Effects of cooling method and top-icing on the quality of peas and sweet corn, *J. Am. Soc. Hortic. Sci.* 75:470 (1957).

14. H. C. Wager, Physiological studies of the storage of green peas, *J. Sci. Food Agric.* 15:245 (1964).

15. R. W. Buescher and H. Brown, Regulation of frozen snap bean quality by postharvest holding in carbon-dioxide enriched atmospheres, *J. Food Sci.* 44:1494 (1979).

16. C. C. Huxsol, Reducing the refrigeration load by partial concentration of foods prior to freezing, *Food Technol.* 36:98 (1982).

17. A. Lenart and J. M. Flink, Osmotic concentration of potato. I. Criteria for the end-point of the osmosis process, *J. Food Technol* 19:45 (1984).

18. A. Lenart and J. M. Flink, Osmotic concentration of potato. II. Spatial distribution of the osmotic effect, *J. Food Technol.* 19:65 (1984).

19. R. N. Biswal, K. Bozorgmerh, F. D. Tompkins, and X. Lin, Osmotic concentration of green beans prior to freezing, *J. Food Sci.* 56:1008 (1991).

20. R. B. Toma, H. K. Leung, J. Augustin, and W. M. Iritani, Quality of french fried potatoes as affected by surface freezing and specific gravity or raw potatoes, *J. Food Sci.* 51:1213 (1986).

21. S. D. Holdsworth, *The Preservation of Fruits and Vegetable Food Products*, Macmillan Press, London, 1983.

22. D. C. Willians, M. H. Lim, A. O. Chen, R. M. Pangborn, and J. R. Whitaker, Blanching of vegetables for freezing—which indicator enzyme to choose, *Food Technol.* 3:130 (1986).

23. E. Winter, Behaviour of peroxidase during blanching of vegetables, *Z. Lebensm. -Unters. Forsch.* 141:201 (1969).

24. H. Delincée and W. Schaefer, Influence of heat treatments of spinach at temperatures up to 100°C on important constituents. Heat inactivation of peroxidase, *Lebensm. -Wiss. Technol.* 8:217 (1975).

25. H. Böttcher, Enzyme activity and quality of frozen vegetables. I. Remaining residual activity of peroxidase, *Nahrung* 19:173 (1975).

26. M. A. Joslyn, Enzyme activity in frozen vegetable tissue, *Adv. Enzymol.* 9:613 (1949).

27. A. T. Lu and J. R. Whitaker, Some factors affecting rates of heat inactivation and reactivation of horseradish peroxidase, *J. Food Sci.* 39:1173 (1974).

28. K. P. Poulsen, Optimization of vegetable blanching, *Food Technol.* 40:122 (1986).

29. J. L. Bomben, W. C. Dietrich, J. S. Hudson, H. K. Hamilton, and D. F. Farkas, Yields and solids loss in steam blanching, cooling and freezing of vegetables, *J. Food Sci.* 40:660 (1975).

30. J. L. Bomben and J. S. Hudson, Cooked weight and solids loss of air-cooled and water cooled frozen vegetables, *J. Food Sci.* 42:1128 (1977).

31. P. A. Carroad, J. B. Swartz, and J. L. Bomben, Yields and solid loss in water and steam blanching, water and air cooling, freezing, and cooking of broccoli spears, *J. Food Sci.* 45:1408 (1980).

32. S. R. Drake and D. M. Carmichael, Frozen vegetable quality as influenced by high temperature short time (HTST) steam blanching, *J. Food Sci.* 51:1378 (1986).

33. Environmental Protection Agency, Vibratory sprial blancher-cooler, EPA-600/2-78-206, Washington DC.

34. E. Steinbuch, The effect of heat shocks on quality retention of green beans during frozen storage, *J. Food Technol.* 15:353 (1980).

35. E. Steinbuch, Heat shock treatment for vegetables to be frozen as an alternative for blanching, *Thermal Processing and Quality of Foods* (P. Zeuthen, J. C. Cheftel, C. Eriksson, M. Jul, H. Leniger, P. Linko, G. Varela, and G. Vos, eds.), Elsevier Applied Science Publ., London, 1984, p. 553.

36. J. B. Adams, Blanching of vegetables, *Nutr. Food Sci.* 73:11 (1981).

37. C. Thomopoulos, Influence de certains sels de sodium sur la durée du blanchiment des haricots verts, *Ind. Aliment. Agric.* 92:531 (1975).

38. P. Harrison and M. Croucher, Packaging of frozen foods, *Frozen Food Technology* (C. P. Mallet, ed.), Blakie Academic and Professional, London, 1993, p. 59.

39. EC Directive 90/128/EEC, Plastic materials and articles intended to come into contact with foodstuffs, European Communities, Brussels, 1990.

40. P. S. Taouckis, B. Fu, and P. Labuza, Time temperature indicators, *Food Technol.* 70 (1991).

41. K. S. Katsaboxakis, The influence of the degree of blanching on the quality of frozen vegetables, *Thermal Processing and Quality of Foods* (P. Zeuthen, J. C. Cheftel, C. Eriksson, M. Jul, H. Leniger, P. Linko, G. Varela, and G. Vods, eds.), Elsevier Applied Science Publ., London, 1984, p. 559.

42. G. C. Walker, Color deterioration in frozen green beans (*Phaseolus vulgaris*), *J. Food Sci.* 29:383 (1964).

44. W. C. Dietrich and H. J. Neumann, Blanching Brussels sprouts, *Food Technol.* 19:1174 (1965).

45. W. C. Dietrich, R. L. Olson, M. D. Nutting, H. J. Neumann, and M. M. Boggs, Time-temperature tolerance of frozen foods. XVIII. Effect of blanching conditions on color stability of frozen beans, *Food Technol.* 13:258 (1959).

46. N. B. Guerrant and N. B. O'Hara, Vitamin retention in peas and lima beans after blanching, freezing, processing in the tin and in glass after storage and after cooking, *Food Technol.* 7:473 (1953).

47. H. L. Hudson, V. J. Sharples, E. Pickford, and N. Leach, Quality of home vegetables. Effects of blanching and/or cooling in various solutions on organoleptic assessments and vitamin C content and on conversion of chlorophyll, *J. Food Technol.* 9:95 (1974).

48. D. Odland and M. Eheart, Ascorbic acid, mineral and quality retention in frozen broccoli blanched in water, steam and ammonia steam, *J. Food Sci.* 40:1004 (1975).

49.  J. Philippon, La congélation des haricots vrts. Effects combinés du temps et de la température pendant la preparation et la congélation sur la qualité finale du produit, *Rev. Gén. Froid.* 60:101 (1969).

50.  S. Mirza and J. D. Morton, Effects of different types of blanching on the color of slices of carrots, *J. Sci. Food Agric.* 28:1035 (1977).

51.  N. I. Krinki, The role of carotenoid pigments as protective agents against photosentized oxidations in chloroplasts, *Biochemistry of Chloroplasts* (I. W. Goodwin, ed.), Academic Press, New York, 1966, p. 423.

52.  L. W. Mapson, T. Swain, and A. W. Tomalin, Influence of variety, cultural conditions and temperature of storage on enzymatic browning of potato tubers, *J. Sci. Food Agric.* 14:673 (1963).

53.  O. Smith and C. O. Davis, Potato Processing, *Potatoes: Production, Storing and Processing* (O. Smith, ed.), AVI Publishing, Westport, CT, 1968, p. 558.

54.  D. Gray and J. C. Hughes, Tuber quality, *The Potato Crop, The Scientific Basis for Improvement* (P. M. Harris, ed.), Chapman and Hall, New York, 1978, p. 504.

55.  A. S. Jaswal, Disodium dihydrogen pyrophosphate blanch and the texture of french fries, *Am. Potato J.* 47:145 (1970).

56.  T. R. Gormley and P. E. Walshe, Reducing shrinkage in canned and frozen mushrooms, *Ir. J. Food Sci. Technol.* 6:165 (1982).

57.  G. Crivelli, E. Sinesi, and A. Maestrelli, Recherches sur le comportement des légumes à la congélation rapide—Note IV. Aptitude variétale du chou-fleur et influence des traitements pré-congélation, *Rev. Gen. Froid.* 67:521 (1976).

58.  C. Y . Lee, M. C. Van Bourne, and J. P. Van Buren, Effect of blanching treatments on the firmness of carrots, *J. Food Sci.* 44:615 (1979).

59.  H. G. Schutz, M. Martens, B. Wilsher, and M. Rodbotten, Consumer perception of vegetable quality, *Acta Hort.* 163:31 (1984).

60.  M. Martens, Sensory and chemical/physical quality criteria of frozen peas studied by multivariate data analysis, *J. Food Sci.* 51:599 (1986).

61.  O. Fennema, Freezing rate and food quality, *Food Technol.* 23:1282 (1969).

62.  W. C. Dietrich, M. D., Nutting, R. L. Olson, F. E. Linquist, M. M. Boggs, G. S. Bohart, H. J. Neumann, and H. J. Morris, Time-temperature tolerance of frozen foods. XVI. Quality retention of frozen green beans in retail packages, *Food Technol.* 13:136 (1959).

63.  M. S. Brown, Texture of frozen vegetables: Effect of freezing rate on green beans, *J. Sci. Food Agric.* 18:77 (1967).

64.  J. P. Van Buren, J. C. Moyer, D. E. Wilson, W. B. Robinson, and D. B. Hand, Influence of blanching conditions on sloughing, splitting and firmness of canned snap beans, *Food Technol.* 14:233 (1960).

65.  R. Ulrich, Modifications de la structure et de la composition des fruits et légumes non blanchis et coséquences du blanchiment, *Rev. Gén. Froid* 73:11 (1983).

66.  R. M. Reeve, Relationships of histological structure to texture of fresh and processed fruits and vegetables, *J. Text. Stud.* 1:247 (1970).

67.  T. E. Weier and C. R. Stocking, Histological changes induced in fruits and vegetables by processing, *Adv. Food Res.* 2:297 (1949).

68.  G. Crivelli and C. Buonocore, Rate of freezing and integrity of vegetable tissues, *J. Text. Stud.* 2:89 (1971).

69. M. S. Brown, Texture of frozen fruits and vegetables, *J. Text. Stud.* 4:391 (1977).

70. A. Monzini, G. Crivelli, C. Buonocore, and M. Bassi, Structural modification in frozen vegetables, *Bull. Inst. Intern. Froid Annexe* 3:239 (1974).

71. J. Van Buren and M. A. Joslyn, Current concepts on the texture of fruits and vegetables, *Crit. Rev. Food Technol.* 1:5 (1970).

72. E. Steinbuch, Improvement of texture of frozen vegetables by stepwise blanching treatments, *J. Fd. Technol.* 11:313 (1976).

73. M. S. Brown, Texture of frozen vegetables: Effect of freezing rate on softening during cooking, *Proc. XIII Int. Congr. Refrig.* 3:491 (1971).

74. M. A. Joslyn, The freezing of fruits and vegetables, *Cryobiology* (H. T. Meryman, ed.), Academic Press, New York, 1966, p. 565.

75. F. A. Lee, The blanching process, *Adv. Food Res.* 8:63 (1958).

76. A. L. Tappel, Haematin compounds and lipoxidase as biocatalysts, *Symposium of Foods: Lipids and Their Oxidation*, (H. W. Schultz, E. A. Day, and R. O. Sinnhuber, eds.), AVI Publishing Co, Westport, CT, 1962, p. 122.

77. P. C. Paul, B. I. Cole, and J. C. Friend, Precooked frozen vegetables, *J. Home Econ.* 44:199 (1952).

78. H. L. Hanson, H. M. Winegarden, M. B. Horton, and H. Lineweaver, Preparation and storage of frozen cooked poultry and vegetables, *Food Technol.* 4:430 (1950).

79. B. E. Halpin and C. Y. Lee, Effect of blanching on enzyme activity and quality changes in green peas, *J. Food Sci.* 52:1002 (1987).

80. J. Philippon and M. A. Rouet-Mayer, Blanchiment et qualité des légumes et des fruits surgelés. Revue. 2. Aspects sensoriels, *Int. J. Refrig.* 2:48 (1985).

81. J. Philippon and M. A. Rouet-Mayer, Incidence du blanchiment et de la dureé de conservation sur l'évolution des caractères organoleptiques, de l' activité peroxydasique et de l' acide ascorbique dans les haricots mange-tout congeèles, *Rev. Gèn. Froid.* 7:685 (1971).

82. H. G. Muller and G. Tobin, *Nutrition and Food Processing*, AVI Publishing Co., Westport, CT, 1980.

83. J. D. Selman and E. J. Rolfe, Effects of water blanching on peas seeds. I. Fresh weight changes and solute loss, *J. Food Technol.* 14:493 (1979).

84. J. D. Selman and E. J. Rolfe, Effects of water blanching on pea seeds. II. Changes in vitamin C content, *J. Food Technol.* 17:219 (1982).

85. S. R. Drake, S. E. Spayd, and J. B. Thompson, The influence of blanch and freezing methods on the quality of selected vegetables, *J. Food Qual.* 4:271 (1981).

86. D. B. Cumming and R. Stark, The development of a new blanching system, *J. Can. Diet. Assoc.* 41:39 (1980).

87. E. G. Gleim, D. K. Tressler, and F. Fenton, Ascorbic acid, thiamine, riboflavin and carotene contents of asparagus and spinach in the fresh, stored and frozen states both before and after cooking, *Food Res.* 9:471 (1944).

88. M. A. Rouet-Mayer and J. Philippon, Etude critique de l'utilization des teneurs en acide ascorbique pour l'appréciation de la qualité organolpetique des haricots congelés, *CR Congrès Intern. Froid Washington.* 3:271 (1971).

89. P. Varoquaux, Les pertes en matières solubles ou cours du blanchiment des pois, *CR Acad. Agric. France* 1:949 (1971).

90. D. Odland and M. S. Eheart, Ascorbic acid, mineral and quality retention in frozen broccoli blanched in water, steam and ammonia-steam, *J. Food Sci.* 40:1004 (1975).

91. A. Lopez and H. L. Willians, Essential elements and cadmium and lead in fresh, canned and frozen green beans (Phaseolus vulgaris L.), *J. Food Sci.* 50:1152 (1985).

92. F. A. Lee, A. C. Wagenknecht, and J. C. Hening, A chemical study of the progressive development of off-flavor in frozen raw vegetables, *Food Res.* 20:289 (1955).

93. C. L. Bedford and H. A. Joslyn, Enzyme activity in frozen vegetables. String beans, *Ind. Eng. Chem.* 31:751 (1939).

94. A. C. Wagenknecht and F. A. Lee, The action of lipoxidase in frozen raw peas, *Food Res.* 21:605 (1956).

95. B. L. Bengtsson and I. Bonsund, Lipid hydrolysis in unblanched frozen peas, *J. Food Sci.* 31:474 (1966).

96. K. S. Rhee and B. M. Watts, Lipid oxidation in frozen vegetables in relation to flavor change, *J. Food Sci.* 31:675 (1966).

97. K. E. Murray, J. Shipton, F. B. Whitfield, and J. H. Last, The volatiles of off-flavoured unblanched green peas, *J. Sci. Food Agric.* 27:1093 (1976).

98. P. A. Buck and M. A. Joslyn, Accumulation of alcohol in underscalded frozen broccoli, *J. Agric. Food Chem.* 1:309 (1953).

99. J. Shipton and J. H. Last, Estimating volatile reducing substances and off-flavor in frozen green peas, *Food Technol.* 22:105 (1968).

100. T. Fuleki and J. J. David, Effect of blanching, storage, temperature and container atmosphere on acetaldehyde and alcohol production and off-flavor formation in frozen snap-beans, *Food Technol.* 17:101 (1963).

101. H. Campbell, Undesirable color changes in frozen peas stored at insufficiently low temperatures, *Food Res.* 2:55 (1937).

102. G. Mackinney and C. A. Weast, Color changes in green vegetables, *Ind. Eng. Chem.* 32:392 (1940).

103. W. C. Dietrich, F. E. Lindquist, J. C. Miers, G. S. Bohart, H. J. Neumann, and W. F. Talburt, Objective tests to measure adverse changes in frozen vegetables, *Food Technol.* 11:109 (1957).

104. A. C. Wagenknecht, F. A. Lee, and F. P. Boyle, The loss of chlorophyll in green peas during frozen storage and analysis, *Food Res.* 17:343 (1952).

105. H. H. Strain, Unsaturated fat oxidase, specificity, occurrence and induced oxidations, *J. Am. Chem. Soc.* 63:3542 (1941).

106. N. E. Tolbert and R. H. Burris, Light activation of the plant enzyme which oxidizes glycolic acid, *J. Biol. Chem.* 186:791 (1950).

107. B. Gini and R. B. Koch, Study of a lipohydroperoxide breakdown factor in soy extracts, *J. Food Sci.* 26:359 (1961).

108. M. Holden, Chlorophyll bleaching by legume seeds, *J. Sci. Food Agric.* 16:312 (1965).

109. K. A. Buckle and R. A. Edwards, Chlorophyll degradation and lipid oxidation in frozen unblanched peas, *J. Sci. Food Agric.* 21:307 (1970).

110. K. L. Simpson, Chemical changes in natural food pigments, *Chemical Changes in Food During Processing* (T. Richardson and J. W. Findey, eds.), AVI Publishing, Co. Westport, CT, 1985, p. 409.

111. A. Lukton, C. O. Chichester, and G. Mackinney, The breakdown of strawberry anthocyanin pigment, *Food Technol.* 10:427 (1956).

112. K. L. Simpson, T. C. Lee, D. B. Rodriguez, and C. O. Chichester, Metabolism in senescent and stored tissues, *Chemistry and Biochemistry of Plant Pigments* (C. T. Goodwin, ed.), Vol. 1 Academic Press, New York, 1976, p. 780.

113. J. B. Adams, Thermal degradation of anthocyanins with particular reference to the 3-glycosides of cyanidin. I. In acidified aqueous solution at 100°C, *J. Sci. Food Agric.* 24:747 (1973).

114. R. E. Wrolstrad, Anthocyanin pigment degradation and nonenzymatic browning reactions in fruit juice concentration, *Oreg. Agric. Exp. Stn., Techn. Bull.*:6234 (1981).

115. M. Karel, Lipid oxidation secondary reactions and water activity in food, *Autoxidation in Food and Biological Systems* (M. G. Simic and M. Karel, eds.), Plenum Press, New York, 1980, p. 34.

116. N. I. Krinski, The role of carotenoid pigments as protective agents against photosensitized oxidations in chloroplasts, *Biochemistry of Chloroplasts* (I. W. Goodwin, ed.), Academic Press, New York, 1966, p. 423.

117. J. R. Whitaker, Mechanisms of oxidoreductases important in food component modification, *Chemical Changes During Processing* (T. Richardson and J. W. Findley, eds.), AVI Publishing Co., Westport, CT, 1985, p. 121.

118. A. Golan-Goldhirsh and J. R. Whitaker, Effect of ascorbic acid, sodium bisulphite and thiol compounds on mushroom polyphenol oxidase, *J. Agric. Food Chem.* 32:1003 (1984).

119. K. P. Dimick, Heat inactivation of polyphenolase in fruit purees, *Food Technol.* 5:237 (1950).

120. J. P. Van Buren, The chemistry of texture in fruits and vegetables, *J. Text. Stud.* 10:1 (1979).

121. M. J. H. Keijberts and W. Pilnik, β-Elimination of pectin in the presence of anions and cations, *Carb. Res.* 33:359 (1974).

122. D. S. Warren and J. S. Woodman, The texture of cooked potatoes: A review, *J. Sci. Food Agric.* 25:129 (1974).

123. L. G. Bartolome and J. E. Hoff, Firming of potatoes: Biochemical effects of preheating, *J. Agric. Food Chem.* 20:266 (1972).

124. C. Hoogzand and J. J. Doesburg, Effect of blanching on the texture and pectin of canned cauliflower, *Food Technol.* 15:160 (1961).

125. K. H. Moledina, M. Haydar, B. Ooraikul, and D. Hadziyev, Pectin changes in the pre-cooking step of dehydrated mashed potato production, *J. Sci. Food Agric.* 32:1091 (1981).

126. J. P. Van Buren, Calcium binding to snap bean water-insoluble solids. Calcium and sodium concentrations, *J. Food Sci.* 45:752 (1980).

127. M. C. Bourne, Effect of blanch temperature on kinetics of thermal softening of carrots and green beans, *J. Food Sci.* 52:667 (1987).

128. J. F. Gregory III, Chemical changes of vitamins during food processing, *Chemical Changes in Food During Processing* (T. Richardson and J. W. Findley, eds.), AVI Publishing Co., Westport, CT, 1985, p. 373.

129. A. T. Ogunlesi and C. Y. Lee, Effect of thermal processing on the stereoisomerization of major carotenoids and vitamin A value of carrots, *Food Chem.* 4:311 (1979).

130. International Institute of Refrigeration, *Recommendations for the Processing and Handling of Frozen Foods*, (I.I.R., eds.), Paris, 1972.

131. S. R. Tannenbaum, Vitamins and minerals, *Principles of Food Science*. Part I. *Food Chemistry* (O. R. Fennema, ed.), Marcel Dekker, New York, 1976.

132. G. G. Birch and T. Pepper, Protection of vitamin C by sugars and their hydrogenated derivatives, *J. Agric. Food Chem.* 31:980 (1983).

133. L. Van der Berg, Physiochemical changes in foods during freezing and subsequent storage, *Low Temperature Biology of Foodstuffs* (J. Hawthorne and E. Rolfe, eds.), Pergamon Press, Oxford, 1968, p. 205.

134. J. W. Holmquist, L. E. Clifcorn, D. G. Heberlein, C. F. Schmidt, and E. C. Rithell, Steam blanching of peas, *Food Technol.* 8:437 (1954).

135. D. F. Splittstoesser and D. A. Corlett, Aerobic plate counts of frozen blanched vegetables processed in the United States, *J. Food Protect.* 43:717 (1980).

136. H. D. Michener and R. P. Elliot, Microbiological conditions affecting frozen food quality, *Quality and Stability in Frozen Foods* (W. B. Van Arsdel, M. J. Copley, and R. L. Olson, eds.) Wiley-Interscience, 1969, p. 43.

137. D. F. Splittstoesser, W. P. Wettergreen, and C. S. Pederson, control of microorganisms during preparation of vegetables for freezing. II. Peas and corn, *Food Technol.* 15:332 (1961).

138. A. C. Pederson and M. F. Gunderson, *The Freezing Preservation of Foods*, Vol. 2 (D. K. Tressler, W. B. Van Arsdel, and M. J. Copley, eds.) AVI Publishing Co., Westport, CT, 1968, p. 289.

139. N. H. Sanderson, Jr., The bacteriology and sanitation of quick frozen foods, *Refrig. Eng.* 42:228 (1941).

140. R. L. Petran, E. A. Zottola, and R. B. Gravani, Incidence of *Listeria monocytogenes* in market samples of fresh and frozen vegetables, *J. Food Sci.* 53:1238 (1988).

141. C. Knight, Crop production, harvesting and storage, *Vegetable Processing* (D. Arthey and C. Dennis, eds.), Blakei Academic and Professional, London, 1991, p. 12.

142. T. Mayes and G. Telling, Product safety from factory to consumer, *Frozen Food Technology* (C. P. Mallet, ed.), Blakie Academic and Professional, London, 1993, p. 93.

143. International Institute of Refrigeration, *Recommendations for the Processing and Handling of Frozen Foods*, 3rd ed. (I.I.R., eds.) Paris, 1986.

144. A. E. Bender, Food manufacture and nutrition. Nutrition and the food industry, *Näringforskning. Suppl.* 19:24 (1981).

145. M. P. Aparicio-Cuesta and C. García-Moreno, Quality of frozen cauliflower during storage, *J. Food Sci.* 53:491 (1988).

146. B. L. Bengtsson, I. Bosund, and I. Rasmussen, Hexanal and ethanol formation in peas in relation to off-flavor development, *Food Technol.* 21:478 (1967).

# 8
# Freezing of Dairy Products

H. Douglas Goff and Michael E. Sahagian

*University of Guelph*
*Guelph, Ontario, Canada*

## I.  INTRODUCTION

Frozen dairy products can be divided into two categories: those that are frozen as a means of increasing their shelf life and will be thawed prior to consumption or further processing and those in which the freezing process is responsible for the development of the desired structure and texture and are consumed in the frozen state. Unlike most other frozen food commodities, the majority of frozen dairy products fall into the latter category, namely, ice cream and related frozen, aerated dairy desserts such as ice milk, sherbet, and frozen yogurt. The amount of dairy products that fall into the first category, those that need thawing prior to further processing or consumption, is very small relative to the frozen dairy dessert industry. Therefore, a major portion of this chapter will overview the technology and manufacture of ice cream and related products. In keeping with the theme of the book, the composition and ingredients used in ice cream and the process of mix manufacture will be covered briefly, while the topics of ice cream freezing, hardening, and stability and quality attributes associated with the ice phase will be covered in more detail. Following a review of ice cream freezing, the freezing of other dairy products will be covered, focusing on stability, shelf life, and quality deterioration that may occur in dairy products upon freezing.

The composition of normal bovine milk is presented in Table 1. The major components include the milkfat fraction in the form of an emulsion, and the milk solids-not-fat (snf) fraction, consisting of the casein proteins suspended in the form of micelles, the whey or serum proteins, the dissolved lactose, and the mineral salts. The first section of this chapter will cover the major effects that freezing has on the two main categories of milk components, the milkfat emulsion, and the milk snf.

**Table 1**  Average Composition of Normal Bovine Milk
and Range Exhibited from Various Species

| Component | Mean (%) | Range (%) |
|---|---|---|
| Milkfat | 3.9 | 3.5 – 5.1 |
| Milk solids-not-fat | 8.7 | 7.9 –10.0 |
| Protein | 3.25 | 2.3 – 4.0 |
| Casein | 2.6 | 2.4 – 2.8 |
| Whey (serum) proteins | 0.45 | 0.3 – 0.7 |
| Lactose | 4.7 | 4.6 – 4.9 |
| Ash | 0.74 | 0.72– 0.77 |
| Water | 87.4 | 85.5 –88.7 |
| Total solids | 12.6 | 12.2 –14.4 |

*Source*: Refs. 2 and 3.

Most dairy products are based on altering the composition by physically separating one of the components, e.g., concentrating the fat phase to produce creams or butter or concentrating the protein phase to produce cheese, or by removal of water, e.g., the production of condensed or dried milk products. Figure 1 presents an illustration of the variety of dairy products that can be manufactured from milk. The amount of water in the various products varies considerably. Ice cream and related products are usually manufactured by combining several dairy ingredients together with sweetening and flavoring agents. In addition to ice cream, many fluid products, cheeses, and butter are also frozen, although few of these products are distributed directly to the consumer in the frozen state. The freezing of dairy products has little, if any, effect on their nutritional properties. During frozen storage, however, the thiamine and ascorbic acid content of milk may decrease [1].

## II.  INFLUENCE OF FREEZING ON MILK COMPONENTS

### A.  Fat Fraction

The fat fraction in milk and most other dairy products exists in the form of an emulsion, with droplet sizes ranging from 0.5 to 5.0 μm [2]. The fat globules are coated with a protein and phospholipid membrane after secretion from the mammary gland. Homogenization greatly increases the number and surface area of the fat globules present, and milk proteins quickly adsorb to a homogenized fat globule to reduce its interfacial tension [3]. During freezing, the physical defect of greatest concern related to the fat phase is the loss of the emulsified state and the separation of the fat phase. Deemulsification and coalescence of the fat during static freezing are caused by the internal pressures set up in the freezing milk or cream by

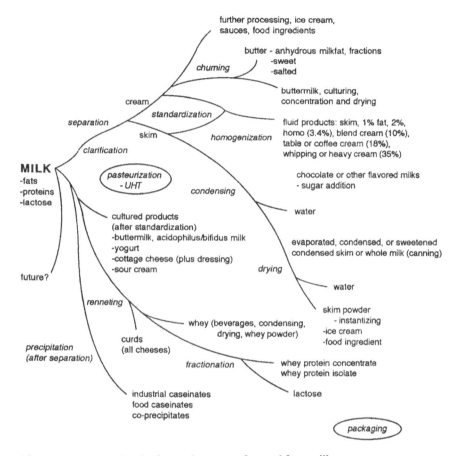

**Figure 1**    The family of dairy products manufactured from milk.

expanding ice crystals, and the degree is closely related to the extent of freezing
[1,4]. Homogenization prior to freezing greatly overcomes the deemulsification of
the fat globules in milk, concentrated milk, or low-fat cream due to their smaller
size and enhanced surface layers, making them less prone to rupture, although some
fat coalescence can still be evident, dependent largely upon the rates of freezing
and thawing. During dynamic freezing of the fat emulsion, in the presence of high
shear rates such as in ice cream manufacture, partial coalescence of the fat emulsion
is desirable to establish optimal structure and textural characteristics in the finished
product. In a study on the process of partial coalescence during freezing, it was
determined that neither shear forces nor ice crystallization alone were sufficient to
induce the rate of partial coalescence seen when both processes occurred simulta-

neously, demonstrating the importance of freeze-concentration of the emulsion, ice crystallization, and mechanical action on emulsion stability [5].

In addition to physical effects on the fat emulsion, chemical and enzymatic changes related to the fat phase are also of interest. While chemical and enzymatic reactions continue in frozen products, albeit at a much slower rate, the enzymatic lipolysis of milk (rancidity) is not usually a problem associated with freezing. Autoxidation, on the other hand, proceeds at varying rates depending on the temperature and availability of oxygen. It has been demonstrated that higher than normal pasteurization temperatures (85°C for 5 min) will have an inhibitory effect on the formation of the oxidized flavor, primarily through the formation of sulfhydral compounds [6]. Low storage temperatures of frozen products inhibit autoxidation mainly due to the reduction in the amount of unfrozen phase [4].

## B.  SNF Fraction

The process of freezing itself has little significant effect on the proteins of milk, but frozen storage can cause the casein micelles to lose their stability and precipitate upon thawing. This manifests itself as either thickened product or as flocs of casein evident either on the sides of thawed glass or plastic containers or as a precipitate on the bottom [4]. In fact, commercial separation of casein from milk serum may be accomplished by freezing followed by storage and thawing to produce a product that has been referred to as cryo-casein [7]. The flocculation of casein from frozen milk is initially reversible with heat and agitation but becomes irreversible with continued storage. Even minor amounts of casein flocculation can lead to the perception of a chalky texture upon consumption. Factors contributing to its instability include a high degree of concentration of the milk, preheating or pasteurization temperatures in excess of 77°C, storage temperatures above –23°C and especially higher than –18°C, cooling and/or holding the product under refrigeration between concentrating and freezing, and lengthening storage periods [3,8]. These factors are all related to the state of the lactose and/or serum proteins in the product, as will be discussed subsequently. Slow freezing has also been reported to result in greater protein stability than fast freezing, possibly related to the effect of rapid freezing in promoting lactose nucleation [8].

The stability of casein in frozen milk depends heavily on the state of the lactose. Due to the freeze-concentration process, very little liquid water remains as a solvent for the lactose and mineral salts. A 10-fold concentration of the salts may occur at –7.5°, and up to 30-fold increase may occur at lower temperatures, especially in the presence of lactose crystallization [1,8]. Considerable precipitation of calcium phosphate may also take place during the freezing of milk, leading to a decrease in the pH from 6.7 to ~6.0 [4,8]. When dissolved, the lactose maintains a lowered freezing point, limits the concentration of the dissolved salts in the unfrozen phase, and contributes to a high viscosity in the unfrozen phase.

Studies have shown that casein remains relatively stable under these conditions [8]. However, crystallization of the lactose leads to an increased freezing point of the unfrozen phase, which in turn leads to a further crystallization of water at constant temperatures. Consequently, further concentration of the salts, increased acidity, a change in the balance of colloidal calcium phosphate, and resultant flocculation of the casein occurs. As little as 40% lactose crystallization may be sufficient for casein instability [8]. Dialysis of milk prior to freezing has led to enhanced protein stability, demonstrating the effects of milk salts on this process [8]. Ultrafiltration of milk has also been demonstrated to increase the storage stability by three times up to an optimum concentration level beyond which removal of soluble phosphate accounted for decreased stability [8]. Lactose does not crystallize readily from freeze-concentrated solutions and remains largely in the supersaturated state, provided no nucleation has occurred prior to freezing [8]. At high solution viscosities resulting from high degrees of supersaturation and low temperatures, an amorphous solid (glass) may easily form in the unfrozen phase, rendering greater stability to the product and demonstrating the importance of low storage temperatures.

## C. Freezing Point Depression

Freezing point is a colligative property that is determined by the molarity of solutes rather than by the percentage by weight or volume. The ideal molal depression constant for water as defined by Raoult's law is 1.86 for dilute solutions (i.e., each mole of solute will decrease the freezing point of water by $1.86°C$) [9]. Freezing point can, therefore, be used to estimate the molecular weight of pure solutes or the average molecular weight of mixed solutes. The freezing point of milk is usually very constant as milk is in osmotic equilibrium with blood when it is synthesized [2]. Lactose accounts for about 55% of freezing point depression of whole or skim milk, chloride accounts for about 25%, and the remaining 20% is due to other soluble components including calcium, potassium, magnesium, lactates, phosphates, and citrates [10]. In the dairy industry, the invariability of the freezing point of raw milk is an important quality control parameter as it is used to determine unintentionally or intentionally added water in milk.

Freezing point determinations may be done by the Hortvet procedure [11], which uses a mercury in glass thermometer or, in more modern instruments, a thermistor cryoscope [12]. For many years most cryoscopes were calibrated in degrees Hortvet, because Hortvet's procedure produced freezing points about 3.7% lower than the actual values in Celsius degrees [13]. A formula given by the Association of Official Analytical Chemists [14] for the conversion of °H to °C gives lower values than true values according to Prentice [13]. An alternate formula published by the International Dairy Federation [10] is:

$$C = 0.96418 \, H - 0.00085 \tag{1}$$

The freezing point of milk is usually in the range of –0.512 to –0.550°C, with an average of about –0.522°C [9]. Freezing points of other dairy products depend on their water content and dissolved lactose and other solids in the water phase (Table 2). Fats, proteins, and colloids in general have a negligible effect on the freezing point of water or solutions in which they are dispersed [1]. Thus, the freezing points of cream and skim milk are identical with those of the milk from which they were separated. Soured or fermented milk is unsuitable for added water testing because the freezing point is lowered by lactic acid and increased concentrations of soluble minerals. Several reports suggest that heat treatment of milk, including UHT and retort sterilization, produces little permenant effect on freezing points [10]. If the freezing point of unwatered milk is known, the relationship between added water and freezing point depression is given in Eq. (2). If the actual freezing point of the unwatered milk is not known, a reference value can be used.

$$W = \frac{(C - D)(100 - S)}{C} \tag{2}$$

where  $W$ = percent (w/w) extraneous water in suspect milk
$C$ = actual or reference freezing point of genuine milk
$D$ = freezing point of suspect milk
$S$ = the % (w/w) of total solids in suspect milk

Numerous references are available on factors affecting freezing points [9,15, 16]. There are small differences in freezing points between breeds (in the order of 0.002–0.007°C), with Holstein milks generally having the lowest freezing points. There is a slight tendency towards lower freezing points in late lactation, but it is not clear whether this effect is independent of feed effects. Similarly, seasonal

**Table 2**  Initial Freezing Point of Various Dairy Products

| Product | Freezing point (°C) |
|---|---|
| Condensed milk | |
|     36% T.S. skim | –3.1 |
|     Sweetened condensed milk | –15 |
| Cheese | |
|     Cottage (78% moisture) | –1.2 |
|     Cheddar (33.8% moisture) | –13 |
|     Processed cheddar (38.8% moisture) | –7 |
|     Brick | –9 |
|     Swiss | –10 |
| Butter | |
|     Unsalted | 0 |
|     Salted | –9 to –20 |

*Source*: Refs. 1 and 4.

differences in freezing points are probably due to feed effects. Larger differences may be observed if the cattle do not have free access to water at all times. Variations in the proportions of grains to roughage and fresh versus dry forage have significant but small effects on freezing point. Underfeeding increases freezing points. Udder health (mastitis) also has small effects on freezing point.

In addition to the initial freezing point of fluid milk or milk products, it is important to consider the process of freeze concentration and its temperature-dependent relationship to water:ice ratio. Water freezes out of solution in the form of relatively pure ice crystals, and as it does, it leaves behind an increasing concentration of solutes in a decreasing amount of solvent. Table 3 illustrates this relationship for skim and condensed skim milk. The significance of freeze concentration with respect to the texture of frozen dairy dessert products will be discussed further in Section III.B.2.

## III. ICE CREAM AND FROZEN DAIRY DESSERT PRODUCTS

Table 4 presents the range of compositional variables found in most ice cream mix formulations. The composition of frozen dairy products in most countries is standardized and regulated. In North America, legal definitions exist for minimum fat content, total solids content, and weight per volume for ice cream and lower-fat versions such as ice milks, low-fat ice creams, and sherbets. These products are all consumed in the frozen state and rely on a concomitant freezing and whipping process to establish their structure and texture. The manufacturing process for most of these products is similar and involves the preparation of a liquid mix, whipping and freezing this mix dynamically under high shear to a soft, semi-frozen slurry, incorporation of flavoring ingredients to this partially frozen mix, packaging the

**Table 3**  Percentage of Ice in Skim Milk and Condensed Skim Milk

| Temperature (°C) | % Ice content | |
| --- | --- | --- |
| | 9.3% T.S. | 26% T.S. |
| −2 | 75.0 | 20.0 |
| −4 | 87.5 | 53.0 |
| −8 | 92.5 | 74.0 |
| −12 | 94.5 | 81.0 |
| −16 | 95.0 | 84.5 |
| −20 | 95.5 | 86.0 |
| −24 | 96.0 | 88.0 |

*Source*: Ref. 6.

**Table 4**  Typical Compositional Range for
Components Used in Ice Cream Mix Formulations

| Component | Range (%) |
|---|---|
| Milkfat | 10–16 |
| Milk solids-not-fat | 9–12 |
| Sucrose | 9–12 |
| Corn syrup solids | 4–6 |
| Stabilizers/emulsifiers | 0–0.5 |
| Total solids | 36–45 |
| Water | 55–64 |

product, and further freezing (hardening) of the product under static, quiescent conditions. Swept-surface freezers are used for the first freezing step, while forced convection freezers, such as air blast tunnels or rooms, or plate-type conduction freezers are used for the second freezing step. Comprehensive reviews of ice cream manufacture are available [17–20].

Ice, air, and fat each occupy distinct but interrelated phases (Fig. 2) and together establish the structure and resulting texture. The ice phase is of critical importance to the quality and shelf life of frozen products. The objective of ice cream

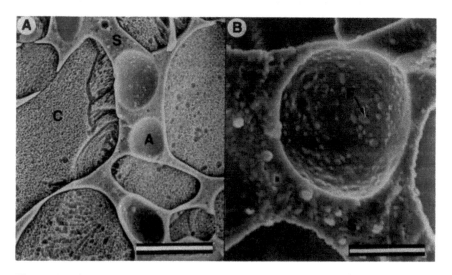

**Figure 2**  The structure of frozen, aerated dairy dessert products showing ice crystals (C), air bubbles (A), fat globules (F), and the freeze-concentrated unfrozen serum phase (S), as evidenced from cryo-scanning electron microscopy A: Bar = 30 µm; B: Bar = 7.5 µm. (From Ref. 21.)

manufacturers is to produce ice crystals below, or at least not significantly above, the threshold of sensory detection at the time of consumption. This threshold has been suggested by Arbuckle [17] to be between 40 and 50 µm. Consequently the freezing steps of the manufacturing process and the temperature profile throughout the distribution system are the critical factors in meeting this objective.

## A. Ingredients

Ice cream and ice milk mix formulations specify the content of milkfat, milk solids-not-fat, sweeteners, stabilizers, emulsifiers, and water that are desired. The role and sources of these components will be discussed in limited detail below, and the reader is directed to the references specified in each section for further information. Usually one mix is used for the production of a variety of flavors. Dairy and other ingredients used to supply the desired components are chosen on the basis of availability, cost, and desired quality. Ingredients available for each of the major components are listed in Table 5. It must be remembered, however, that in addition to supplying the major component, most of the dairy ingredients also supply minor amounts of the other components. Thus, a detailed calculation is necessary to convert the desired formula into a recipe, based on the ingredients chosen. Algebraic methods for the solution of these problems are presented in detail in Arbuckle [17] or in Jimenez-Flores et al. [20], and computer programs to solve mix calculation problems, based on the solution of simultaneous equations, are in widespread use in many commercial operations.

**Table 5**  Ingredients of an Ice Cream Mix and Components Supplied to Its Formulation

| Component | Ingredients to supply (also supplies) |
|---|---|
| Milkfat | Cream (SNF, water) |
| | Butter (SNF, water) |
| Milk solids-not-fat | Skim powder (water) |
| | Condensed skim (water) |
| | Condensed milk (water, fat) |
| | Sweetened condensed skim (water, sugar) |
| | Sweetened condensed milk (water, sugar, fat) |
| | Whey powder (water) |
| Water | Skim milk (msnf) |
| | Milk (fat, msnf) |
| | Water |
| Sweetener | Sucrose |
| | Corn syrup solids |
| | Liquid sugars (water) |
| Stabilizers/emulsifiers | |

## 1. Milkfat

Milkfat increases the richness of flavor in ice cream, produces a characteristic smooth texture by lubricating the palate, helps to give body, and aids in producing desirable melting properties. The fat content of a mix also aids in lubricating the freezer barrel while the ice cream is being manufactured. Limitations on excessive use of butterfat in a mix include cost, a hindered whipping ability, decreased consumption due to excessive richness, and high caloric value. Fat contributes 9 kcal/g to the diet, regardless of its source. The triglycerides in milkfat have a wide melting range: +40 to –40°C. Consequently, there is always a combination of liquid and crystalline fat [22]. Alteration of this solid:liquid ratio at freezer barrel temperatures may affect the ice cream structure formed. Duplicating structure with other sources of fat is difficult [19]. During freezing of ice cream, the fat emulsion that exists in the mix will partially destabilize or churn as a result of air incorporation, ice crystallization, and high shear forces of the blades. This partial churning is necessary to set up the structure and texture in ice cream, which is very similar to the structure in whipped cream [23]. Emulsifiers help to promote this destabilization process, and will be discussed below.

The best source of butterfat in ice cream for high-quality flavor is fresh sweet cream from fresh sweet milk. Other sources of butterfat include sweet (unsalted) butter, frozen cream, or condensed milk blends. Whey creams have also been used but may lead to flavor or texture problems. The fat content is an indicator of the perceived quality and/or value of the ice cream. Ice cream must have a minimum fat content of 10% in most legal jurisdictions. Premium ice creams generally have fat contents of 14–18%. It has become desirable, however, to create light ice creams (<10% fat) with the same perceived quality. In addition to structure formation, fat does contribute a considerable amount of flavor to ice cream, which is difficult to reproduce in low-fat ice creams. Fat content must be altered by at least 1% before any noticeable difference appears in the taste or texture [17].

## 2. Milk Solids-Not-Fat

The snf or serum solids improve the texture of ice cream, aid in giving body and chew resistance to the finished product, are capable of allowing a higher overrun without the characteristic snowy or flaky textures associated with high overruns, and may be a cheap source of total solids [24]. The snf contain the lactose, caseins, whey proteins, minerals (ash), vitamins, acids, enzymes, and gases of the milk or milk products from which they were derived. The content of serum solids used in a mix can vary from 10 to 14% or more. Limitations on their use include off-flavors from certain products and an excess of lactose, which may lead to numerous problems, as discussed below [8].

Proteins contribute much to the development of structure in ice cream, including emulsification, whipping, and water-holding capacity. Emulsification properties of proteins in the mix arise from their adsorption to fat globules at the time of homogenization [25]. Whipping properties of proteins in ice cream contribute to the formation of the initial air bubbles in the mix [26,27]. The water-holding capacity of proteins leads to enhanced viscosity in the mix [28], which imparts a beneficial body to the ice cream, increases the meltdown time of ice cream, and contributes to reduced iciness [29]. The protein content of a mix is usually about 4%. When replacing traditional milk snf sources in ice cream with newer sources with altered protein, it is important to ensure these functional roles are being met.

Lactose is a disaccharide of glucose and galactose. Lactose does not contribute much to sweetness; it is only one fifth to one sixth as sweet as sucrose [2]. Lactose is relatively insoluble and crystallizes in two forms ($\alpha$ and $\beta$). These insoluble crystals (particularly $\alpha$, which takes on a characteristic tomahawk shape) can produce the defect known as sandiness when they are allowed to grow to a large size. Lactose content of ice cream mix is about 6% if no whey powder has been used in the formulation. Excessive levels of lactose in ice cream mix leads to reduced freezing point, causing a softening of the ice cream and the potential for development of iciness, a greater potential for lactose crystallization or sandiness, and salty flavor [30]. The lactose solubility in water at room temperature is about 11% [2]. During freezing, this concentration is exceeded as a result of freeze concentration (water removal in the form of ice). When 75% of the water is frozen in a mix consisting originally of 11% snf (6% lactose), the lactose content in the unfrozen phase corresponds to 39.3%. Probably much of the lactose in ice cream exists in a supersaturated, amorphous (noncrystalline) state, however, due to extreme viscosity [21]. Stabilizers help to hold lactose in a supersaturated state due to viscosity enhancement.

Traditionally, the best sources of serum solids for high-quality products are fresh concentrated skim milk or spray-dried low-heat skim milk powder. Others include sweetened condensed whole or skim milk, superheated condensed skim milk, buttermilk powder or condensed buttermilk, or dried or condensed whey [24]. The production of superheated condensed skim milk leads to a high degree of whey protein denaturation and cooked flavor but creates a product that is capable of holding more water of hydration leading to a higher viscosity and thus may be used to reduce or eliminate the need for stabilizers [17]. Buttermilk powder is a by-product of the butter industry. It contains a higher concentration of fat globule membrane phospholipids than skim milk. Thus it can be used for its emulsifying properties to reduce or eliminate the need for emulsifiers [17]. Whey is a by-product of the cheese industry and contains fat, lactose, whey proteins, and water but no casein. While skim milk powder contains 54.5% lactose and 36% protein, whey powder contains 72–73% lactose and only about 10–12% protein [2]. Thus it can promote some of the problems associated with high lactose as described above.

There are an increasing number of available whey protein products, many of which possess much higher quality than the traditional products [29]. The protein component of whey can be increased through ultrafiltration, which also leads to reduced lactose content. Subsequent drying can produce whey protein concentrates (WPC) with varying protein contents, from 36 to 75%. However, this is all whey protein, rather than the 80% casein/20% whey protein blend found in skim milk. WPC may contribute some added functional properties to a food system, which whey powder cannot, due to the increased concentration of protein. In some cases, use of concentrated whey protein blends in ice cream mixes can lead to more desirable texture in the ice cream [29]. It is also possible to produce concentrated protein products from the casein portion of milk proteins, which have potential for use in ice cream [29,31]. The standards of identity for ice cream allow for the use of up to 1% added edible caseinates in the mix. Caseinates are acid-precipitated protein preparations from milk, which are neutralized by the addition of mineral hydroxides to produce a variety of products such as sodium caseinate or calcium caseinate. These caseinates are formulated to give different functional properties, such as degrees of emulsification, foaming, gelation, or water holding [28].

## 3. Sweeteners

Sweet ice cream is usually desired by the consumer. As a result, sweetening agents are usually added to ice cream mix at a rate of 12–17% by weight. Sweeteners improve the texture and palatability of the ice cream, enhance flavors, and are usually the cheapest source of total solids [17]. In determining the proper blend of sweeteners for an ice cream mix, the total solids required from the sweeteners, the sweetness factor of each sugar, and the combined freezing point depression of all sugars in solution must be calculated to achieve the proper solids content, the appropriate sweetness level, and a satisfactory degree of hardness [19]. The most common sweetening agent used is sucrose, alone or in combination with other sugars. Sucrose and lactose are most commonly present in ice cream in the supersaturated or glassy state, with few crystals being present [19,21]. It has become common practice in the industry to substitute sweeteners derived from corn syrup for all or a portion of the sucrose [17,20]. A typical sweetener blend for an ice cream mix usually includes 10–12% sucrose and 4–5% corn syrup solids (glucose solids) [17,20].

The use of corn starch hydrolysis products (corn syrups or glucose solids) in ice cream is generally perceived to provide greater smoothness by contributing to a firmer and more chewy body, to provide better meltdown characteristics, to bring out and accentuate fruit flavors, to reduce heat-shock potential which improves the shelf life of the finished product, and to provide an economical source of solids [32–34]. The higher molecular weight saccharides (dextrins) are effective stabilizers and provide maximum prevention against coarse ice crystal formation, which is reflected in improved meltdown and heat-shock resistance. They also improve cohesive and adhesive textural properties. Smaller sugars provide smoothness, sweetness, and flavor enhancement. Dextrose offers sweetness synergism with sucrose but has the greatest freezing point depression of all the sugars. The ratio of higher to lower molecular weight fractions can be estimated from the dextrose equivalent (DE) of the syrup. Ice cream manufacturers usually use a 28–42 DE syrup, either liquid or dry [17,33]. When dextrose is converted to fructose, the resultant syrup is much sweeter than sucrose, although it has half the molecular weight and thus contributes more to freezing point depression than sucrose [34].

## 4. Stabilizers

Ice cream stabilizers are a group of ingredients (usually polysaccharides) commonly used in ice cream formulations. The primary purposes for using stabilizers in ice cream are to produce smoothness in body and texture, retard or reduce ice and lactose crystal growth during storage, and provide uniformity to the product and resistance to melting [17,20]. They also increase mix viscosity, stabilize the mix to prevent wheying off (e.g., carrageenan), aid in suspension of flavoring particles, produce a stable foam with easy cut-off and stiffness at the barrel freezer for packaging, slow down moisture migration from the product to the package or the air, and help to prevent shrinkage of the product volume during storage [35]. Stabilizers must also have a clean, neutral flavor, not bind to other ice cream flavors, contribute to acceptable meltdown of the ice cream, and provide desirable texture and mouthfeel upon consumption [35]. Limitations on their use include production of undesirable melting characteristics, excessive mix viscosity, and contribution to a heavy, soggy body. Although stabilizers increase mix viscosity, they have little or no impact on freezing point depression.

Gelatin, a protein of animal origin, was used almost exclusively in the ice cream industry as a stabilizer but has gradually been replaced with polysaccharides of plant origin due to their increased effectiveness and reduced cost [17]. Stabilizers currently in use include:

1. Carboxymethyl cellulose, derived from the bulky components or soluble fibre of plant material
2. Locust bean gum (carob bean gum), which is derived from the beans of the tree *Ceratonia siliqua* grown mostly in the Mediterranean
3. Guar gum, from the guar bush, *Cyamoposis tetragonolba*, a member of the legume family grown in India and Pakistan for centuries and now grown to a limited extent in Texas
4. Carrageenan, an extract of *Chondus crispis*, Irish moss, a red algae, originally harvested from the coast of Ireland, near the village of Carragheen
5. Sodium alginate, an extract of another seaweed, kelp, a brown algae.

Each stabilizer has its own characteristics, and often two or more of these stabilizers are used in combination to lend synergistic properties to each other and improve their overall effectiveness. Guar, for example, is more soluble than locust bean gum at cold temperatures, thus it finds more application in high-temperature, short-time (HTST) pasteurization systems. Carrageenan is a secondary colloid used to prevent the wheying off of mix, which is usually promoted by one of the other stabilizers [17,19,20].

The mechanism of action of stabilizers in enhancing frozen stability is related primarily to their effect on the ice phase. Stabilized ice creams have been observed from electron microscopy techniques to have smaller ice crystals than unstabilized ice cream both before and after storage at fluctuating temperatures, as demonstrated in Figure 3 [36]. The control of iciness by stabilizers has been examined by several researchers [37–44] and has been reviewed by Goff et al. [45]. While it was originally thought that stabilizers bound water, rendering it unfreezable, thereby exerting their stabilizing mechanism by reducing the quantity of ice present [17], recent research has failed to demonstrate this effect [39–44]. The stabilizing effect of the polysaccharides presently appears to result from an alteration of the diffusion properties of water and solutes within the unfrozen phase [45,46].

Polysaccharides in solution have the ability to form gels (implying chemical bonds) or physical entanglements in water, which limit or restrict the diffusion characteristics of water and solutes within their networks. They also hold free water as water of hydration around the polysaccharide structure. At a critical concentration, dependent upon the structure of each polysaccharide, these molecules occupy a volume that exceeds the volume of the solvent causing them to entangle or interact [47]. At this critical concentration, there is an abrupt increase in the serum viscosity. Stabilizers are widely accepted to have an effect on both the microscopic rheological properties and on the macroscopic structure/texture in the frozen ice cream. This can easily be demonstrated by the erratic behavior of unstabilized ice cream mixes during extrusion from the barrel freezer. The action of the polysaccharides in ice cream is probably related to their concentration in the unfrozen serum phase,

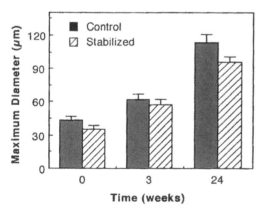

**Figure 3**  The relationship between size of ice crystals and time in the presence and absence of ice cream mix stabilizers. (From Ref. 36.)

resulting from freeze concentration [48]. There is a large discontinuity in the apparent viscosity of unfrozen ice cream mix when the concentration of stabilizer is increased. The concentration at which this occurs is similar to the value that would be obtained from freeze concentration (unpublished data from the author's laboratory). It is well known that polysaccharides increase mix viscosity prior to freezing. After the dynamic freezing process, it is likely that their concentration exceeds the critical concentration in the freeze concentrated unfrozen phase, which leads to a large restriction in solute and solvent motion due to this entanglement process [45,46]. The free volume of the continuous phase is probably lower than the critical volume necessary for transport of solutes beyond this critical concentration, which leads to their protective effect against ice recrystallization and lactose crystallization.

Stabilizer action in ice cream has been examined in a number of recent papers, which focus on thermodynamic, rheological, and mechanical properties [45,46,49]. Thermodynamic properties such as glass transition, heat capacity, and ice content determined by melting endotherm have been shown by DSC to be similar in systems with and without the presence of stabilizer [42,46]. However, it was shown from thermomechanical analyses of stabilized and unstabilized ice cream mixes that apparent viscosity of the stabilized mix increased dramatically and to a much greater extent than the unstabilized mix at temperatures below $-8°C$ [45]. Stabilizers have also been shown by thermomechanical analysis to decrease the rate of thermal deformation, increase apparent viscosity, and decrease compliance at $-26°C$ in frozen 20% sucrose solutions (proposed as model ice cream mixes since sucrose has the greatest influence on subzero rheological properties) [46]. Using $^1$H-nuclear magnetic resonance, it has been shown that stabilizers decrease the molecular relaxation properties of the same solutions [46]. Further, it has been shown that stabilizers increased storage and loss moduli in frozen ice cream mixes when compared to unstabilized mixes of the same composition [49]. It appears from the above results that stabilizers modify the kinetic properties of the unfrozen phase, rather than any thermodynamic properties associated with water:ice equilibrium. Therefore it was postulated that highly concentrated polysaccharides in the viscoelastic liquid surrounding ice crystals exceeded their critical concentration and consequently became entangled. As a result they may restrict diffusion of the water to existing ice crystals during fluctuations in temperature. It should be noted, however, that there may also be differences in stabilizer action related to the specific structure of each polysaccharide or protein hydrocolloid in solution (e.g., gelatin, xanthan), which may be important to ice cream stabilization [45,46].

## 5. Emulsifiers

Emulsifiers have been used in ice cream mix manufacture for many years. They are usually integrated with the stabilizers in proprietary blends, but their function and action is very different than the stabilizers. They are used for (1) improvement of the whipping quality of the mix, (2) production of a drier ice cream to facilitate molding, fancy extrusion, and sandwich manufacture, (3) smoother body and texture in the finished product, (4) superior drawing qualities at the freezer to produce a product with good stand-up properties and melt resistance, and (5) more exact control of the product during freezing and packaging operations [17,50].

Egg yolk was formerly commonly used as an ice cream emulsifier. Emulsifiers used in ice cream manufacture today are of two main types: the mono- and diglycerides and the sorbitan esters. Mono- and diglycerides are derived from the partial hydrolysis of fats of animal or vegetable origin. The sorbitan esters are similar to the monoglycerides in that they have a fatty acid molecule such as stearate or oleate attached to a sorbitol molecule, whereas the monoglycerides have a fatty acid molecule attached to a glycerol molecule. To make the sorbitan esters water soluble, polyoxyethylene groups are attached to the glucose molecule. Thus Polysorbate 80, the most common of these sorbitan esters, is actually polyoxyethylene sorbitan monooleate. Polysorbate 80 is a very active drying agent in ice cream [5] and is used in practically all commercial stabilizer/emulsifier blends.

The emulsifier molecules move instantly to and penetrate the interface between fat and water in an emulsion, thus reducing the interfacial tension between the two phases. An ice cream mix contains many surface-active molecules, such as the constituent milk proteins. Immediately after homogenization, these surface-active molecules compete for adsorption to the fat surface. Due to their ability to lower the interfacial tension, the emulsifiers displace proteins from the surface of the fat during the aging period of the mix [5,25]. During the ice cream–manufacturing process, the finely divided fat emulsion created by homogenization is subjected to great shear forces from freezing, agitation, and whipping in of air in the barrel freezer. This causes the emulsion to partially coalesce with the formation of fat globule clumps or clusters [51,52] as the emulsifiers do not produce sufficient coating of the fat to inhibit partial coalescence. These clusters are responsible for surrounding and stabilizing the air cells and creating a semicontinuous network or matrix of fat throughout the product [50]. As a result, the product is drier in appearance, possesses a characteristic smooth texture, and resists meltdown [5]. Hence the fat destabilization process is critical to the formation of the desired structure in ice cream [19,48,53].

## B.  Processing

### 1.  Mix Manufacture

Ice cream–processing operations can be divided into two distinct stages: mix manufacture and freezing operations. Ice cream mix manufacture consists of the following unit operations: combination and blending of ingredients, batch or continuous pasteurization, homogenization, and mix aging [18]. Ingredients are usually preblended prior to pasteurization, regardless of the type of pasteurization system used. Blending of ingredients is relatively simple if all ingredients are in the liquid form, as automated metering pumps or tanks on load cells can be used. When dry ingredients are used, powders are added either through a pumping system under high velocity or through a liquifier—a large centrifugal pump with rotating knife blades that chops all ingredients as they are mixed with the liquid [18].

Pasteurization is the biological control point in the system, designed for the destruction of pathogenic bacteria. In addition, it serves a useful role in reducing the bacterial load and in the solubilization of some of the components (proteins and stabilizers). Both batch and continuous (HTST) systems are in common use [18]. In a batch pasteurization system, blending of the proper ingredient amounts is done in large jacketed vats equipped with some means of heating, usually saturated steam or hot water. The product is then heated in the vat to at least 69°C (155°F) and held for 30 minutes to satisfy legal requirements for pasteurization, necessary for the destruction of pathogenic bacteria [54]. Various time-temperature combinations can be used, depending on the legal jurisdiction. The heat treatment must be severe enough to ensure destruction of pathogens and to reduce the bacterial count to a maximum (e.g., 100,000/g), depending on the legal jurisdiction [54]. Following pasteurization, the mix is homogenized using high pressures and then is passed across some type of heat exchanger (plate or double or triple tube) for the purpose of cooling the mix to refrigerated temperatures (4°C).

Continuous pasteurization is usually performed in an HTST heat exchanger following the blending of ingredients in a large, insulated feed tank. Some preheating, to 30–40°C, may be necessary for solubilization of the components. The HTST system is equipped with a heating section, a cooling section, and a regeneration section (Fig. 4). The mix first enters the raw regeneration section, where cold or preheated mix is heated as much as possible on one side of a plate heat exchanger while the pasteurized hot mix is cooled as much as possible running countercurrent on the opposite sides of the plates. Following the raw regeneration section, the mix enters the heating section where a minimum temperature of 80°C is obtained. The mix is held at this temperature for 25 seconds by flowing either through a series of holding tubes or through an additional set of plates in the HTST unit. Holding times much longer than the minimum can be accomplished with longer holding tubes. Holding times of 2 or 3 minutes may produce superior mixes, which retain many of the advantages of batch pasteurization [19,20]. After leaving the holding tube, the mix enters the homogenizer, depending upon the particular configuration, then flows back through the pasteurized side of the regeneration section and enters the cooling plates where a chilled brine solution or chilled water bring the mix down to around 4°C. Cooling sections of ice cream mix HTST presses are usually larger than milk HTST presses [54]. Due to the preheating of the mix, regeneration is lost and the mix entering the cooling section is still quite warm.

Homogenization is responsible for the formation of the fat emulsion by forcing the hot mix through a small orifice under pressures of 15.5–18.9 MPa (2000–3000

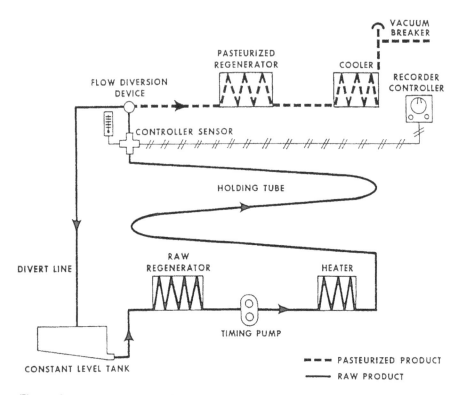

**Figure 4**  A schematic illustration of the equipment used for HTST pasteurization of milk, frozen dairy dessert mixes, and other fluid dairy products. (From Ref. 54.)

psig), depending on the mix composition, producing fat globules with an average diameter of <1 μm [19]. Homogenization is also responsible for the formation of the fat globule membrane, comprised of adsorbed proteins and emulsifier [25]. At the time of homogenization, many molecules (proteins, phospholipids, added surfactants, etc.) compete for space at the newly created fat surface and rapidly adsorb to the fat globule [55]. However, these molecules rearrange over time to their lowest energy state by reducing interfacial tension between the phases, thereby producing the fat globule membrane. With single-stage homogenizers, fat globules tend to cluster as bare fat surfaces come together. Therefore, a second homogenizing valve is frequently placed immediately after the first with applied back pressures of 3.4 MPa (500 psig) [18]. With two-stage homogenization, more time is allowed for adsorption of the membrane by physically separating clusters in the second stage. The net effects of homogenization are in the production of a smoother, more uniform product with a greater apparent richness and palatability and better whipping ability [17]. Homogenization also decreases the danger of churning the fat in the freezer and makes it possible to use products that could not otherwise be used, such as butter and frozen cream.

An aging time of 4 hours or more is recommended following mix processing prior to freezing to allow for hydration of milk proteins and stabilizers, crystallization of the fat globules, and a membrane rearrangement to produce a smoother texture and better quality product [56]. Emulsifiers generally displace milk proteins from the fat surface during the aging period [5,25]. The whipping qualities of the mix are usually improved with aging. Aging is performed in insulated or refrigerated storage tanks, silos, etc. Mix temperature should be kept as low as possible (at or below 5°C) without freezing.

## 2. Ice Cream Freezing

Ice cream freezing also consists of two distinct stages: passing mix through a swept-surface heat exchanger under high shear conditions to promote extensive ice crystal nucleation and air incorporation and freezing the packaged ice cream under conditions that promote rapid freezing and small ice crystal sizes [17,18]. The freezing and whipping process is one of the most important unit operations for the development of quality, palatability, and yield of finished product. Flavoring and coloring can be added as desired to the mix prior to passing through the barrel freezer, and particulate flavoring ingredients, such as nuts, fruits, candy pieces, or ripple sauces, can be added to the semi-frozen product at the exit from the barrel freezer prior to packaging and hardening.

*Equipment and Air Incorporation.* Continuous freezers dominate the ice cream industry [18]. In this type of process, mix is drawn from the flavoring tank into a swept surface heat exchanger, which is jacketed with a liquid, boiling refrigerant [usually ammonia, a chlorinated, fluorinated hydrocarbon (CFC) such as R-12, R-22, or R-502 or one of the newly developed CFC substitutes]. Incorporation of air into ice cream, termed the overrun, is a necessity to produce desirable body and texture. Overrun is calculated as follows [17]:

$$\% \text{ Overrun} = \frac{\text{Vol. of ice cream produced} - \text{Vol. of mix used}}{\text{Vol. of mix used}} \times 100\% \qquad (3)$$

As a guide, maximum overrun should be 2.5–3 times the total solids content to avoid possible defects in the finished ice cream [17].

Three types of air incorporation systems are used in continuous freezers. In older systems, the pump configuration resulted in a vacuum either at the pump itself or on the mix line entering the pump. Air was then incorporated through a spring-loaded, controllable needle valve. Newer types of freezers utilize compressed air, which is injected into the mix. This type of air-handling system allows for filtration of the air prior to entering into the mix [18]. Additionally, recent experimentation with preaeration systems in which the mix is whipped independently prior to entering into the barrel freezer has demonstrated that much smaller air bubble sizes and improved body and texture can be attained [57]. Following aeration, the water in the mix is partially frozen as the mix and air combination passes through the barrel of the heat exchanger. Rotating knife blades and dashers keep the product agitated and prevent freezing on the side of the barrel. Residence time for mix through the annulus of the freezer varies from 0.4 to 2 minutes, freezing rates can vary from 5 to 27°C per minute, and draw temperatures of –6°C can easily be achieved [19].

Batch freezing processes differ slightly from the continuous systems just described. The barrel of a batch swept surface heat exchanger is jacketed with refrigerant and contains a set of dashers and scraper blades inside the barrel. It is filled to about one-half volume with the liquid mix. Barrel volumes usually range from 2 to 12 liters. The freezing unit and agitators are then activated, and the product remains in the barrel for sufficient time to achieve the desired degree of overrun and stiffness. Whipping increases with time and cannot exceed the amount that will fill the barrel with product (i.e., 100% overrun when starting half full). Batch freezers are used in smaller operations where it is desirable to run individual flavored mixes on a small scale or to retain an element of the "homemade"-style manufacturing process. They are also operated in a semi-continuous mode for the production of soft-serve desserts. A hopper containing the mix feeds the barrel as product is removed.

The refrigerant in the continuous freezer must be sufficient to remove all the latent heat of fusion from the crystallizing water. It is extremely important to maintain freezers in excellent condition to obtain rapid freezing by keeping the heat-exchange surfaces free of oil or debris on the refrigerant side and keeping the blades sharp and true on the mix side. Moving the ice cream through the ingredient feeders, filling pipelines, and packaging machines should be done rapidly, so that the tiny preformed crystals do not have a chance to melt. Measures like precooling ingredients prior to addition to the fruit feeder, insulating the pipelines to the filling machines, and keeping pipelines short are essential to small, uniform ice crystal size distributions. Shrink wrapping should not cause any temperature increase in the product [17].

*Physical Chemistry of the Phase Change.*    Ice cream mix contains approximately 60% water (100 – TS%). The lower the total solids, the more water present. When ice cream is frozen, the water in the mix loses its latent heat of fusion and converts into ice, which freezes out of the mix in the form of fairly pure ice crystals. The crystallization of water to ice involves three major steps: undercooling, nucleation, and crystal growth [58,59]. (These processes have been more fully explored in Chapter 1.) Undercooling occurs when liquids such as water, milk, or mix are cooled several degrees below their initial freezing point through the removal of sensible heat but without a phase change (i.e., no formation of ice). The mix contains a significant amount of dissolved sugars, lactose, and salts and is frozen with rapid agitation, both of which tend to inhibit undercooling in the barrel freezer. However, ice crystallization is a kinetically controlled process and as such reaches a maximum rate, limited by diffusion of water to a growing crystal surface and diffusion of solutes away from the crystal surface [60] (see also Chapter 1). This maximum rate of crystallization dictates the time required to approach maximal ice formation and is influenced largely by the temperature difference between the freezing medium and the product and by the heat-transfer characteristics of both the product and equipment. There is some evidence that at the time of drawing of ice cream from the barrel, considerably less ice is present than the equilibrium freezing curve would dictate [21].

The dissolved sugars, lactose, and salts result in an initial freezing temperature in the mix of less than 0°C, usually about –2.5°, depending on their concentration. From Raoult's law, the lower the molecular weight, the greater the ability of a molecule to depress the freezing point at constant concentrations due to the increased numbers of molecules present [9]. Thus monosaccharides such as fructose or glucose produce a much softer ice cream than disaccharides such as sucrose [20]. This limits the amount and type of sugar that can successfully be incorporated into the formulation [30]. At the initial freezing temperature (heterogeneous nucleation temperature), water molecules begin to take on an ordered structure (in the form of a hexagonal cluster), and after a critical size of molecular cluster is surpassed, the cluster becomes a stable nucleus [59]. This process of exceeding a critical size of molecular cluster is known as nucleation [43]. Enthalpy associated with the energy state of the molecules and their phase change is released in the form of latent heat during nucleation. The process of crystal growth refers to the addition of further water molecules to the surface of a nucleus [44]. Together, nucleation and crystal growth constitute the process of crystallization. Rapid heat removal, which results from low temperatures (large $\Delta T$) in the freezing medium, and rapid agitation, which are both present in the barrel freezer, promote extensive nucleation and create numerous, tiny ice crystals. Further temperature reduction during hardening accounts for continued growth of the preformed crystals. Crystal growth is more rapid and nucleation rates are reduced, however, at higher freezing medium temperatures (smaller $\Delta T$s) during both dynamic freezing and hardening. These phenomena are related to the rates of diffusion of water within the freezing product compared to the rates of ice crystallization [59].

Another phenomenon occurring as water turns into ice is freeze concentration [58]. As was mentioned earlier, ice crystals are relatively pure. Thus the formation of ice acts to concentrate the dissolved sugars, lactose, milk proteins, salts, and hydrocolloids in an ever-decreasing amount of water. Water and its dissolved components are referred to as the serum in the mix [17]. Because the freezing point of the serum is a function of the concentration of dissolved solids, the formation of more ice concentrates the serum and results in an ever-decreasing freezing temperature for the remaining serum. Thus at temperatures several degrees below the initial freezing temperature, there is always an unfrozen phase present. Bradley [61,62] presented a method of calculation of initial freezing points and equilibrium freezing curves for sweetened dairy mixes based on their composition. An equilibrium freezing curve for an ice cream mix, which plots the amount of water frozen as a function of temperature, is shown in Figure 5. This represents the maximum amount of ice that can form at any given temperature, based on freezing point considerations [60]. Ice cream hardness is a function of temperature due to its effect on this conversion from unfrozen water to ice and further concentration of the serum phase surrounding the ice crystals, which helps to give ice cream its ability to be scooped and chewed at freezer temperatures.

At the draw temperature, only a portion of the water will be frozen, while the rest of the water remains in a concentrated solution. The equilibrium freezing curve shown in Figure 5 indicates that at a draw temperature of –5°C about 50% of the water would be frozen. However, probably much of the water remains undercooled at this time, thus the system is far from equilibrium [36]. During hardening, water continues to freeze but crystal growth takes over from nucleation. It is also very probable that maximal formation of ice does not occur during hardening, consequently, more unfrozen phase exists than the equilibrium freezing curve would predict. Ice crystal formation is an event that takes time to occur, and thus may be under kinetic control rather than thermodynamic control (i.e., the amount of ice present at any temperature is time dependent and therefore freezing rate dependent, so that it approaches but only reaches its maximum amount after infinite time)

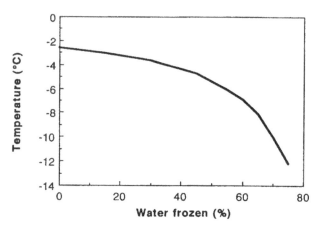

**Figure 5**  An equilibrium freezing curve for a typical ice cream mix as calculated by the method of Bradley [61,62] illustrating the relationship between temperature and unfrozen water as determined by freezing point depression characteristics. This relationship illustrates the maximum amount of ice that may form at any temperature, but due to kinetic considerations the amount of ice may be considerably less as a result of freezing rate conditions.

[60,63]. As was mentioned earlier, rapid initial freezing is essential to set up as many crystal nuclei as possible so that during the maturation or growth stage, their size stays small. In the same context, rapid hardening is also necessary to keep ice crystal size small. Desirable ice cream texture results from having many ice crystals of small size. Therefore, it is of utmost importance that freezing be done properly at this stage. Continuous freezers are superior to batch freezers due to the freezing time involved and the extensive nucleation that results. The goal is to produce an ice crystal size distribution that is undetectable during consumption [17].

As freeze concentration progresses, various solutes eventually reach and exceed their saturation concentrations, consequently crystallization may occur. The temperature at which crystallization occurs is the eutectic temperature. For example, the eutectic temperature is $-5°C$ for solutions of pure glucose, $-14°C$ for sucrose, $-21.1°C$ for sodium chloride, and $-55°C$ for calcium chloride [19]. However, crystallization of the sugar at its eutectic temperature is extremely unlikely due to the high viscosity and low temperature of the concentrated, unfrozen solution in a mix [60]. The amount of unfrozen phase is both temperature and freezing rate dependent due to both thermodynamic and kinetic considerations. This unfrozen phase is then capable of forming a glass, a noncrystalline solid with a high viscosity $(10^{13}–10^{15}$ poise) [63–67]. In the glassy region the serum phase exists as an unreactive amorphous solid, consequently no ice crystal growth can occur. However, in the rubbery state above the glass transition temperature, the serum phase is mobile and reactive so that recrystallization and ice crystal growth can occur in time frames significant to food storage [66]. Recrystallization is discussed further in Section III.C. (The reader is also referred to Chapter 1 for further details on the stability and formation of the glassy state in frozen foods.) A technologically significant transition temperature is apparent in ice cream at $\sim-30°C$, dependent upon its composition, which affects its stability [45,64]. It is uncertain, however, whether this is a true glass transition temperature or the onset of ice melting [67–70].

*Formation of the Fat Structure.* The dynamic whipping and freezing process is also responsible for the formation of a fat network or structure in the product. Ice cream is both an emulsion and a foam. The milkfat exists in tiny globules that have been formed by the homogenizer. There are many proteins, which act as emulsifiers and give the fat emulsion its needed stability. The emulsifiers discussed in the ingredients section, when added to ice cream, actually reduce the stability of this fat emulsion, because they replace proteins on the fat surface. When the mix is subjected to the whipping action of the barrel freezer, the fat emulsion begins to partially coalesce and the fat globules begin to flocculate [5,51]. The air bubbles being beaten into the mix are stabilized by this partially coalesced fat [21]. This process is similar to that which occurs during the whipping of heavy cream as the liquid is converted to a semi-solid with desirable stand-up qualities and mouthfeel.

In ice cream, the emulsion destabilization phenomenon also results in desirable textural qualities both at the time of draw from the barrel freezer and during consumption, resulting in a smoother, creamier product with a slower meltdown.

*Hardening.* Ice cream, following dynamic freezing, ingredient addition, and packaging, is immediately transferred to a hardening chamber (–30°C or colder, either forced convection or plate-type conduction freezers) where the majority of the remaining water freezes [18]. Rapid hardening is necessary for product quality, as it helps to create a small ice crystal size distribution due to the speed of ice crystal formation and the increase in the nucleation rate of crystals. When hardening is slow, there is too great an opportunity for water still remaining in the ice cream to migrate to crystal centers already formed, resulting in large ice crystals. Many factors need to be considered during the hardening process. The main factors affecting heat transfer are the temperature difference between the product and the freezing medium, the area of product being exposed to the freezing medium, and the heat-transfer coefficient for the particular operation [18]. The temperature of the ice cream when placed in the hardening room should be as cold as possible [17]. Draw temperatures from the barrel freezer are limited by the necessity of packaging the product. The addition of ingredients and the packaging operation should not increase the temperature of the ice cream as it is drawn from the barrel freezer any more than necessary. The temperature of the hardening chamber is also critical for rapid freezing and smooth product. The surface area of the ice cream also needs to be considered and is especially important when packaging in large packages or in shrink-wrapping product bundles. Palletizing or stacking of product should not interfere with rapid air circulation and fast freezing. Convective heat-transfer coefficients are greatly increased through the use of forced convection systems. The evaporator should be free of frost from the outside and oil from the inside of the coils as these act to reduce the heat-transfer coefficients. Following rapid hardening, ice cream storage should occur at low, constant temperatures, usually –25°C [71].

## C. Factors Affecting Storage Stability

During manufacture, a great deal of emphasis is placed on producing an optimal size distribution of ice crystals, which is as small and numerous as possible. Control of ice crystal size during storage and distribution would be relatively easy if crystals were stable. Unfortunately, they are relatively unstable and undergo morphological changes in number, size, and shape during frozen storage. This is known collectively as recrystallization and leads to a coarse, icy texture and the defect of iciness [60,72]. Recrystallization is probably the most important change producing quality losses and limitations in shelf life [71]. It also probably accounts for countless lost sales through customer dissatisfaction with quality. Some recrystallization, due to a process known as Ostwald ripening, occurs naturally at constant temperatures.

This is related to the differences in vapor pressures and relative free energy between small and larger ice crystals, resulting in a driving force for a reduction in the number of small crystals and the growth of large crystals to minimize free energy [60]. However, the majority of recrystallization is stimulated by fluctuating temperatures (known as heat shock) and can be minimized by maintaining a low and constant storage temperature [66].

Heat shock occurs readily in frozen foods. If the temperature during the frozen storage increases, some of the ice crystals melt, particularly the smaller ones as they have the highest free energy and lowest melting point [60]. Consequently, the amount of unfrozen water in the serum phase increases. Conversely, as temperatures decrease, water does not renucleate but rather refreezes on the surface of larger crystals, resulting in the total number of crystals diminishing and the mean crystal size increasing [59,60]. Figure 6 illustrates at least four independent ice crystals

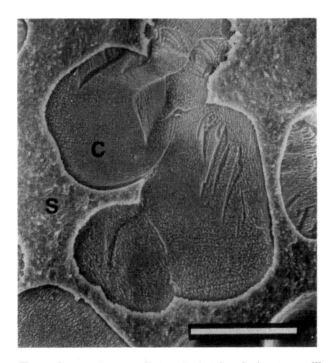

**Figure 6** An electron micrograph showing the ice recrystallization process as demonstrated by at least four independent ice crystals (C), which have migrated within and grown from the serum phase (S) and fused following 3 weeks of temperature-abusive storage. Bar = 27 μm. (From Ref. 36.)

that have merged as a result of heat shock. Temperature fluctuations are common during frozen storage, due to the cyclic nature of refrigeration systems and the need for automatic defrost. However, mishandling of product is probably the biggest culprit [73]. This growth in ice crystals is detrimental to the texture of all frozen foods and is usually the limiting factor in their storage stability [66]. If one were to track the temperature history of ice cream during distribution, retailing, and finally consumption, one would find a great number of temperature fluctuations [48]. Each time the temperature changes, the ice to serum content changes, and the smaller ice crystals decrease in number or disappear while the larger ones grow even larger, leading to a change in the mean crystal size distribution.

Processors have known for a long time how to prevent iciness: formulate the ice cream properly to begin with, freeze the ice cream quickly in a well-maintained barrel freezer, harden the ice cream rapidly, and avoid temperature fluctuations during storage and distribution to the extent possible [71,74]. Proper formulation with stabilizers designed to combat heat shock is an essential defense against the inevitable growth of ice crystals. As discussed in Section III.A.4, the most important function of the polysaccharide stabilizer is to limit the growth of ice crystals during storage. Figure 3 demonstrates the effectiveness of stabilizers during this process. Mixes with low total solids are also more difficult to effectively stabilize, since the increased content of water leads to more ice at any given temperature. Also, high concentrations of sucrose or lactose change the ratio of water to ice and lead to increased problems of recrystallization. Education of people involved in the retailing sector and the consumer regarding the causes of iciness and preventative action to maintain a smooth-textured ice cream is also effective in minimizing heat shock [71].

## D.  Factors Affecting Quality

The main factors affecting ice cream quality can be divided into two categories: compositional factors and processing factors. The reader is referred to an excellent review on ice cream defects by Bodyfelt et al. [75] for further details. Defects in ice cream quality associated with compositional factors are related to flavor, body and texture, meltdown, color, and appearance. Dairy ingredients may contribute to rancid or oxidized flavors associated with fat content, stale or old ingredient flavors associated especially with the use of dried milk ingredients, salty flavors associated with the use of high levels of whey, or sandiness, a textural defect associated with excessive lactose crystallization. The sweetening system may also contribute to either a lack of or too much sweetness or to a syrupy flavor from the corn sweeteners. Stabilizers can produce gummy types of defects if used in excess. The flavoring system may likewise produce many flavor defects, associated with either too high or not enough flavor or an unnatural flavor. Quality factors associated with processing parameters are mostly related to problems in the ice phase, as discussed in Section III.B.2.

## IV.  FREEZING OF OTHER DAIRY PRODUCTS

Dairy products vary considerably in their moisture content, from 87 to 91% for whole and skim milk to 3% for milk powders. As mentioned previously, freezing refers to the conversion of water to ice. Obviously, products with higher water contents will involve significantly more physical alterations during freezing than those with lower water contents. Although milk can be frozen for preservation or extended shelf life, its high water content makes this process somewhat economically unattractive and suggests the need for concentration prior to freezing [76]. Many products that have very low moisture contents or freezing points can be stored at subzero temperatures to inhibit microbiological and chemical changes [76], but they are not frozen since they have not undergone a phase change and no ice is present. For example, storage of milk powders is often conducted at subzero temperatures but will not be considered in this chapter on frozen dairy products.

The flavor of dairy products after freezing and thawing generally are comparable to their fresh counterparts under normal circumstances, with lipid oxidation being the greatest concern. However, the physical effects of freezing may be quite noticeable upon thawing [4,76]. Sen and Gupta [77] reviewed a number of textural defects occurring in frozen dairy products as a result of freezing and thawing, but suggested that the flavor of frozen products remained high. While flavor deterioration, if present, cannot be rectified, physical changes that occurred during storage may be of little consequence if the thawed product is to be pasteurized and homogenized for further processing into dairy or other food products, since the additional heat and homogenization can melt and resuspend the casein and the fat, respectively [1].

### A.  Fluid Milk and Condensed Milks

Very little commercial freezing of fluid milk occurs in North America, primarily due to the widespread and constant supply of fresh milk on a year-round basis and the unfavorable economics of preservation of milk in this form. There are a few technological problems associated with the freezing of milk, namely, protein coagulation, fat separation, and fat autoxidation [4,6], but these can be overcome if commercial interest develops in application of the process. During freezing of homogenized milk, a considerable degree of phase separation usually occurs, since the solids and the unfrozen phase drain or migrate away from the frozen ice, which may produce a distinct layering of the product during quiescent thawing. However, the phases are usually redispersed quickly with agitation. Tulloch and Cheney [78] observed considerable textural deterioration in milk when stored at −10°C for 2 months. However, much improved texture was observed when milk was stored at −18°C for the same time period. The flavor of all samples was acceptable. The microbiological implications of milk freezing have also been studied. El-Kest and Marth [79,80] examined the influence of freezing and frozen storage of milk on the survival of *Listeria monocytogenes* and observed that death and injury after 4 weeks of storage at –18°C ranged from 11 to 67%, depending on the strain. Cells were protected from death and injury by the presence of milkfat, casein, and lactose. Therefore, survival rates were higher in milk than in buffer solutions.

Milk could be sold in the frozen, concentrated form, in a fashion similar to the widespread distribution of orange juice, if not for the instability of the casein micelle system, as discussed in Section II.B. The stability of these products can be enhanced by removing calcium ions through dialysis or ultrafiltration of milk or by adding calcium ion–complexing chemicals, such as hexametaphosphate or sodium polyphosphate [76] at 0.2% [6] to form more soluble casein aggregates and submicelles [81]. Lonergan et al. [82] examined the use of electrodialysis to remove calcium and thereby dissociate micellar casein prior to freezing of skim milk at concentrations of 1–3.3 times. Removal of 40% of the total calcium resulted in protein stability greater than 17 weeks at –8°C, compared to less than 1 week for the control sample. A pasteurization treatment just prior to freezing, presumably to solubilize all lactose crystals, or removal of calcium to 70% extended storage stability up to 1 year at –8°C. Other techniques to stabilize the casein focus on the crystallization of lactose, as discussed in Section II.B. These include the crystallization and removal of lactose from concentrated products prior to freezing [76], the addition of hydrocolloids to increase viscosity and suppress crystallization of lactose, the addition of sugar to suppress lactose crystallization, the use of heat treatments postconcentration to dissolve lactose nuclei, and the enzymatic hydrolysis of lactose by β-galactosidase (lactase), which results in a slightly sweeter flavor in the reconstituted milk [8,83].

## B.  Cream and Butter

The freezing of cream in bulk containers increases its shelf life greatly but can also lead to gross separation of fat and serum solids upon thawing. Cream processed in this manner is suitable for reprocessing into cream soups, recombined milk, butter, or ice cream, where pasteurization and homogenization will restore the original fat emulsion or churning will continue to induce fat (butter) separation [84]. Bulk frozen cream of desirable quality can be prepared with fresh, low-acid cream, with desirable flavor and 40–60% fat, 52% being optimal [76]. Manufacturing steps include addition of sugar at a rate of 10% prior to pasteurization, if desired, pasteurizing at 88°C for 5 minutes, and freezing the cream in stainless steel cans or plastic containers to –23°C or lower. Cream for further processing can be stored for periods of 6–10 months at temperatures below –18°C if handled properly [73]. Creams of 25% fat or more freeze relatively homogeneously with little separation of the fat phase. For this reason, it is easier to sample frozen cream by removing only small samples [1]. The addition of sugar to cream considerably retards fat coalescence, due primarily to the lowering of the freezing point by the dissolved sugar. Usually 5–20% of the weight of the cream is added. Frozen sweetened cream melts much more readily and is therefore easier to handle than unsweetened cream [76]. The same result can be accomplished by increasing the milk solids-not-fat content or by rapid freezing through the use of low temperatures, thin films, or small containers [1]. Rapid freezing will not damage the fat emulsion to the same extent as slower freezing, even after long periods of frozen storage.

Rotary drum freezing provides a means for the continuous production of rapidly frozen "chipped" or flaked cream of high quality [84]. The packaged flakes have a lower density than bulk frozen cream and may be convenient to use in further processing applications. Storage temperatures of –18°C or lower are required for such a product. Consumer products of this nature (e.g., whipping cream) can also be prepared utilizing rapid freezing techniques. These products are normally distributed in bags, similar to frozen vegetables. Additionally, packaged cream in small, consumer packages can be frozen rapidly using cryogenic techniques (e.g., liquid nitrogen tunnels) to produce a high-quality product upon thawing [84]. It is essential that the fat emulsion be very stable prior to freezing to prevent fat coalescence upon thawing. Emulsifiers and stabilizers may also assist in the freeze-thaw stability of cream if rapid freezing is not possible. Low, constant storage temperatures are equally important for the maintenance of quality. Temperature cycling may lead to the formation of large crystals with resultant damage to the fat globules [84]. However, the development of overrun in whipped cream may be impaired by a previous freezing step [76].

Freezing has no observable effect of the characteristics or quality of butter. The water content of butter is less than 20%, which is evenly distributed in the form of tiny droplets throughout the product. The freezing point of the water phase is determined primarily by the addition of salt and the degree of residual lactose and dissolved milk salts after washing. Temperatures of –20°C for a few months or –30°C for storage periods of one year or more are normally employed during the storage of butter to minimize flavor deterioration [73,76]. When butter is to be frozen, freezing immediately after processing has shown to produce a better quality product than freezing after chilled storage (4°C) for several days. Butter is notorious for the absorption of flavors from its storage environment. Therefore, butter should be properly wrapped to prevent penetration of air and off-odors [76].

## C. Starter Cultures

Frozen starter cultures for the manufacture of a variety of cultured dairy products have been widely accepted for industrial use [4,76,85,86]. This section will not review their production aspects in detail but will simply serve to give the reader an appreciation of the freezing parameters associated with their production by the culture manufacturer and their handling by the culture user. An excellent review by Gilliland [85] offers more details. Commercial production of starter cultures now involves the following steps: growth of the desired culture in a suitable medium, concentration of the cells by centrifugation, resuspension at a desired concentration, based on activity, in a suitable medium, packaging to provide the appropriate amount of starter for inoculation of a fixed batch size of product, and cryogenically frozen, usually in liquid nitrogen (–196°C). If transportation and storage is carried out at suitable temperatures, usually less than –40° (closer to –80°C) [86], frozen cultures are almost instantly active at the appropriate concentration. The advantages of this approach to inoculation of cultured products include convenience, culture reliability, improved daily performance and strain balance, greater flexibility, better control of phage, and possible improvement in quality [86]. The disadvantages, however, include the need for handling of cryogens, higher cost, and dependence on starter suppliers. Thawing becomes particularly crucial. Consequently, suppliers' recommendations must be followed carefully. Thawing is usually conducted in water containing 100–200 ppm hypochlorite at 20°C for 10 minutes, followed by the addition to milk for processing when the contents are thawed enough to be loosened from the can. Rapid thawing (rather than at 4°C) is most often preferred [86].

Appropriate media for frozen culture preparation may include 10% skim milk, 5% sucrose, fresh cream, and 0.9% sodium chloride or 1% gelatin. Concentrated cells in the presence of certain mixtures of other ingredients (sodium citrate, glycerol, sodium-b-glycerophosphate, yeast extract, sucrose, cream, sterile skim milk, Tween 80, sodium oleate, peptone, or lactose) have also been prepared with excellent results [86]. For freezing in liquid nitrogen, media comprising only 16% sterile, reconstituted skim milk solids with no added cryoprotectants are adequate. Storage life of more than 2 years in liquid $N_2$ has been reported [86].

Considerable interest in the production of cultures that retain suitable activity when stored at temperatures between –20 and –40°C has developed to avoid the use of cryogens such as liquid nitrogen or dry ice ($CO_2$) in their distribution and handling. Initial growth and preparation of the microorganism has proven to affect the biological activity after thawing. An example is the growth of lactic streptococci on tryptone-lactose agar and resuspension after initial harvest in 1:1 solutions of glycerol:water. These preparations did not lose viability after 8 months of storage at –30°C and were suitable for direct use as an inoculum in cheese making [87]. Another example is the growth of cultures in papain-digested milk enriched with yeast extract and lactose, followed by resuspension of the harvested cells in a glycerol–skim milk mixture and storage at –30°C [87].

## D.  Fermented Products and Cheese

One of the parameters of interest in the freezing of cultured products is the retention of microbiological activity after thawing. Foschino et al. [88] reported that plate counts decreased less than 1 log cycle and fermentation activity was 40–70% when cultures of *Lactobacillus delbruchii* sbsp. *bulgaricus* were stored at –80°C for 1 year. However, fermentation activity was less than 10% when cultures were stored for 1 year at –30°C. The fermentation activity of *Streptococcus salavarius* sbsp. *thermophilus* was similarly reduced to 10–60% after 1 year of storage at –30°C. Sheu et al. [89] reported a 40% survival of *L. bulgaricus* cells in ice milk from a continuous freezer but suggested this could be increased to 90% through entrapment of cells in beads of calcium alginate of greater than 30 μm diameter with no adverse effects on the overrun or sensory characteristics of the product. Unpublished results from the authors' laboratory have shown that a strawberry-flavored frozen yogurt mix stored at –10°C for a period of 75 days prior to thawing, whipping, and freezing showed considerable deterioration in texture, flavor, and odor compared to similar product stored at –15°C for the same time period, which showed excellent sensory attributes.

Interest in the freezing of cheese has focused on two distinct processes: the freezing of curd for further processing and the freezing of fresh or aged cheese for extended storage. Le Jaouen [90] has reviewed the process of freezing cheese curd from goat's milk for deferred utilization. The following discussion is taken from this report. It was considered most desirable to freeze the curd rapidly in thin layers or flat blocks less than 10 cm thick and wrapped tightly to avoid oxygen exposure. Constant storage temperatures at less than –20°C are needed prior to utilization of the curd. Thawing can be accomplished at 4°C for 24–48 hours prior to warming the curd to 22°C for further processing, molding, inoculation, ripening, etc. Flavor and texture defects of products manufactured from frozen curd, particularly lipid oxidation and crumbly, grainy textures, may readily occur. The lactic acid microflora may be reduced to about one tenth of the original numbers after prolonged storage of curd, and yeasts and coliforms may completely disappear, while other microorganisms may undergo relatively little change in their numbers. These changes in microflora may have an effect on the ripening process after curd utilization. Salt contributes greatly to the formation of oxidized flavor. Only unsalted curd can be stored for deferred utilization. During ripening of cheeses made from frozen curd, a distinctly higher noncasein nitrogen is found, presumably due to the increased susceptibility of the casein fraction to degradation. A lack of characteristic flavor may also result from a loss of carbonyl compounds during frozen storage of the curd. Similarly, dry uncreamed cottage cheese curd is sometimes frozen for extended storage prior to creaming, packaging, and retail distribution. Such curd should be less than 80% moisture, salted lightly, fast frozen to –30°C or lower in air-tight packaging, and stored at –23°C or lower for periods of no longer than 3–6 months.

During the freezing of cheese, the flavor usually remains good, unless lipid oxidation occurs. However, the body and texture become more crumbly and mealy after thawing, especially after extensive freezing [1,76,91]. Fontecha et al. [92] reported that there was an increase in the unordered structure of the protein in frozen cheese, particularly in slowly frozen samples, consistent with greater damage to the microstructure observed by scanning electron microscopy and greater proteolysis. Oberg et al. [91] have reported on the freezing of mozzarella cheese. Their results suggest similar change in protein structure as evident by an alteration in the stretch and melting properties of shredded or bulk mozzarella after thawing.

## V. CONCLUSIONS

The freezing of dairy products presents an interesting study in contrasts from the freezing of many other commodity groups. First, milk is readily available both seasonally and geographically, making its long-term preservation less critical than that of seasonal fruits and vegetables. Second, many dairy products from a structural viewpoint are carbohydrate solutions, protein suspensions, and fat emulsions and thus present very different physical challenges than freezing of tissue-based systems such as plant and animal products. Third, the vast majority of frozen dairy products, namely, the dessert category, are eaten in the frozen form, unlike other categories of frozen foods. The dairy industry has made a tremendous success of the frozen dessert category but has used freezing to a much smaller extent in other dairy products. Thus other frozen dairy products generally have little commercial significance. Frozen dairy starter cultures are the most important exception, but the volume of milk utilization in this case is relatively insignificant to the dairy industry as a whole.

One must also distinguish the freezing of dairy products for further processing and the freezing of dairy products for the consumer market. In the latter category, the greatest opportunities may lie in two areas: the frozen pelleted or chipped cream product described above and the distribution of frozen concentrated milk in sizes suitable for easy reconstitution, analogous to the tremendous success in the frozen orange juice market. Although there is little use of these products in the North American consumer market at present, evidence of commercial interest in these and other frozen dairy products exists.

## REFERENCES

1. P. G. Keeney and M. Kroger, Frozen dairy products, *Fundamentals of Dairy Chemistry*, 2nd ed. (B. H. Webb, A. H. Johnson, and J. A. Alford, ed.), AVI Publishing Co., Inc., Westport, CT, 1974, p. 873.
2. P. Walstra and R. Jenness, *Dairy Chemistry and Physics*, John Wiley and Sons, New York, 1984, p. 1.
3. H. D. Goff and A. R. Hill, Dairy chemistry and physics, *Dairy Science and Technology Handbook*, Vol 1, *Principles and Properties* (Y. H. Hui, ed.), VCH Publishers, New York, 1993, p. 1.
4. B. H. Webb, Characteristics and quality changes in dairy products during freezing and storage, *The Freezing Preservation of Foods*. Vol. 2. *Factors affecting quality in*

*frozen foods*, 4th ed. (D. K. Tressler, W. B. Van Arsdel, and M. J. Copley, ed.), AVI Publishing Co. Inc., Westport, CT, 1968, p. 224.

5. H. D. Goff and W. K. Jordan, The action of emulsifiers in promoting fat destabilization during the manufacture of ice cream, *J. Dairy Sci.* 72:18 (1989).

6. W. D. Powrie, Characteristics of fluid foods, *Low-Temperature Preservation of Foods and Living Matter* (O. R. Fennema, W. D. Powrie, and E. H. Marth, ed.), Marcel Dekker, Inc., New York, 1973, p. 242.

7. P. Walstra, On the stability of casein micelles, *J. Dairy Sci.* 73:1965 (1990).

8. P. A. Morrisey, Lactose: Chemical and physicochemical properties, *Developments in Dairy Chemistry -3- Lactose and Minor Constituents*, (P. F. Fox, ed.), Elsevier Applied Science, New York, 1985, p. 1.

9. J. W. Sherbon, Physical properties of milk, *Fundamentals of Dairy Chemistry*, 3rd ed. (N. P. Wong, R. Jenness, M. Keeney, and E. H. Marth, ed.), Van Nostrand Reinhold Company, New York, 1988, p. 409.

10. F. Harding, *Measurement of Extraneous Water by the Freezing Point Test*, Int. Dairy Fed. Bulletin 154, Brussels, Belgium, 1983, p. 1.

11. J. Hortvet, The cryoscropic examination of milk, *J. Assoc. Off. Agric. Chem.* 5:470 (1923).

12. W. F. Shipe, The use of thermistors for freezing point determinations, *J. Dairy Sci.* 39:916 (1956).

13. J. H. Prentice, Freezing point data on aqueous solutions of sucrose and sodium chloride and the Hortvet test: A reappraisal, *Analyst* 103:1269 (1978).

14. Association of Official Analytical Chemists, *Methods of Analysis*, 15th ed., Washington, DC, 1990, p. 819.

15. G. Brathen, Milk constituents and freezing point depression, *Measurement of Extraneous Water by the Freezing Point Test* (F. Harding, ed.), Int. Dairy Fed. Bulletin 154, Brussels, Belgium, 1983, p. 5.

16. W. F. Shipe, The freezing point of milk. A review, *J. Dairy Sci.* 42:1745 (1959).

17. W. S. Arbuckle, *Ice Cream*, 4th ed., AVI Publishing Co., Westport, CT, 1986, p. 1.

18. H. L. Mitten and J. M. Neirinckx, Developments in frozen products manufacture, *Modern Dairy Technology*, Vol 2, *Advances in Milk Products* (R. K. Robinson, ed.), Elsevier Applied Science Publishers, New York, 1986, p. 215.

19. K. G. Berger, Ice cream, *Food Emulsions*, 2nd ed. (K. Larsson and S. Friberg, eds.), Marcel Dekker, New York, 1990, p. 367.

20. R. Jimenez-Flores, N. J. Klipfel, and J. Tobias, Ice cream and frozen desserts, *Dairy Science and Technology Handbook*. Vol. 2. *Product Manufacturing* (Y. H. Hui, ed.), VCH Publishers, Inc., New York, 1993, p. 57.

21. K. B. Caldwell, H. D. Goff, and D. W. Stanley, A low-temperature SEM study of ice cream. I. Techniques and general microstructure, *Food Struct.* 11:1 (1992).

22. P. Walstra, Physical chemistry of milkfat globules, *Developments in Dairy Chemistry—2. Lipids* (P. F. Fox, ed.), Applied Science Publishers, New York, 1983, p. 119.

23. B. E. Brooker, M. Anderson, and A. T. Andrews, The development of structure in whipped cream, *Food Microstruct.* 5:277 (1986).

24. H. D. Goff, Examining the milk solids-not-fat in frozen dairy desserts, *Modern Dairy* 71(3):16 (1992).

25. H. D. Goff, M. Liboff, W. K. Jordan, and J. E. Kinsella, The effects of Polysorbate

80 on the fat emulsion in ice cream mix: Evidence from transmission electron microscopy studies, *Food Microstruct.* 6:193 (1987).

26.  P. J. Halling, Protein stabilized foams and emulsions, *CRC Crit. Rev. Food Sci. Nutr.* 15:155 (1981).

27.  B. E. Brooker, Observations on the air serum interface of milk foams, *Food Microstruct.* 4:289 (1985).

28.  J. E. Kinsella, Milk proteins: Physicochemical and functional properties, *CRC Crit. Rev. Food Sci. Nutr.* 21(3):197 (1984).

29.  H. D. Goff, W. K. Jordan, and J. E. Kinsella, The influence of various milk protein isolates on ice cream emulsion stability, *J. Dairy Sci.* 72:385 (1989).

30.  D. E. Smith, A. S. Bakshi, and C. J. Lomauro, Changes in freezing point and rheological properties of ice cream mix as a function of sweetener system and whey substitution, *Milchwissenschaft* 39:455 (1984).

31.  J. G. Parsons, S. T. Dybing, D. S. Coder, K. R. Spurgeon, and S. W. Seas, Acceptability of ice cream made with processed whey and sodium caseinate, *J. Dairy Sci.* 68:2880 (1985).

32.  H. D. Goff and A. M. Pearson, Aspartame and corn syrup solids as sweeteners for ice cream, *Modern Dairy* 62(3):11 and (4):14 (1983).

33.  H. D. Goff, R. D. McCurdy, and G. N. Fulford, Advances in corn sweeteners for ice cream, *Modern Dairy* 69(3):17 (1990).

34.  H. D. Goff, R. D. McCurdy, and E. A. Gullett, Replacement of carbon-refined corn syrups with ion-exchanged corn syrups in ice cream formulations, *J. Food Sci.* 55:827 (1990).

35.  H. D. Goff and K. B. Caldwell, Stabilizers in ice cream. How do they work?, *Modern Dairy* 70(3):14 (1991).

36.  K. B. Caldwell, H. D. Goff, and D. W. Stanley, A low-temperature SEM study of ice cream. II. Influence of ingredients and processes, *Food Struct.* 11:11 (1992).

37.  E. K. Harper and C. F. Shoemaker, Effect of locust bean gum and selected sweetening agents on ice recrystallization rates, *J. Food Sci.* 48:1801 (1983).

38.  Y. Shirai, K. Nakanishi, R. Matsuno, and T. Kamikubo, Effects of polymers on secondary nucleation of ice crystals, *J. Food Sci.* 50:401 (1985).

39.  E. R. Budiaman and O. Fennema, Linear rate of water crystallization as influenced by temperature of hydrocolloid suspensions, *J. Dairy Sci.* 70:534 (1987).

40.  E. R. Budiaman and O. Fennema, Linear rate of water crystallization as influenced by viscosity of hydrocolloid suspensions, *J. Dairy Sci.* 70:547 (1987).

41.  N. Buyong and O. Fennema, Amount and size of ice crystals in frozen samples as influenced by hydrocolloids, *J. Dairy Sci.* 71:2630 (1988).

42.  A. H. Muhr and J. M. V. Blanshard, The effect of polysaccharide stabilizers on ice crystal formation, *Gums and Stabilisers for the Food Industry. 2. Applications of Hydrocolloids* (G. O. Phillips, D. J. Wedlock, and P. A. Williams, ed.), Pergamon Press, New York, 1983, p. 321.

43.  A. H. Muhr, J. M. V. Blanshard, and S. J. Sheard, Effects of polysaccharide stabilizers on the nucleation of ice, *J. Food Technol.* 21:587 (1986).

44.  A. H. Muhr, and J. M. V. Blanshard, Effect of polysaccharide stabilizers on the rate of growth of ice, *J. Food Technol.* 21:683 (1986).

45.  H. D. Goff, K. B. Caldwell, D. W. Stanley, and T. J. Maurice, The influence of

polysaccharides on the glass transition in frozen sucrose solutions and ice cream, *J. Dairy Sci.* 76:1268 (1993).

46.  M. E. Sahagian and H. D. Goff, Thermal, mechanical and molecular relaxation properties of stabilized sucrose solutions at sub-zero temperatures, *Food Res. Int.* 28:1 (1995).

47.  E. Morris, Polysaccharide solution properties: Origin, rheological characterization, and implications for food systems, *Frontiers in Carbohydrate Research: Food Applications*, Elsevier Applied Science Publishers, London, 1989, p. 132.

48.  P. G. Keeney, Development of frozen emulsions, *Food Technol.* 36(11):65 (1982).

49.  H. D. Goff, B. Freslon, M. E. Sahagian, T. D. Hauber, A. P. Stone, and D. W. Stanley, Structure development in ice cream—dynamic rheological measurements, *J. Texture Stud.* (in press).

50.  H. D. Goff, Emulsifiers in ice cream: How do they work?, *Modern Dairy* 67(3):15 (1988).

51.  D. F. Darling, Recent advances in the destabilization of dairy emulsions, *J. Dairy Res.* 49:695 (1982).

52.  M. Kalab, Microstructure of dairy foods. 2. Milk products based on fat, *J. Dairy Sci.* 68:3234 (1985).

53.  P. M. Lin and J. G. Leeder, Mechanism of emulsifier action in an ice cream system, *J. Food Sci.* 39:108 (1974).

54.  Food and Drug Administration, *Milk pasteurization controls and tests*, U.S. Department of Health and Human Services, Cincinnati, OH, 1981, p. 25.

55.  P. Walstra and H. Oortwijn, The membranes of recombined fat globules. 3. Mode of formation, *Neth. Milk Dairy J.* 36:103 (1982).

56.  E. K. Iversen and K. S. Pedersen, *Aging of Ice Cream*, Grinsted Technical Paper TP 33-1e, Brabrand, Denmark, 1982, p. 1.

57.  B. W. Tharp, T. V. Gottemoller, and A. Kilara, *The Role of Processing in Achieving Desirable Properties in Health-Responsive Frozen Desserts*, Germantown Manufacturing Co. Technical Paper 100, Broomall, PA, 1992.

58.  F. Franks, *Biophysics and Biochemistry at Low Temperatures*, Cambridge University Press, New York, 1985, pp. 21, 37, 167.

59.  F. Franks, Complex aqueous systems at subzero temperatures, *Properties of Water in Foods* (D. Simatos and J. L. Multon, ed.), Martinus Nijhoff Publishers, Dordrecht, Netherlands, 1985, p. 497.

60.  J. M. V. Blanshard and F. Franks, Ice crystallization and its control in frozen food systems, *Food Structure and Behavior*, (J. M. V. Blanshard and P. Lillford, ed.), Academic Press, New York, 1987, p. 51.

61.  R. Bradley and K. Smith, Finding the freezing point of frozen desserts, *Dairy Record* 84(6):114 (1983).

62.  R. Bradley, Plotting freezing curves for frozen desserts, *Dairy Record* 85(7):86 (1984).

63.  H. Levine and L. Slade, Principles of cryo-stabilization technology from structure/property relationships of carbohydrate/water systems. A Review, *Cryo-Letters* 9:21 (1988).

64.  H. Levine and L. Slade, Cryostabilization technology: Thermoanalytical evaluation of food ingredients and systems, *Thermal Analysis of Foods* (C. Y. Ma and V. R. Harwalker, ed.), Elsevier Applied Science, London, 1988, p. 221.

65.  L. Slade and H. Levine, Beyond water activity: Recent advances based on an alternative approach to the assessment of food quality and safety, *Crit. Rev. Food Sci. Nutr.* 30:115 (1991).

66.  H. D. Goff, Low-temperature stability and the glassy state in frozen foods, *Food Res. Int.* 25:317 (1992).

67.  M. E. Sahagian and H. D. Goff, Effect of freezing rate on the thermal, mechanical and physical aging properties of the glassy state in frozen sucrose solutions, *Thermochimica Acta* 246:271 (1994).

68.  Y. Roos and M. Karel, Phase transitions of amorphous sucrose and frozen sucrose solutions, *J. Food Sci.* 56:266 (1991).

69.  Y. Roos and M. Karel, Nonequilibrium ice formation in carbohydrate solutions, *Cryo-Letters* 12:367 (1991).

70.  Y. Roos and M. Karel, Amorphous state and delayed ice formation in sucrose solutions, *Int. J. Food Sci. Technol.* 26:553 (1991).

71.  H. D. Goff, Heat shock revisited, *Modern Dairy.* 72(3):24 (1993).

72.  D. P. Donhowe, R. W. Hartel, and R. L. Bradley, Jr., Determination of ice crystal size distributions in frozen desserts, *J. Dairy Sci.* 74:3334 (1991).

73.  International Institute of Refrigeration, *Recommendations for the Processing and Handling of Frozen Foods*, 3rd ed., IIR, Paris, 1986, p. 342.

74.  R. Bradley, Protecting ice cream from heat shock, *Dairy Record* 85(10):120 (1984).

75.  F. W. Bodyfelt, J. Tobias, and G. M. Trout, *The Sensory Evaluation of Dairy Products*, Van Nostrand Reinhold, New York, 1988, p. 166.

76.  B. H. Webb, Preparation for freezing and freezing of dairy products, *The Freezing Preservation of Foods. Vol. 3. Commercial Food Freezing Operations*, 4th ed. (D. K. Tressler, W. B. Van Arsdel, and M. J. Copley, ed.), AVI Publishing Co. Inc., Westport, CT, 1968, p. 295.

77.  D. C. Sen and S. K. Gupta, Effect of freezing and thawing on sensory qualities of dairy products, *Ind. Dairyman* 39(5):231 (1987).

78.  D. Tulloch and M. Cheney, Keeping quality of Western Australian frozen milk, *Aust. J. Dairy Technol.* 39(2):85 (1984).

79.  S. E. El-Kest and E. H. Marth, Injury and death of frozen *Listeria monocytogenes* as affected by glycerol and milk components, *J. Dairy Sci.* 74:1201 (1991).

80.  S. E. El-Kest and E. H. Marth, Strains and suspending menstrua as factors affecting death and injury of *Listeria monocytogenes* during freezing and frozen storage, *J. Dairy Sci.* 74:1209 (1991).

81.  C. V. Morr and R. L. Richter, Chemistry of Processing, *Fundamentals of Dairy Chemistry*, 3rd ed. (N. P. Wong, R. Jennes, M. Keeney, and E. H. Marth, ed.), Van Nostrand Reinhold Co., New York, 1988, p. 739.

82.  D. A. Lonergan, O. Fennema, and C. H. Amundson, Use of electrodialysis to improve the protein stability of frozen skim milks and milk concentrates, *J. Food Sci.* 47:1429 (1982).

83.  R. R. Mahoney, Modification of lactose and lactose-containing dairy products with b-galactosidase, *Developments in Dairy Chemistry -3- Lactose and Minor Constituents* (P. F. Fox, ed.), Elsevier Applied Science, New York, 1985, p. 69.

84.  C. Towler, Developments in cream separation and processing, *Modern Dairy Tech-*

nology, Vol. 1. *Advances in Milk Processing* (R. K. Robinson, ed.), Elsevier Applied Science Publishers, New York, 1986, p. 51.

85. S. E. Gilliland, *Bacterial Starter Cultures for Foods*, CRC Press, Inc., Boca Raton, FL, 1985, p. 145.

86. A. Y. Tamime, Microbiology of starter cultures, *Dairy Microbiology* Vol. 2, *The Microbiology of Milk Products* (R. K. Robnson, ed.), Elsevier Applied Science, New York, 1990, p. 164.

87. J. F. Frank and E. H. Marth, Fermentations, *Fundamentals of Dairy Chemistry*, 3rd ed. (N. P. Wong, R. Jennes, M. Keeney, and E. H. Marth, ed.), Van Nostrand Reinhold Co., New York, 1988, p. 655.

88. R. Foschino, C. Beretta, and G. Ottogalli, Study of optimal conditions in freezing and thawing for thermophilic lactic cultures, *Industria del Latte* 28(2):49 (1992).

89. T.-Y. Sheu, R. T. Marshall, and H. Heymann, Improving survival of culture bacteria in frozen desserts by microentrapment, *J. Dairy Sci.* 76:1902 (1993).

90. J. C. Le Jaouen, Deferred utilization of curd, *Cheesemaking Science and Technology* (A. Eck, ed.), Lavoisier Publishing Inc., New York, 1987, p. 249.

91. C. J. Oberg, R. K. Merrill, R. J. Brown, and G. H. Richardson, Effects of freezing, thawing, and shredding on low moisture, part-skim mozzarella cheese, *J. Dairy Sci.* 75:1161 (1992).

92. J. Fontecha, J. Bellanato, and M. Juarez, Infrared and Raman spectroscopic study of casein in cheese: Effect of freezing and frozen storage, *J. Dairy Sci.* 76:3303 (1993).

# 9

# Effects of Freezing, Frozen Storage, and Thawing on Eggs and Egg Products

Paul L. Dawson

*Clemson University*
*Clemson, South Carolina*

## I. INTRODUCTION

The U.S. National Agricultural Statistics Service report on liquid egg production [1] shows a fairly level production rate for liquid frozen egg, while liquid egg for immediate use has shown an increase since 1982 (Fig. 1). Mandatory inspection of egg products began in 1971, therefore, values prior to 1972 are not shown due to lack of accuracy. The use of nonfrozen liquid egg in quick-service restaurants has probably contributed to the increase in liquid egg consumption. Advances in continuous-process pasteurization technology have contributed to the improved shelf life and quality of liquid egg and, therefore, its increased usage.

Commercial poultry eggs are composed of approximately 9.5% shell, 63% albumen, and 27.5% yolk. Raw egg products can be divided into three general forms: whole egg (WE), egg albumen or white (EA), and egg yolk (EY). The egg can also be subdivided into fractions used in food applications. Lecithin and lysozyme are two such components that can be extracted from egg and may be frozen or dried for extended storage. WE, EA, and EY are often frozen from the liquid state. Cooked and dried egg products are also frozen to improve their quality and safety during extended storage. EA and EY differ in chemical composition, which has a dramatic effect on the properties of these components during freezing and thawing. EA is composed primarily of water (88%) and protein (10%), while EY contains mostly water (48%) and lipid (33%).

The effects of freezing on the quality of egg products differ from the effects of freezing on the quality of meats and vegetables, since eggs lack the physical cell structure and tissue associated with meats and vegetables [2]. Freezing egg is also

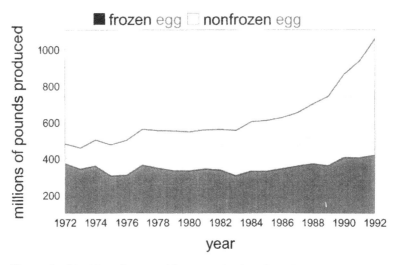

**Figure 1** Liquid nonfrozen and frozen production of eggs in the United States from 1972 to 1992. (From Ref. 1.)

different from freezing milk and juices since egg contains a higher solute concentration than most other fluid foods. Freezing rate, storage time, storage temperature, thawing rate, and additives can all influence the quality of the final product. This chapter will discuss the freezing process and freezing methods for eggs, frozen egg products and their uses, and the effects of freezing on egg functionality. Each function will be discussed in relation to how freezing, storage, and thawing affects WE, EA, and EY from the liquid, dried, and cooked state.

Freezing liquid foods (which contain solutes) usually results in a concentration of the solid phase and a crystallization of water into ice. If the liquid food (solution) contains different solutes (lipid, protein, carbohydrate, and salts), freezing will occur in phases with the successive separation of single components. During ice crystal growth, the water separates from the solution and the colloidal substances at different temperature levels. The temperature at which separation occurs is inversely proportional to the strength of the bonds between each component and water. This phenomenon is a sequence of specific equilibrium situations, characteristic of each solution due to the concentration of salts and various organic compounds (proteins, lipids, sugars, salts), each of which has a binding potential for water [3]. It is this separation of solutes that affects the functionality of eggs. Increased ionic concentrations as freezing progresses can result in protein denaturation [2]. The egg is comprised of the basic food components of protein, fat, carbohydrate, minerals, and water (Fig. 2). The first three components are intimately associated with water and give the egg its unique functional properties. Due

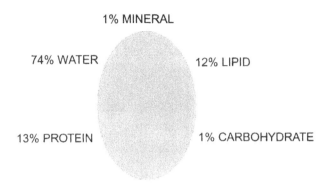

**Figure 2**   Composition of whole egg, excluding shell.

to the compositional differences in whole egg and whole egg fractions (yolk and albumen), freezing has dramatically different effects on these fractions.

## II.   THE FREEZING PROCESS

The freezing process generally involves both a lowering of the temperature and change of phase from liquid to solid. In foods this phase change rarely involves all of the water associated with the food. Biological materials such as eggs contain water that is bound to the solid components to varying degrees. A small part of this water is so tightly bound to the food components that the water will not freeze under commercial conditions. On a macro scale the product will appear to have changed phase to a solid, but on a molecular level, bound liquid water will still exist. Using egg yolk as an example, about 5.8% of the original total water is unfreezable (Fig. 3) [3]. As the temperature is lowered, the "freezable" water becomes crystallized and the solutes (protein, lipid, carbohydrate, minerals and salts) become more concentrated in the remaining water. The increased concentration of solids in the remaining nonbound water causes the remaining water to be more difficult to freeze. The concentration of these solutes in the remaining water affects the quality of the thawed product. This description of the freezing process can be applied to all foods since all contain some "bound" water. Even cooked products contain bound water, but they have a lower "free" water content when the freezing process is initiated.

## III.   USDA REGULATIONS

The United States Department of Agriculture (USDA) Regulations Governing Inspection of Eggs and Egg products (Title 7 of the Code of Federal Regulations

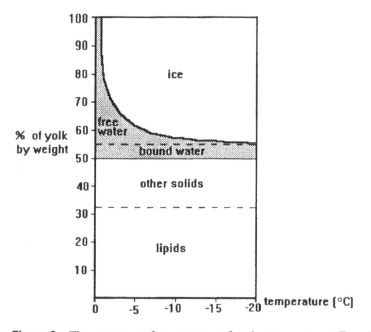

**Figure 3**   The percentage of components at freezing temperatures. (From Ref. 3.)

Part 59) has detailed requirements for the handling and freezing of egg products. This information is available from the USDA Agricultural Marketing Service Poultry Division in Washington, D.C. Part 59.534 describes freezing facility requirements:

(a) Freezing rooms, either on or off the premises, shall be capable of freezing all liquid egg products in accordance with the freezing requirements as set forth in part 59.536. Use of off-premise freezing facilities is permitted only when prior approval in writing from the National Supervisor is on file.

(b) Adequate air circulation shall be provided in all freezing rooms.

Part 59.536 describes freezing operations:

(a) Freezing rooms shall be kept clean and free of objectionable odors.

(b) Requirements. (1) Nonpasteurized egg products which are to be frozen shall be solidly frozen or reduced to a temperature of 10°F or lower within 60 hours from time of breaking. (2) Pasteurized egg products which are to be frozen shall be solidly frozen or reduced to a temperature of 10°F or lower within 60 hours of the time of pasteurization. (3) The temperature of the products not solidly frozen shall be taken at the container to determine compliance with this section.

(c) Containers shall be stacked so as to permit circulation of air around the containers.

(d) The outside of the liquid egg containers shall be clean and free of evidence of liquid egg.

(e) Frozen egg products shall be examined by organoleptic examination after freezing to determine their fitness for human food. Any such products which are found to be unfit for human food shall be denatured and any official identification mark which appears on any container thereof shall be removed or completely obliterated and the containers identified as required in Parts 59.840 and 58.860 (name and address of the processor) with the words "Inedible Egg Products—Not to Be Used as Human Food."

## IV. PRODUCTS AND USES

Egg products are frozen primarily to extend their shelf life, quality, and safety. Commercial egg products are frozen in two forms: pasteurized-liquid egg and cooked egg. The uses and specifications of some liquid-frozen egg products appear in Table 1.

Liquid-frozen egg products are usually packaged in 30-lb cans, plastic containers, or in 4-, 5-, 8-, and 10-lb plastic pouches. Other special liquid-frozen mixtures include salted whole egg, whole egg with added corn syrup, whole eggs with added yolk, and blends of yolk and albumen with and without salt or sugars.

Cooked egg products are sold primarily in two forms: cooked whole egg (scrambled) and hard-cooked. Cooked-frozen, scrambled egg is marketed in breakfast dishes and entrees and in breakfast sandwiches. Frozen, hard-cooked eggs are often used in the form of yolk and albumen pieces as found in restaurant salad bars as a topping. Some specialty cooked-frozen egg products include egg patties, fried eggs, quiches, omelets, and French toast.

## V. FREEZING METHODS

### A. Blast/Slow Freezing

Liquid egg is most often frozen and held in plastic or metal containers using conventional-type air-convection freezers. These conventional freezers can freeze large volumes of liquid egg and can be of either the "blast" or "slow" variety. Blast freezers use fans to increase the air flow around the containers, thus increasing the cooling rate. This type of freezer operates at around –40°C (–40°F). An advantage of the blast freezer compared to the slow freezer is the rapid cooling rate. A slow freezer with no forced-air convection may be used to hold the egg at below freezing temperatures, usually below –12°C (10°F).

**Table 1**  Specifications and Uses of Frozen Egg Products

| Product | Specifications | Uses |
|---|---|---|
| Whole egg | Whole egg in natural proportions; no additives with a minimum of 25.5% egg solids | In baked goods (cakes, cookies, sweet doughts, and pastries) |
| Fortified whole egg | Whole eggs with extra yolk and sugar, salt or syrup added according to user's formula | Same as whole egg |
| Standard albumen | Albumen with a minimum of 11.5% egg solids and a maximum fat content of 0.03% | In baked goods (angel food and white cakes, meringues, icings, and candy) |
| Quick-whipping albumen | Specially processed albumen to promote quicker whipping | In angel food cakes |
| Plain yolk | Yolks with a minimum of 45% solids with no additives | In egg noodles and baby foods |
| Sugared yolk | Yolks with a minimum of 43% solids with 10% sugar added | In baked goods (cakes, pastries, ice cream, and baby food |
| Salted yolk | Yolks with a minimum of 43% solids with 10% salt added | In mayonnaise, salad dressings, and condiments |

*Source*: Ref. 4.

## B.  Quick Freezing

A variety of liquid carbon dioxide ($CO_2$) and liquid nitrogen ($N_2$) freezers are used to freeze cooked egg products. Different versions of these "quick" freezers include tunnel, spiral, or cabinet (Figs. 4–7). The cooked egg product is sprayed with liquid $CO_2$ or liquid $N_2$ and the food is almost instantaneously frozen. Liquid $N_2$ tunnel freezers and cabinet freezers expose the egg to temperatures as low as –157°C (–250°F) and –101°C (–150°F), respectively. Only the thickness of the food material itself limits the freezing rate.

## C.  Concentrate Then Quick Freeze

A liquid $CO_2$ drum freezer (Fig. 8) is available to quickly freeze liquids into frozen flakes; however, this is not currently used by the egg industry. Researchers are investigating the quick freezing of liquid egg that has been concentrated by vacuum

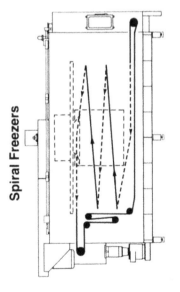

The Model JE-U6 Compact Height Spiral Freezer

**Spiral Freezers**

**Figure 4** Spiral freezer. (Courtesy of Liquid Carbonic.)

**Figure 5**   Spiral freezer. (Courtesy of Frigoscandia.)

evaporation. The concentrated egg is formed into pellets by quick freezing with liquid $CO_2$ or $N_2$. An excellent description of freezing equipment can be found in Refs. 2 and 5.

## VI.   EFFECTS ON FUNCTIONALITY

The major functions of eggs in food products include coagulation, foaming, emulsification, and nutritional contributions. Several analytical tests are used to measure functionality, including emulsification capacity, emulsion stability, viscosity, cake or foam height (when used as an ingredient), water-holding ability (cooked egg), and texture profile characteristics of cooked egg.

## A.   Liquid Frozen Eggs

### 1.   Egg Albumen

The functional properties of liquid EA are generally stable to freezing and thawing. Researchers report a decrease in the sulfhydryl flavor groups during 10 days of frozen storage [6]. This may be related to a slight thinning of the thick albumen fraction. Almost no loss in volume has been observed in cakes made from frozen EA compared to those made from fresh egg albumen [7]. While EA is relatively stable during freezing, slower freezing rates, higher thawing temperatures, and longer storage times increase the loss of denaturation enthalpy of EA [8]. Con-

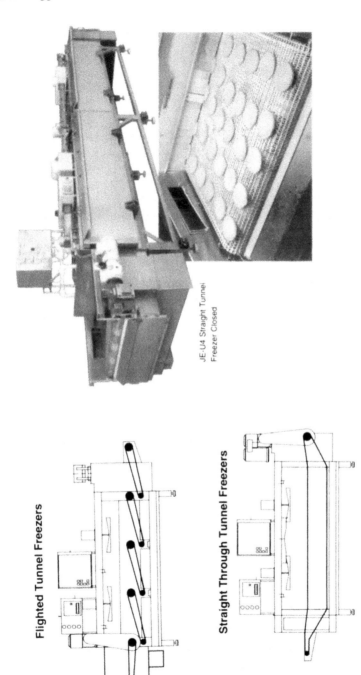

JE-U4 Straight Tunnel
Freezer Closed

**Flighted Tunnel Freezers**

**Straight Through Tunnel Freezers**

**Figure 6** Tunnel freezer. (Courtesy of Liquid Carbonic.)

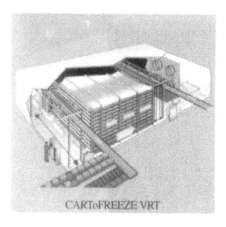 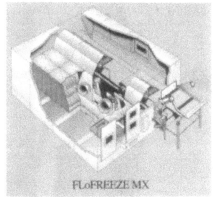

**Figure 7**   Tunnel freezers. (Courtesy of Frigoscandia.)

albumin suffered the greatest loss, while ovalbumin had a smaller loss in enthalpy. The slower freezing rates resulted in the greatest decrease in denaturation enthalpy (Fig. 9). Thus, some denaturation of the EA proteins does occur during freezing, but this has only a slight effect on the egg albumen functionality. These functional changes are detectable in tests for viscosity and foam stability (Fig. 10). The EA viscosity and foam instability is reduced by slow freezing rates, higher thawing rates, longer storage times, and lower storage temperatures. Egg albumen displays pseudoplastic, non-Newtonian flow behavior. Shear stress will decrease with increased shear rates.

Heat set gels formed from liquid EA that have been frozen and thawed show increased gel strength and elasticity from the addition of 5 and 10% sucrose to the liquid albumen before freezing [9]. The viscosity index of the cooked albumen with added sucrose is similar to that of plain albumen. The addition of 5 and 10% sodium chloride to liquid EA will reduce the gel strength, elasticity, and viscosity index of cooked EA gels [9].

## 2.  Egg Yolk

Unlike EA, significant changes in EY texture occur during freezing and thawing. Freezing below –6°C and subsequent thawing causes an increase in viscosity and gelling of the yolk [10]. The longer EY is held under frozen storage conditions, the greater the degree of gelation and increase in viscosity. Like EA, EY is a pseudoplastic fluid (Fig. 11). The pseudoplasticity of yolk is caused by the presence of granules, proteins, and low-density lipoproteins (LDL). The pseudoplastic behavior of the remaining yolk plasma is reduced when the granules are removed by ultracentrifugation [11] (Fig. 11).

Under laboratory conditions, EY will freeze at –1°C but does not gel until –6°C.

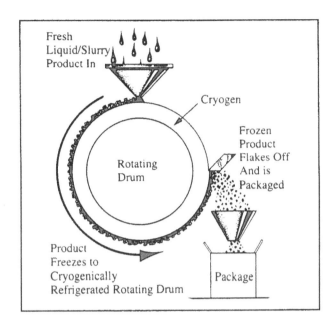

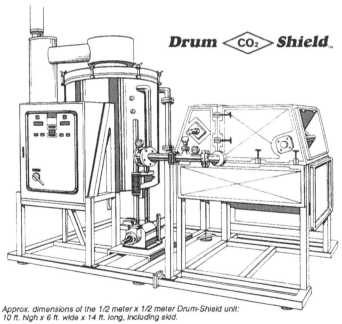

Approx. dimensions of the 1/2 meter x 1/2 meter Drum-Shield unit:
10 ft. high x 6 ft. wide x 14 ft. long, including skid.

**Figure 8** Quick-freezing liquid $CO_2$ drum freezer. (Courtesy of Liquid Carbonic.)

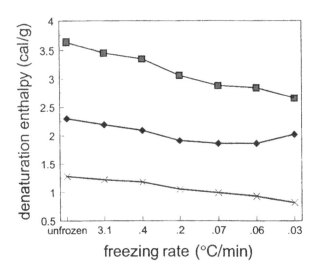

**Figure 9**  Effect of freezing and freezing rates on the denaturation enthalpy of egg albumen, ovalbumin, and conalbumin. (■), Egg albumen; (♦), ovalbumin; (✕), conalbumin. (Adapted from Ref. 8.)

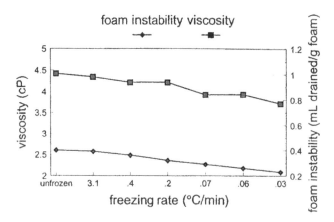

**Figure 10**  Effect of freezing and freezing rate on the viscosity and foam instability of egg albumen. (Adapted from Ref. 8.)

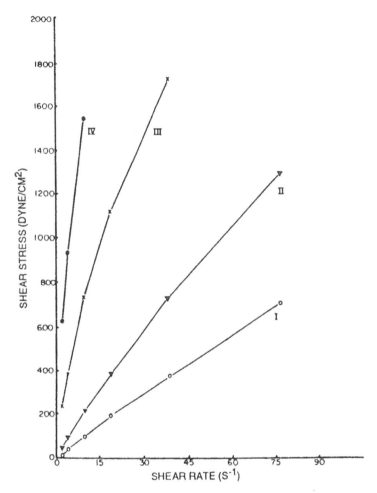

**Figure 11** The shear stress and shear rate relationships of nonfrozen and frozen-thawed samples of egg yolk and yolk plasma at 25°C. I = native yolk plasma; II = native yolk; III = frozen plasma for 5 h at –14°C; IV = yolk frozen for 3 h at –14°C. (From Ref. 11.)

The maximum rate of gelation occurs at –18°C. Egg yolk stored at –0.5 and –3°C was more viscous than nonfrozen yolk but was still less viscous than 10% salted yolk stored at –20°C [12]. However, commercial freezers do not have adequate temperature control to take advantage of this small window between freezing and gelation temperatures. Thus, all frozen EY will essentially gel under commercial freezing applications.

Egg yolk is a complicated system during the freezing process when compared to other fluid foods. Yolk contains stable and unstable particles suspended in a protein solution. These particles consist of yolk granules (stable) and micelles of lipoproteins (unstable) with varying degrees of stability and solubility (Fig. 12).

The unstable particles have both yellow and white types. The yellow and white particles are relatively large (50-100μm) and unstable compared to the granules (1-5μm) that can withstand centrifugation. The yellow particles are packed with smaller particles believed to contain lipoproteins. The granules also contain lipoproteins. The yellow and white yolk particles are damaged during the freeze-thaw cycle and contribute to gelation by the release and denaturation of lipoproteins. Due to the relatively high solids content, EY can be lowered to –6°C without the separation of water from the solutes into ice crystals. Once separation begins, the yolk will freeze and gelation occurs. The egg yolk becomes "pasty" upon thawing. Thus, yolk gelation is induced by the separation of ice resulting in physical damage to egg components and the subsequent denaturation of the lipoprotein fraction. At –6°C, 91% of the "freezable" water is crystallized into ice. The damage to the yolk due to freezing can be seen as the formation of large pores under electron microscopic examination [13] (Fig. 13).

The degree of gelation is affected by the freezing rate, storage temperature, duration of storage, and thawing rate. The relative importance of each of these factors on EY gelation is somewhat vague due to their interaction. However, the effect of the individual factors on gelation has been thoroughly researched. Very rapid freezing will reduce gelation, while a lower storage temperature will increase gelation. The dramatic changes in yolk texture due to freezing are theorized to be caused by protein denaturation as a result of the high salt concentrations found as water is transformed into ice and/or aggregation of the LDL fraction with the yolk granules [13–17]. The granule-LDL interaction is implied, since when the plasma was separated from the yolk by centrifugation, the remaining granules showed minimal gelation during freezing-thawing [17,18]. However, researchers have reported that the yolk plasma will gel after removal of the granules by centrifugation, thus implicating the LDL in the yolk plasma as the primary reason for

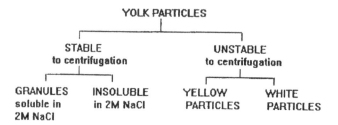

**Figure 12**    The particles found in egg yolk.

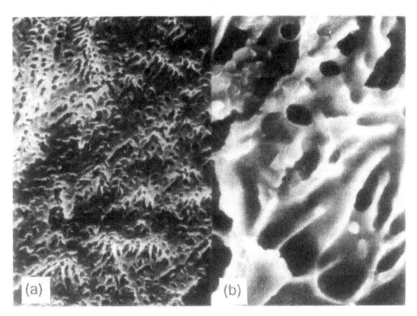

**Figure 13** Scanning electron micrographs (750×) of control yolk: (a) nonfrozen; (b) frozen-thawed. (From Ref. 13.)

freeze-thaw gelation [14,18]. The LDL in yolk (density = 0.962 g/ml) is comprised of 11% protein, 60.5% triglycerides, 25% phospholipid, and about 4% cholesterol or cholesterol esters [19]. The unfreezable water in the LDL is less than 10%. The general structure of yolk LDL is a liquidlike lipid region (triglycerides) with little ordered structure anchored by the polar head of lecithin (phospholipid) attached to the peptide groups (Fig. 14) [19].

Both yolk and yolk plasma show increased viscosity after a freeze-thaw cycle (Fig. 15). The yolk plasma (with granules removed) has a reduced degree of gelation shown by a smaller increase in viscosity (Fig. 15). After freezing and thawing, yolk plasma has a pasty consistency. About 15% of the lipoproteins are insoluble in 10% NaCl [14]. Aggregation of lipoproteins has been observed in frozen liquid whole egg via transmission electron microscopy [20]. This aggregation was observed more clearly in thawed egg magma than in freeze-dried egg, suggesting that aggregation continues during and/or after thawing. Under examination by electron microscopy it was observed that freezing induced LDL aggregation by the alteration of water associated with the LDL [13]. The LDL of nonfrozen yolk migrates under certain paper electrophoretic conditions, while much of the LDL from frozen yolk remains at the origin [11,15,16] (Fig. 16). The LDL are transformed into high molecular weight lipoproteins by a direct aggrega-

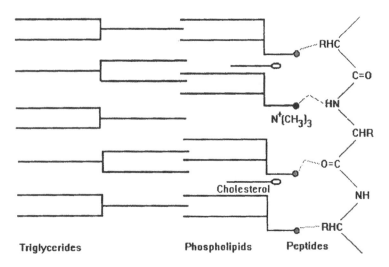

**Figure 14**   Proposed structure of egg yolk low-density lipoprotein. (From Ref. 19.)


tion with other yolk lipoproteins of high molecular weight [21,22]. The loss of a large part of the LDL fraction and an increase in the high-density fraction has been verified by gel filtration [21]. These findings verify that the LDL is denatured during the freeze-thaw cycle.

Alteration in LDL is caused by the concentration of salts [11] or change in pH [15] due to freezing. This concentration in the nonfrozen phase results in precipitation of the LDL, a disruption of the yolk granules, and/or aggregation of the LDL [23], as stated earlier. Several other points can be made concerning freeze-thaw–induced yolk gelation. The transformation of a minimum amount of liquid water to ice is required for gelation. The LDL aggregation may be related to pH changes in the unfrozen phase as well as a concentration of salts. The protein and phospholipid moieties may participate in the formation of a LDL–water–sodium chloride complex [23]. However, alteration in the salt concentration and pH have little effect on the very low-density lipoprotein (VLDL) fraction [24]. Several researchers have suggested that gelation was caused by the gradual removal of water from the LDL and aggregation by the protein moiety. Proteolytic enzymes inhibited yolk gelation when added prior to freezing [19,25,26]. It is likely that freezing causes damage to the hydrogen bond linkages between the phospholipid and the peptide chain (Fig. 14). Theoretically, once the proteins have been cleaved from the LDL by freezing and thawing, their intrinsic property of aggregation results in the formation of a gel [27]. Three proteins (apoprotein A, B, and C) can be isolated from the LDL of eggs, with apoprotein A having the greatest tendency to gel under cold storage [27]. Apoprotein B gels at concentrations above 10 mg/ml, while apoprotein C shows no

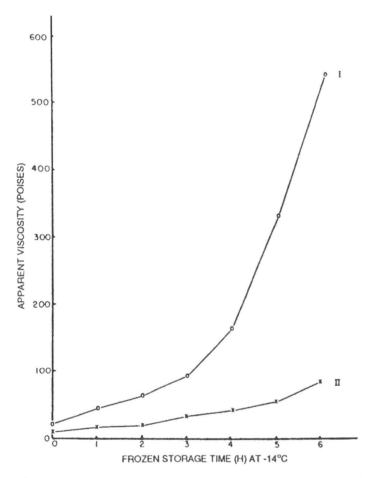

**Figure 15**  Influence of frozen storage at –10°C on the apparent viscosities at 25°C of yolk and yolk plasma. I = yolk; II = yolk plasma. (From Ref. 14.)

tendency to gel. Apoproteins A and B have similar amino acid compositions, with the exception that B has more than two times the isoleucine content [27]. Aggregation of LDL micelles may also be due to electrostatic bonding between positively and negatively charged peptide groups [26]. Yolk VLDL loosens components that stabilize the molecule at the surface, leading to the formation of a meshlike structure [28]. Differential scanning calorimetry can be used to detect the lipid thermal transition patterns related to denaturation and aggregation. These patterns are indicative of similarities or differences in structure. Three distinguishable patterns have been obtained for native (type 1), freeze-damaged (type 2), and enzymatically

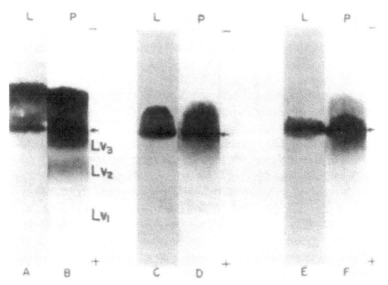

**Figure 16** Paper electrophoretograms of yolk, native and frozen for 30 days at −10 and −14°C: A+B, native yolk; C+D, yolk frozen for 30 days at −10°C; E+F, yolk frozen for 30 days at −14°C. L = Lipid stain; P = protein stain. (From Ref. 11.)

protein-depleted fractions (type 3) from VLDL [29]. Treatments used to prevent gelation (addition of salt) produced lipid patterns intermediate to types 1 and 2, indicating that while VLDL aggregation may be inhibited, additives do not prevent changes in the lipid fraction of these particles [29].

Yolk gelation can be inhibited by chemical additives, proteolytic enzymes, and mechanical shearing. Egg yolk gelation during freezing is controlled easily by the addition of low molecular weight compounds such as sucrose (8% level) and sodium chloride (2% level). Ten percent solutions of sugar and salt are commonly used commercially to retard gelation, even though lower levels are effective. Other additives that also retard gelation include syrups, glycerin, phosphates, propylene glycol, glycerol, sorbitol, and sugars other than sucrose. The mechanism of gelation inhibition by these additives is theorized to be due to either the reduction of freezable water, the reduction of the reactivity of the LDL fraction, and/or the formation of a salt-granule complex to prevent the granule-LDL interaction [30]. A reduction in the charged peptide groups produced by binding with additives may be responsible for a reduction in gelation [26]. Mechanical shearing such as homogenization will reduce the viscosity of frozen yolk, but this is an unacceptable method to reverse gelation for most commercial operations [31]. The addition of proteolytic enzymes (papian, trypsin, and rhozyme) inhibits yolk gelation during freezing and thawing, but only papian does not result in unacceptable product flavor [25]. Enzymatic hydrolysis reduces the number of negatively charged sites [18]. Although adding 10% sodium chloride to egg yolk controls gelation, freezing still significantly increases the viscosity of salted yolk [32]. Freeze-drying results in an even more viscous yolk material than conventional freezing of salted yolk [32]. In addition, freezing near −20°C and extended frozen storage increases the viscosity of salted yolk [33]. Temperature fluctuations between −18 and −29°C approximately doubles the viscosity of 10% salted yolk [33]. Thus, control of the holding temperature is very important in maintaining the yolk's fluid consistency.

## 3.  Whole Egg

Freezing alters the rheological and functional properties of liquid whole egg (LWE) [34]. Liquid whole egg gelatinizes during freezing and thawing, but gelling is not as extensive in LWE as in pure yolk. Reduction in the degree of LWE gelation compared to EY gelation attributable to freezing is most probably due to dilution of the pure yolk with albumen in the LWE. Gelation increases the viscosity and decreases the foam height of LWE upon thawing [35]. The freezing and holding temperature as well as storage time determines the extent to which the functional properties of LWE are altered [20]. Storage times of less than 4 hours were found to have no effect on the viscosity of frozen-thawed LWE [36].

The viscosity of fresh egg is 5-6 cp (Fig. 17). A freeze/hold temperature of 5°C has little effect on LWE viscosity. Freezing and holding at −20°C increases the

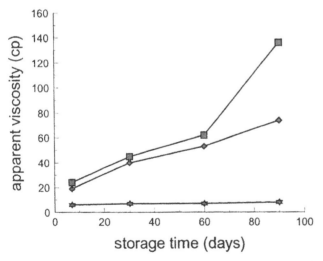

**Figure 17**   The effect of freezing and holding temperature on the viscosity of whole egg. Freezing/holding temperatures (°C) were as follows: (■), -5/-20; (♦), -20/-20; (★), -5/-5. (Adapted from Ref. 20.)

viscosity to 20 cp when held 5 days and to 60 cp after 90 days. The increase in LWE viscosity is directly related to freezing temperature and length of frozen storage (Fig. 17). Freezing at –5°C followed by holding at –20°C increases LWE viscosity to a greater extent than freezing at –20°C and holding at –20°C. The viscosity of LWE increases most during the first day of frozen storage [21]. Frozen storage has been found to double the viscosity of LWE. Thawed egg behaves as a pseudoplastic fluid [37]. More recent research has shown that thawed and non-frozen LWE behave as either a Newtonian fluid (no yield stress) or a Bingham plastic (with a yield stress) [21]. For both, a linear relationship exists between shear stress and shear rate$^{-1}$ [21]. Nonpasteurized LWE, frozen up to 80 days at –10°C, did not develop a yield stress. However, higher pasteurization temperatures required shorter frozen storage times to produce a yield stress [21]. Maximum gelation occurs between –18 and –23°C. Blast freezing at –29°C limited the viscosity to three quarters of that of "slow"-frozen whole egg (–20°C) [34].

Egg foam height is not affected by the length of frozen storage (Fig. 18). However, LWE foam height is decreased from 90 mm in fresh egg to 73–79 mm (–5°C freeze/–5°C hold) and to 50–79 mm (–5 or –20°C freeze/–20°C hold) in frozen-thawed egg. Different freezing and holding temperatures below –5°C do not

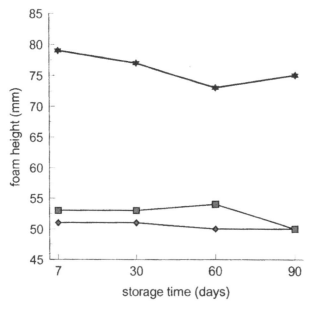

**Figure 18**   The effect of freezing and holding temperature on the foam height of whole egg. Freezing/holding temperatures (°C) were as follows: (★), -5/-5; (◆), -20/-20; (■), -5/-20. (Adapted from Ref. 20.)

change the foaming ability of LWE. Whipping properties and volume of sponge cakes are not decreased by the use of frozen LWE [18,38].

Long-term freezing generally decreases the solubility of protein, however, freezing at 24°C for one day produced a 5–8% increase in soluble protein content of whole egg [39]. The decrease of soluble protein correlates to pie filling viscosity and yield stress in frozen whole egg [40]. Functionality changes in LWE have been linked to the denaturation and insolubilization of egg proteins [40].

## B.  Cooked Frozen Eggs

### 1.  Whole Egg

Freezing and thawing of cooked WE results in syneresis and a loss of water. Using a centrifuge method, the expressible moisture of cooked nonfrozen WE was approximately 38% [42]. Freezing and thawing increases the expressible moisture of WE to 59% [43]. The volume of expressed serum from cooked-frozen-thawed-reheated (CFTR) scrambled eggs decreases as the pH is increased from 6.0 to 7.0 [44]. The solids, protein, and lipid content of this serum is maximum at pH 7.0. The pH of CFTR whole egg increased above the initial pH, particularly in the 6.8–8.4 initial pH range [44]. An initial pH of 6.8 increased to 8.2 after CFTR; about half of this increase was observed after cooking only. A clear explanation is not available for this increase in pH, but it may be related to some loss of buffering capacity due to protein denaturation [44]. The increase in expressible moisture in cooked-frozen-thawed WE can be reduced through the addition of sodium caseinate to the liquid egg prior to cooking [43]. Adding 4.5% sodium caseinate also maintained the cooked egg's texture attributes during the freeze-thaw cycle and resulted in similar springiness and cohesiveness when compared to nonfrozen cooked egg. Other additives such as corn starch and modified corn starch did not produce an increase in expressible moisture during the freeze-thawing of cooked WE [43]. An increased rate of freezing reduces the size of the ice crystals formed and, thus, reduces the expressible moisture and increases the sensory scores [45] (Fig. 19).

### 2.  Hard-Cooked Whole Egg, Cooked Yolk, and Cooked Albumen

Research on the freezing of hard-cooked eggs can be included with information on freezing cooked egg yolk and egg albumen since these components are essentially separated in the hard-cooked egg during cooking. Cooked egg albumen does not normally maintain its texture and water-binding properties after freezing and thawing. Cooked albumen will develop a porous texture that "weeps." Freezing causes the water held within the gel to migrate toward ice nuclea, forming relatively large ice crystals within the albumen protein structure. Thawing causes the ice crystals to melt, leaving pores. The protein gel surrounding the pores is hardened due to the migration of water during freezing [46]. The damage of freezing is produced by the development of the physical formation of ice crystals and by the

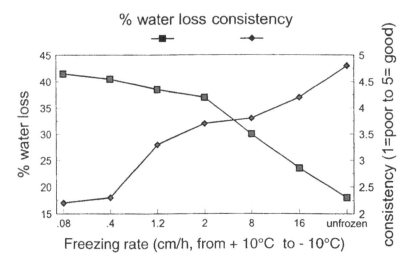

**Figure 19**   Effect of freezing rate on the water loss and texture score of cooked, frozen-thawed egg albumen. (Adapted from Ref. 45.)

loss of bound water located within the gel. The degree of damage produced by freezing can therefore be lessened by reducing the size of the ice crystals. Factors used to reduce ice crystal size include extremely high rates of freezing produced by dipping in liquid Freon or by using compressed liquid nitrogen gas [46]. Syneresis can also be reduced by the addition of water-binding carbohydrates, such as algin, carrageenan, agar, and starch, at the 2–4% level [47].

Gelatin may facilitate supercooling of egg white and, therefore, reduce ice crystal size [48]. Calcium carbonate particles provide sites for formation of ice crystals, thereby reducing their average size [48].

The effect of freezing rate on the microstructure of cooked egg albumen, cooked egg yolk, and cooked whole egg can be seen in Figures 20, 21, and 22. These micrographs were produced during a procedural optimization for examining pH and other effects on egg gel microstructure. The lower freezing temperature (faster rate) of –95°C (Fig. 20B,C) produced pores (0.1–2.0 μm across) 90% smaller than those in egg albumen frozen at –35°C (15–20 μm across) (Fig. 20A). The pores are surrounded by strands of heat-set protein. The microstructure of egg yolk was similarly affected by freezing rate. The size of the fat globules in yolk paralleled the pore size in egg albumen under similar freezing rates [48] (Fig. 21). Yolk frozen at –35°C (Fig. 21A) displayed particles 1–15 μm in diameter, while the –95°C frozen yolk (Fig. 21C,D) had particles in the range of 0.3-2.0 μm in diameter, indicating that some yolk gelation can occur in cooked yolk. Some loosely bound water may produce ice crystals, which damage the structure holding the lipoprotein

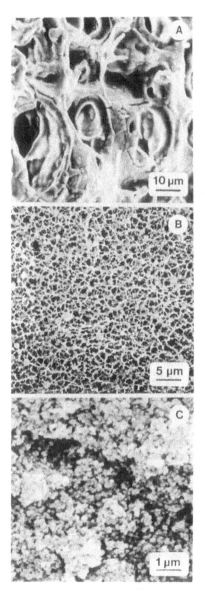

**Figure 20** Micrographs of cooked egg albumen gels prepared for scanning electron microscopy by various methods. (A) Frozen at –35°C and freeze-dried; (B) frozen in liquid hexane (–95°C), immersed in liquid nitrogen (–196°C), and freeze-dried; (C) higher magnification of B. (From Ref. 48.)

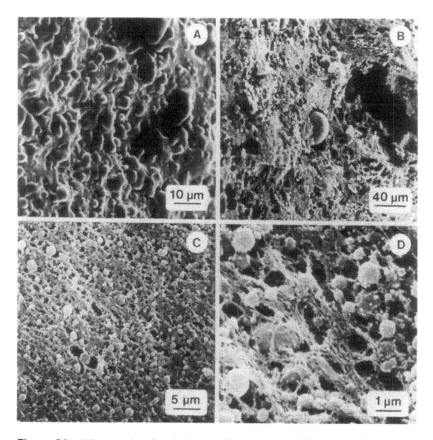

**Figure 21** Micrographs of cooked egg yolk gels prepared for scanning electron microscopy by various methods. (A) Frozen at –35°C, freeze-dried; (B) frozen at –35°C, freeze-dried, defatted and critical point dried; (C) frozen in liquid hexane (–95°C), immersed in liquid nitrogen (–196°C), freeze-dried, defatted, and critical point dried; (D) higher magnification of (C). (From Ref. 48.)

structure together. Gummy sections of cooked, frozen-thawed yolk can sometimes be seen in yolk preparations for salad bars. Freezing WE at –35°C produced 10- to 20- μm-diameter pores surrounded with protein strands coated with a lipid layer (Fig. 22A). The WE frozen at 95°C had pores 90% smaller than the slow-frozen WE with an open network structure (Fig. 22B).

## C. Egg as a Component

Sponge cakes made from freeze-dried eggs have a reduced volume and decreased tenderness compared to frozen, foam-spray-frozen, and spray-dried eggs. In chif-

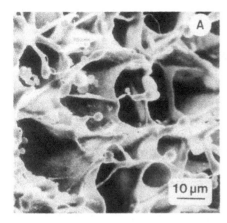
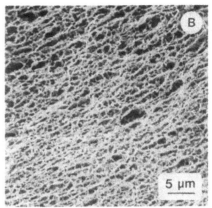

**Figure 22** Micrographs of cooked whole egg gels prepared for scanning electron micros-copy by various methods. (A) Frozen at –35°C and freeze-dried; (B) frozen in liquid hexane (–95°C), immersed in liquid nitrogen (–196°C), freeze-dried, defatted, and critical point dried. (From Ref. 48.)

fon cakes prepared from frozen and spray-dried eggs, the foam produced from the frozen and spray-dried eggs albumen is more stable in supporting the weight of other cake ingredients compared to the foam from foam-spray-dried eggs [49]. However, freeze- and foam-spray-dried yolks produced more stable foams, but the frozen yolk would not form a foam due to high viscosity [50].

Whole egg, EY, and EA are commonly used as a component in other foods which are cooked prior to freezing. Souffles and meringues can be frozen for several months without loss of emulsion stability. However, increasing the concen-tration of flour and methylcellulose reduces the loss of souffle height during 1 to 3 months of frozen storage [51]. The optimal storage temperature for souffles is –18 to –23°C. Off flavors develop at higher freezing temperatures and greater loss of height occurs at lower temperatures [51].

Mayonnaise made from salted-frozen yolks is less stable and more viscous than mayonnaise made from fresh yolk [32]. The emulsifying capacity of salted yolk, however, is not affected by freezing, but freezing does decrease the stability and stiffness of mayonnaise made from 10% salted yolk. Freeze-drying of salted yolk further decreases the mayonnaise stability and stiffness compared to frozen salted yolk [52].

Egg yolks act to stabilize fat in cream puffs. Freezing salted yolks results in cream puffs 2 to 6% smaller than puffs made from non-frozen salted yolks [33].

## VII.  MICROBIOLOGICAL QUALITY AND SAFETY

Freezing reduces bacterial counts in egg products but will not destroy all strains. Gram-negative bacteria (*Aeromonas, Enterobacter, Flavobacterium* and *Pseudomonas* spp.) are predominant in nonpasteurized egg products [53–55]. However, gram-positive bacteria (*Bacillus, Micrococcus, Staphylococcus,* and *Coryneform* spp.) are predominant in pasteurized egg products [56–58]. The levels of both in nonpasteurized and pasteurized WE remained constant or decreased when these products were stored at –3°C [12]. *Bacillus* and *Streptococcus* strains are particularly resistant to freezing at –3°C [12]. No bacteria were recovered from pasteurized EA frozen at –3°C, but several strains survived pasteurization and freezing in EY [12]. The numbers of coliforms remained constant in frozen WE stored up to 50 days at –3°C [12].

*Listeria monocytogenes* survived frozen storage at 0 and 18°C in both the dried

**Table 2**  General Effects of Freezing Rate, Storage Time, Storage Temperature, Thawing Rate, and Additives on Liquid Egg Products Quality

| Influencing Factor | Affect on quality | | |
|---|---|---|---|
| | Egg albumen[a] | Egg yolk[b] | Whole egg[c] |
| Freezing rate | Slower rate causes reduced viscosity and increased foam stability | Slower rate causes increased viscosity and gelation | Same as liquid EY but less severe |
| Storage time | Longer time causes reduced viscosity and increased foam stability | Longer time causes increased viscosity and gelation | Same as liquid EY but less severe |
| Storage temperature | Lower temperature causes reduced viscosity and increased foam stability | –18°C results in maximum increase in viscosity and gelation | Same as liquid EY but less severe |
| Thawing rate | Faster rate causes some protein denaturation | Slower rate causes increased viscosity and gelation | Same as liquid EY but less severe |
| Additives | None normally needed | 2% NaCl and 8% sucrose inhibits gelation; 10% used commercially | None normally needed |

[a]freezing usually has only a slight effect on egg albumen properties.
[b]freezing often has a drastic effect on egg yolk viscosity
[c]freezing has a greater effect on whole egg properties than egg albumen but less than the effect on egg yolk.

**Table 3**  General Effects of Freezing Rate, Storage Time, Storage Temperature, Thawing Rate, and Additives on Cooked Egg Product Quality

| Influencing factor | Affect on quality | | |
|---|---|---|---|
| | Egg albumen | Egg yolk | Whole egg |
| Freezing rate | Slower rate causes greater syneresis | Little affect on functionality | Same as albumen but less severe |
| Storage time | Longer time causes greater syneresis | Little effect on functionality | Same as albumen but less severe |
| Storage temperature | Lower temperature slightly increases syneresis | Little effect on functionality | Same as albumen but less severe |
| Thawing rate | Faster rate increases syneresis | Little effect on functionality | Same as albumen but less severe |
| Additives | Starches, algin carrageenan, agar inhibit syneresis | None normally needed | Sodium caseinate, gums, and starches at 2–5% inhibit syneresis |

and liquid state [59]. A slight reduction (1–1.5 logs) was observed with the dried-then-frozen treatments, but no loss of inoculated viable cells was observed in whole egg frozen from the liquid form [59].

## VIII.  SUMMARY

Some general conclusions of five influences on egg product quality can be drawn and used as guidelines when considering freezing effects (Tables 2 and 3). Of course, exceptions are usually found, thus, these guidelines may not hold in all cases. In an oversimplification of the general effects of freezing, liquid albumen and cooked yolk are stable, while liquid yolk and cooked albumen undergo drastic changes in viscosity and texture, respectively.

## REFERENCES

1. USDA, *U.S. Egg and Poultry Statistical Series, 1960-1992* (M. Madison and A. Perez, eds.), 1994.
2. D. S. Reid, Physical phenomena in the freezing process, *Frozen Food Technology* (C. P. Mallett, ed.), Blackie Academic and Professional, London, 1993, pp. 14–15, 25–50.
3. A. Monzini and E. Maltini, Structural modification by freezing in liquid foods, *Prog. Food Eng: Proc European Symp. Food Working Party of the E.F.C.E.* 7:435–469 (1983).

4.  O. Feet, T. J. Hedrick, L. E. Dawson, and H. E. Larzelere, Frozen egg products specifications and uses. *Mich. Agri. Expt. Sta. Bull.* 46(80): (1963).

5.  W. R. Woolrich, Refrigerating systems used in cold and freezer storage, *The Freezing Preservation of Foods* (D. K. Tressler, W. B. Van Arsdel, and M. J. Copley, eds.), AVI Publishing, Inc., 1968, pp. 49–73.

6.  S. A. Hussaini and F. Alm, Denaturation of proteins of egg white and fish and its relation to the liberation of sulfhydryl groups on frozen storage, *Food Res.* 20:264–272 (1955).

7.  C. A. Clinger, A. Young, I. Prudent, and A. R. Winter, The influence of pasteurization, freezing and frozen storage on the functional properties of egg white, *Food Technol.* 5:166–170 (1951).

8.  M. Wootton, N. T. Hong, and H. L. Phan Thi, A study on the denaturation of egg white proteins during freezing using differential scanning calorimetry, *J. Food Sci.* 46:1336–1338 (1981).

9.  C. W. Dill, J. Brough, E. S. Alford, F. A. Gardner, R. L. Edwards, R. L. Richter, and K. C. Diehl, Rheological properties of heat-induced gels from egg albumen subjected to freeze-thaw, *J. Food Sci.* 56:764–768 (1991).

10. T. Moran, The effect of low temperature on hen's eggs, *Proc. Royal Soc. London* B98:436–456 (1925).

11. C. H. Chang, W. D. Powrie, and O. Fennema, Studies on the gelation of egg yolk and plasma upon freezing and thawing, *J. Food Sci.* 41:1658–1665 (1977).

12. C. Imai, J. Saito, and M. Ishikawa, Storage stability of liquid egg products below 0°C, *Poultry Sci.* 65:1679–1686 (1986).

13. R. J. Hasiak, D. V. Vadehra, R. C. Baker, and L. Hood, Effect of certain physical and chemical treatments on the microstructure of egg yolk, *J. Food Sci.* 37:913–917 (1972).

14. A. Saari, W. D. Powrie, and O. Fennema, Influence of freezing egg yolk plasma on the properties of low density lipoproteins, *J. Food Sci.* 29:762–765 (1964).

15. W. D. Powrie, H. Little, and A. Lopez, Gelation of egg yolk, *J. Food Sci.* 28:38–46 (1963).

16. D. D. Meyer and M. Woodburn, Gelation of frozen-defrosted egg yolk as affected by selective additives: Viscosity and electrophoretic findings, *Poultry Sci.* 44:437–446 (1965).

17. S. Mahadevan, T. Satyanarayana, and S. A. Kumar, Physical and chemical studies on the gelation of hen's egg yolk. Separation of gelling protein components from yolk plasma, *J. Agric. Food Chem.* 17:767–770 (1969).

18. S. T. McCready and O. J. Cotterill, Centrifuged liquid whole egg. 3. Functional performance of frozen supernatant and precipitate fractions, *Poultry Sci.* 51:877–881 (1972).

19. V. B. Kamat, G. A. Lawrence, M. D. Barrat, A. Darke, S. S. Leslie, G. G. Shipley, and J. M. Stubbs, Physical studies of egg yolk low density lipoprotein, *Chem. Phys. Lipids* 9:1–25 (1972).

20. Y. Nonami, M. Akasawa, and M. Saito, Effect of frozen storage on functional properties of commercial frozen whole egg, *Nippon Shokuhin Kogyo Gakkaishi* 39:49–54 (1992).

21. T. J. Herald, F. A. Osori, and D. M. Smith, Rheological properties of pasteurized liquid whole egg during frozen storage, *J. Food Sci.* 54:35–38 (1989).

22. F. S. Soliman and L. Van den Berg, Factors affecting freeze aggregation of lipoprotein, *Cryobiology* 8:265–270 (1971).

23. T. Wakamatu, Y. Sato, and Y. Saito, On sodium chloride action in the gelation process of low density lipoprotein (LDL) from hen egg yolk, *J. Food Sci.* 48:507–516 (1983).

24. Y. Sato and Y. Aoki, Influence of various salts on gelation of low density lipoprotein (egg yolk) during its freezing and thawing, *Agric. Biol. Chem.* 39:29–35 (1975).

25. A. Lopez, C. R. Ferrers, and W. D. Powrie, Enzymatic inhibition of gelation in frozen egg yolk, *J. Milk Food Technol.* 18:77–80 (1955).

26. C. M. Nowak, W. D. Powrie, and O. Fennema, Interaction of low density lipoprotein (LDL) from yolk plasma with methyl orange, *Food Res.* 31:812–818 (1966).

27. K. S. Raju and S. Mahadevan, Protein components in the very low density lipoproteins of hen's egg yolk. Identification of highly aggregating (gelling) and less aggregation (nongelling) proteins, *Biochim. Biophys. Acta* 446:387–398 (1976).

28. J. I. Kurisaki, S. Kaminogawa, and K. Yamauchi, Studies on freeze-thaw gelation of very low density lipoprotein from hen's egg yolk, *J. Food Sci.* 45:463–466 (1980).

29. S. Mahadevan, Differential scanning calorimetry studies of native and freeze-damaged very low density lipoproteins in hens' egg yolk plasma, *J. Biosci.* 11(1-4):299–310 (1987).

30. R. Jordan and E. S. Whitlock, A note on the effect of salt (NaCl) upon the apparent viscosity of egg yolk, egg white and whole egg magma, *Poultry Sci.* 34:566–571 (1955).

31. J. E. Marion, J. G. Woodroof, and R. E. Cook, Some physical and chemical properties of eggs from hens of five different stocks, *Poultry Sci.* 44:529–534 (1965).

32. S. S. Yang and O. J. Cotterill, Physical and functional properties of 10% salted egg yolk in mayonnaise, *J. Food Sci.* 54:210–213 (1989).

33. H. H. Palmer, K. Ijichi, S. L. Cimono, and H. Roff, Salted egg yolks. I. Viscosity and performance of pasteurized and frozen samples, *Food Technol.* 23(2):1480–1485 (1969).

34. K. Ijichi, H. H. Palmer, and C. Lineweaver, Frozen whole eggs for scrambling, *J. Food Sci.* 35:695–698 (1970).

35. C. Miller and A. R. Winter, The functional properties and bacterial content of pasteurized and frozen whole egg, *Poultry Sci.* 50:1810–1817 (1950).

36. J. A. Pearce and C. G. Lavers, Liquid and frozen egg. V. Viscosity, baking quality and other measurements on frozen egg products, *Can. J. Res.* 27:231–240 (1949).

37. J. Torten and H. Eisenberg, Studies on colloidal properties of whole egg magma, *J. Food Sci.* 47:1423–1428 (1982).

38. S. T. McCready, M. E. Norris, M. Sebring, and O. J. Cotterill, Centrifuged liquid whole egg. 1. Effect of pasteurization on the composition and performance of the supernatant fraction, *Poultry Sci.* 50:1810–1817 (1971).

39. T. L. Parkinson, The effect of pasteurization and freezing on the low density lipoproteins of egg, *J. Food Sci. Agric.* 28:806–810 (1977).

40. T. J. Herald and D. M. Smith, Functional properties and composition of liquid whole egg proteins as influenced by pasteurization and frozen storage, *Poultry Sci.* 68:1461–1469 (1989).

41. T. L. Parkinson, Effect of pasteurization on the chemical composition of liquid whole

egg. III. Effect of staling and freezing on the protein fractionization patterns of raw and pasteurized whole eggs. *J. Food Sci. Food Agric.* 19:590–596 (1968).

42. C. Jauregui, J. Regenstein, and R. Baker, A simple centrifugal method for measuring expressible moisture, a water binding property of muscle foods, *J. Food Sci.* 46:1271–1273 (1981).

43. G. Poole, Freeze-thaw determination for precooked whole egg, M.S. thesis, Clemson University, 1992.

44. G. E. Feiser and O. J. Cotterill, Composition of serum from cooked-frozen-thawed-reheated scrambled eggs at various pH levels, *J. Food Sci.* 47:1333–1337 (1982).

45. N. Bengtsson, Ultrafast freezing of cooked egg white, *Food Technol.* 21:1259–1261 (1967).

46. J. Davis, H. Hanson, and H. Lineweaver, Characterization of the effect of freezing on cooked egg white, *Food Res.* 17(4):393–401 (1952).

47. R. L. Hawley, Egg product (egg roll proteolytic enzyme added to yolk and starch to white to reduce rubberiness and syneresis), U.S. Patent No. 3,510,315 (1970).

48. S. A. Woodward and O. J. Cotterill, Preparation of cooked egg white, egg yolk and whole egg gels for scanning electron microscopy, *J. Food Sci.* 50:1624–1628 (1985).

49. M. E. Zabik and S. L. Brown, Comparison of frozen, foam-spray-dried, freeze-dried and spray-dried eggs 4. Foaming ability of whole eggs and yolks with corn syrup solids and albumen, *Food Technol.* 23:262–266 (1969).

50. M. E. Zabik, C. M. Anderson, E. M. Davey, and N. J. Wolfe, Comparison of frozen, foam-spray-dried, freeze-dried and spray-dried eggs 5. Sponge and chiffon cakes, *Food Technol.* 23:359–364 (1969).

51. S. Cimino, L. Elliott, and H. Palmer, The stability of souffles and meringues subjected to frozen storage, *Food Technol.* 21:1149–1152 (1967).

52. L. J. Harrison and F. E. Cunningham, Influence of frozen storage time on properties of salted yolk and its functionality in mayonnaise, *Food Qual.* 9:167–173 (1986).

53. E. M. Sevior and R. C. Board, The behavior of mixed infection in the shell membranes of the hen's egg, *J. Food Hyg.* 13:33–44 (1972).

54. N. Sashihara, H. Mizutani, S. Takayama, H. Konuma, A. Suzuki, and C. Imai, Bacteriological survey on raw materials of liquid whole egg, its products and manufacturing processes, *J. Food Hyg. Soc. Jpn.* 20:127–136 (1979).

55. A. Suzuki, H. Konuma, S. Takayama, C. Imai, and N. Sashihara, Preservability of unpasteurized liquid whole eggs by chilling, *J. Food Hyg. Soc. Jpn.* 20:442–449 (1979).

56. R. Shafi, O. J. Cotterill, and M. L. Nichols, Microbial flora of commercially pasteurized egg products, *Poultry Sci.* 49:578–585 (1970).

57. J. Payne and J. E. Gooch, Heat resistant bacteria in pasteurized whole egg, *J. Appl. Bacteriol.* 46:601–613 (1979).

58. J. Payne and J. E. Gooch, Survival of faecal streptococci in raw and pasteurized egg products, *Br. Poultry Sci.* 21:61–70 (1980).

59. R. E. Brackett and L. R. Beuchat, Survival of *Listeria monocytogenes* in commercially dried and frozen liquid eggs, *Abstr. Gen. Meet. Am. Soc. Microbiol.* 3:270 (1991).

# 10
# Effects of Freezing, Frozen Storage, and Thawing on Dough and Baked Goods

**Yoshifumi Inoue**

*Japan School of Baking*
*Edogawa-ku, Tokyo, Japan*

**Walter Bushuk**

*University of Manitoba*
*Winnipeg, Manitoba, Canada*

## I. GENERAL INSTRUCTIONS

Freshly baked goods possess a desirable flavor and texture. These highly desirable attributes diminish rapidly with a lapse of time after baking. For example, a crispy crust develops a moist leathery texture, and a soft crumb becomes firm and dry. Fresh flavor dissipates within a few hours after baking and is generally referred to as staling. This problem and its accompanying economic losses have traditionally forced bakers to do midnight or early morning baking to provide the consumer with fresh bread on a daily basis. These factors have also limited the distance over which baked products can be transported from a large automated bakery.

Some of these problems can be overcome by using mixed and molded frozen dough, which can be quickly transformed into fresh-baked product in a small local bakery. Many such bakeries are now located in large grocery stores and contribute significantly to overall sales by attracting customers to the store.

The use of frozen doughs for the production of baked goods has several advantages over conventional processing. It eliminates night or early morning labor, decreases the need for highly skilled bakers, reduces processing space and equipment, and increases the variety of baked items that can be produced in a single bakery. These advantages constitute the main reason for the rapid growth in the number of in-store bakeries since the early 1970s.

This chapter discusses the technology used for the production of bread from frozen dough and the factors that contribute to the quality of the baked product.

## II. EFFECTS OF FREEZING, FROZEN STORAGE, AND THAWING ON GAS PRODUCTION IN DOUGH

### A. Introduction

Frozen doughs encounter processing stresses during freezing, frozen storage, and thawing to which conventional nonfrozen doughs are not subjected. It is common knowledge that these stresses cause a deterioration in product quality. Accordingly, for optimum product quality from frozen dough, special attention must be paid to the selection of ingredients, optimization of formulas and processing conditions, especially conditions during freezing, frozen storage, and thawing. These aspects of frozen dough technology will be discussed in the sections that follow. This section will cover yeast, the ingredient mainly responsible for gas production in a fermenting dough.

### B. Effect of Fermentation Before Freezing

Whether a thawed dough expands properly in the oven depends on two interacting factors: gas production and the gas retention required for a optimum loaf volume and crumb grain. Gas production in a fermenting dough is a function of yeast activity and availability of fermentable sugars.

The changes in yeast activity in frozen doughs will be discussed from three perspectives. This section will consider the effects of fermentation before freezing. Subsequent sections will cover the effects of rate of freezing and freezing temperature and the effect of frozen storage.

Fermentation in mixed doughs prior to freezing reduces subsequent freeze tolerance of the yeast [1,2]. When dough pieces were made up and frozen immediately after mixing, yeast activity remained stable after prolonged storage periods (Fig. 1). However, when fermentation was allowed to proceed after mixing before freezing, the yeast became less tolerant of freezing temperature and activity declined. The reason for this loss of activity is not known. It is presumed that with onset of fermentation, the yeast cell membrane becomes more sensitive to damage by freezing than the membrane of dormant yeast cells [1,3]. The products of fermentation, ethanol and other volatile organics, have been demonstrated to mitigate the tolerance of the yeast to damage during freezing and frozen storage [3,4]. To minimize loss of yeast activity resulting from prefreezing fermentation, the dough should be mixed at as low a temperature as possible and frozen as quickly as possible after mixing and dough make-up.

### C. Effect of Freezing Rate and Temperature

Rapid freezing is generally recommended for foods to minimize damage due to ice crystallization. However, this does not appear to apply to frozen doughs, where rapid freezing seems to have a detrimental effect on yeast activity [5-7]. This effect was shown quite explicitly by Neyreneuf and Delpuech [7]. Yeast activity decreased significantly when the rate of cooling was increased to 1.565°C min from 0.978°C min (Table 1).

Although the cause of the detrimental effect of rapid freezing is not clearly understood, the following explanation has been postulated by Mazur [8] and appears plausible. During rapid cooling, the formation of intracellular ice crystals, invariably lethal to yeast cell membranes, is unavoidable. During slow cooling, the supercooled intracellular water may be transferred to external ice crystals due to the vapor pressure differential. The resulting dehydration may be sufficient to render the small quantity of residual internal water incapable of freezing due to interaction with the solid components of the yeast cell (i.e., bound water).

Hsu et al. [5] reported that the final freezing temperature also affected the activity of yeast in frozen dough after thawing. Doughs that were frozen to –35°C or lower gave inferior results, regardless of cooling rates. In baking practice, doughs are usually frozen at –20 to –30°C.

## D. Effect of Frozen Storage and Thawing

During storage of frozen dough, yeast activity gradually decreases [2, 3]. The loss of yeast activity can be measured directly by determining the number of viable yeast cells, as shown in Figure 1, and indirectly by measuring the gassing power of thawed doughs or the proof time required to raise the dough to a constant height [2].

The mechanism for the loss of yeast activity during frozen storage and thawing is not fully understood. However, it is believed that the major cause of loss of activity is the disruption of the yeast membrane by ice crystals and by the diffusion of intracellular moisture when the water activity outside the yeast cell decreases with freezing. Recently, new strains of yeast have been developed which have considerably higher tolerance of freezing [9–11]. Further improvements in freeze tolerance of yeast for frozen dough applications are needed.

## III. EFFECTS OF FREEZING, FROZEN STORAGE, AND THAWING ON GAS-RETENTION PROPERTIES OF DOUGH

### A. Changes in Rheological Properties of Thawed Doughs

Dough pieces that have been subjected to freezing, frozen storage, and thawing retain less of the gas produced by fermentation when compared with analogous nonfrozen doughs. This effect has been demonstrated by indirect measurement of gas-retention properties using the modified extensigraph procedure [12–14]. Ex-

**Table 1** Effects of Freezing Temperatures on the Thermal Characteristics and Baking Performance of Dough Slabs

| Freezing temperatures (°C) | Thermal characteristics | | | | Loaf volumes (ml × 2) after storage | | | |
|---|---|---|---|---|---|---|---|---|
| | Freezing time[a] | | Cooling velocity at surface (°C/min) | Cooling velocity at core (°C/min) | 0 days | 30 days | 60 days | 90 days |
| | min | s | | | | | | |
| -40 | 100 | 34 | 0.363 | 0.348 | 4,350 | 4,265 | 4,285 | 4,275 |
| -40 | 43 | 54 | 0.919 | 0.797 | 4,133 | 4,056 | 4,090 | 4,070 |
| -60 | 35 | 45 | 1.267 | 0.978 | 3,963 | 3,985 | 4,010 | 4,020 |
| -80 | 22 | 21 | 2.080 | 1.565 | 3,390 | 3,353 | 3,380 | 3,370 |
| -100 | 17 | 40 | 2.933 | 1.980 | 2,751 | 2,731 | 2,750 | 2,760 |
| -120 | 13 | 51 | 3.936 | 2.528 | 2,481 | 2,280 | 2,261 | 2,186 |

[a]To reach a core temperature of -15°C
*Source:* Ref. 7.

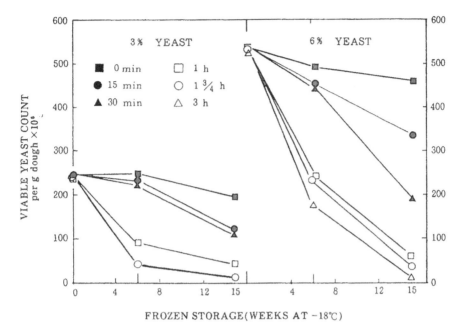

**Figure 1** Changes in viable yeast count of straight doughs during frozen storage at –18°C. Effect of initial yeast level and fermentation time prior to molding and freezing. (From Ref. 2.)

perimental dough pieces subjected to a freeze-thaw cycle showed a significant decrease in extensigraph maximum resistance to extension (Table 2). This resistance decreased further after an additional three freeze-thaw cycles or during prolonged frozen storage for up to 10 weeks. This decrease in maximum resistance was significantly correlated with increase in proof time and with decrease in loaf volume [12,13]. Such observations indicate that the decrease in gas-retention capacity that occurs during freezing, frozen storage, and thawing is partially responsible for the deterioration of baking quality (i.e., loaf volume) of frozen dough.

## B. Factors Involved in Changes in Rheological Properties

### 1. Effect of Ice Crystals

Factors involved in the wakening of frozen doughs are not fully understood. It is generally known that the faster the freezing rate, the lower the degree of structural damage in frozen foods due to ice crystallization and to growth of ice crystals.

**Table 2** Extensigraph Properties and Gassing Power of Nonfrozen and Thawed Frozen Doughs[a]

| Frozen storage time | Extensigraph | | | | |
|---|---|---|---|---|---|
| | Maximum resistance[b] (BU) | | Extensibility[b] (mm) | | Gassing power[c] |
| | | | | (mm Hg) | (%) |
| Nonfrozen | 627 ± 6 | a | 121 ± 3  b | 459 ± 6  a | 100 |
| 1 day | 530 ± 10 | b | 122 ± 3  b | 447 ± 7  a | 97 |
| 7 days | 523 ± 12 | b | 123 ± 4  b | 451 ± 16  a | 98 |
| 7 days | 407 ± 6 | c | 121 ± 4  b | 378 ± 9  b | 82 |
| 70 days | 360 ± 20 | d | 136 ± 4  a | 254 ± 10  c | 55 |

[a]Values are means ± standard deviations (SD). Means with different letters within a column are significantly different ($p < 0.05$).
[b]Mean ± SD of three replicates.
[c]Mean ± SD of two replicates.
*Source*: Ref. 14.

However, freezing of yeasted doughs must occur at a rate that minimizes damage to yeast cells. In commercial practice, the rate of freezing is optimized to minimize the detrimental effects on dough structure and yeast activity. Under such optimized conditions, considerable disruption of the gluten network has been observed by scanning electron microscopy [15,16]. Consequently, it is generally concluded that formation and growth of ice crystals during freezing, storage, and thawing is one of the main causes of dough weakening, which results in a loss of gas-retaining capacity.

## 2. Effect of Dead Yeast Cells

Extensigraph maximum resistance of thawed doughs decreased significantly when dough was subjected to freeze-thaw cycles and during frozen storage (Table 2). On the other hand, extensigraph extensibility did not change after an additional three freeze-thaw cycles but significantly increased after 70 days of frozen storage. This change in extensibility was attributed to the loss of gassing power [14]. Therefore, it seems reasonable to conclude that dead yeast cells contribute to observed weakening of fermenting doughs.

Kline and Sugihara [2] postulated that some dough weakening may be caused by reducing substances from dead yeast cells. A recent study [14] of changes in rheological properties and protein solubility suggests that the effects of reducing substances from dead yeast cells become significant only when a substantial proportion of yeast activity is lost. In addition, dough weakening during frozen storage may occur by mechanism(s) other than the reduction caused by substances from dead yeast cells [13].

## 3. Effect of Fermentation Before Freezing

Doughs that have undergone some fermentation during mixing and dough make-up are more sensitive to damage during freezing, storage, and thawing [12]. However, the deterioration of baking quality can be minimized by using a lower mixing temperature and freezing quickly after mixing so as to limit the extent of fermentation. This appears to be the main reason why proofed doughs cannot be frozen without a significant loss in baking quality.

## IV. FROZEN DOUGH INGREDIENTS AND FORMULATION

### A. Flour

Until recently regular bread flours have been used for the production of frozen doughs for bread. It had been assumed that the deterioration of breadmaking quality of frozen doughs was caused by the loss of yeast. It is now recognized that the deterioration of quality during frozen storage can be mitigated by using an appropriate flour.

Neyreneuf and Van Der Plaat [9] compared the stability of frozen French bread doughs made from two flours of different protein content (11.1 and 12.8%). Doughs from the higher-protein flour had greater stability than those from the flour with the lower protein content (Fig. 2).

Inoue and Bushuk [13] used four different flours to investigate separate effects of protein content and protein quality or mixing strength (Table 3). Frozen doughs were stored for 10 weeks. Baking results (Fig. 3) showed that the strongest flour (cv. Glenlea), which in Canada is considered unsuitable for conventional bread product, gave the best results for the frozen-dough procedure. The superior performance of the extra-strong flour appeared to be due to the ability of its doughs to sustain a higher oven spring even after losing some of its intrinsic strength upon freezing and frozen storage (Fig. 4). Accordingly, to obtain the highest bread quality from frozen doughs, flours that produce doughs with very strong mixing properties should be used. Protein content in the range covered by the study of Inoue and Bushuk [13] appears to be less important.

Dough strength can be increased by adding vital wheat gluten to the flour. The level of vital gluten addition must not be too high, otherwise the color, flavor, and texture of the baked products will suffer.

Intrinsic amylase activity of the flour also affects its performance in frozen-dough processing. Neyreneuf and Van Der Plaat [9] showed that flours with a lower falling number value (higher amylase activity) did not perform in frozen French bread doughs as well as flours with a higher falling number value. Flour for frozen dough should be milled from sound wheat to obviate any problems due to excessive enzyme activity.

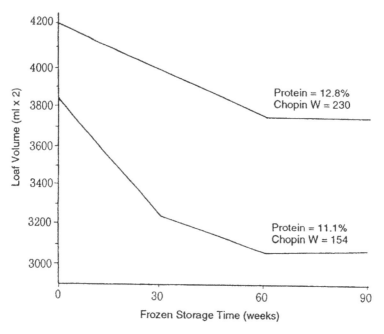

**Figure 2**   Effect of baking quality of flours on frozen dough performance. (From Ref. 9.)

## B.   Water Absorption

Since dough weakening during freezing and frozen storage (due to e.g., ice crystallization) is a major cause of quality deterioration of frozen doughs, doughs used for freezing should contain somewhat less water than doughs used for conventional baking. Generally, water absorption is decreased 3-5% for frozen doughs compared with conventional doughs.

## C.   Yeast

Three types of yeast are available for baking: compressed, dry, and instant dry. As mentioned earlier, yeast loses some of its freeze-thaw tolerance due to activation by fermentation before freezing. Therefore, it had been presumed that dry yeast, which possesses a longer induction period, would minimize the extent of fermentation before freezing and would have a higher freeze-thaw tolerance in a frozen dought system [1]. However, it has been demonstrated that compressed yeast possesses the highest freeze-thaw tolerance among the three yeast types [2,17]. Neyreneuf and Van Der Plaat [9] found that the drying process used in the production of instant dry yeast remarkably decreased the freeze-thaw tolerance of yeast (Fig. 5). Based on this result, they concluded that the drying condition affects

**Table 3** Characteristics of Flour Samples[a]

| | Flours | | | |
|---|---|---|---|---|
| Quality tests | A | B | C | D |
| Protein content, % | 14.4 | 13.9 | 13.7 | 13.7 |
| Ash content, % | 0.52 | 0.54 | 0.58 | 0.60 |
| Starch damage (Farrand unit) | 21 | 23 | 11 | 25 |
| Falling number value | 538 | 566 | 337 | 500 |
| Farinograph | | | | |
|   Absorption, % | 65.1 | 63.5 | 58.6 | 58.9 |
|   Dough development time, min | 7.0 | 6.0 | 5.5 | 30.5 |
| Gassing power,[bc] mm Hg | 459 ± 5 | 461 ± 8 | 519 ± 11 | 477 ± 5 |
| Standard proofing height,[bd] cm | 10.0 ± 0 | 10.0 ± 15 | 10.0 ± 0 | 9.8 ± 0.1 |
| Extensigraph | | | | |
|   Maximum resistance,[bd] BU | 627 ± 6 | 623 ± 15 | 680 ± 10 | 1,273 ± 21 |
|   Extensibility,[ba] mm | 121 ± 3 | 120 ± 2 | 137 ± 2 | 102 ± 3 |
| Loaf volume,[bd] cm$^3$ | 792 ± 8 | 780 ± 5 | 838 ± 5 | 863 ± 8 |

[a]Reported on 14.0% moisture basis.
[b]For nonfrozen dough by frozen dough procedure.
[c]Mean and standard deviation of duplicates.
[d]Mean and standard deviation of three replicates.
*Source*: Ref. 13.

the structure and functional integrity of the yeast cell membrane and thereby increases the sensitivity to damage during freezing.

Neyreneuf and Van Der Plaat [9] compared rapid yeast and reduced-activity yeast for suitability in frozen doughs. The rapid yeast was immediately active and had a much faster fermentation rate before freezing than the reduced-activity yeast (Fig. 6). Figure 7 shows the stabilities of frozen doughs containing the two types of yeast. The dough with reduced-activity yeast was much more stable than the dough with the rapid yeast. Hsu et al. [4] reported that acceptable yeast performance after freezing was associated only with yeasts with protein contents higher than 57%. However, reduced-activity yeast contained a rather low protein level (46%). Neyreneuf and Van Der Platt [9] further indicated that freeze-thaw tolerant yeast should have a high trehalose content in addition to reduced activity. Trehalose has been reported to perform a cryoprotectant function in the yeast cell [10].

A new freeze-tolerant yeast has been developed and is now available commercially in Japan [11]. This yeast can be frozen in a dough for prolonged periods of time without significantly losing any of its activity even after considerable fermentation prior to freezing (Fig. 8). However, this unique characteristic of the yeast cannot be used for made-up frozen doughs because fermentation before freezing decreases the freeze-thaw tolerance of partially developed doughs (see above). This type of yeast can be used to prepare fully fermented doughs for freezing only if the

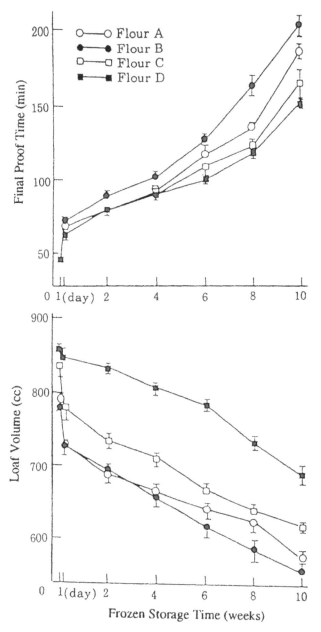

**Figure 3**  Effect of frozen storage time on final proof time and loaf volume of doughs. (From Ref. 13.)

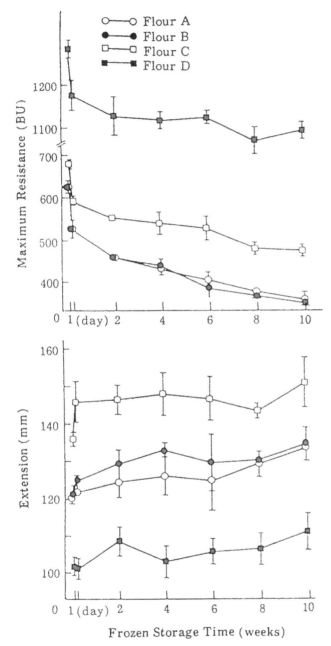

**Figure 4**  Effect of frozen storage time on extensigraph maximum resistance and extension of doughs. (From Ref. 13.)

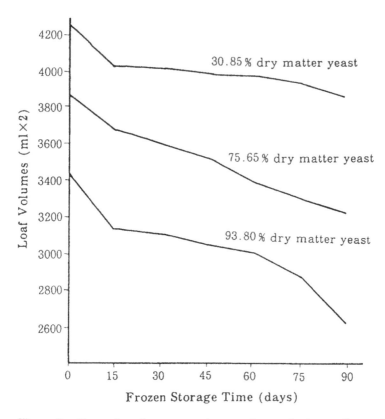

**Figure 5**   Conversion of a compressed yeast to instant dry forms-effect on frozen dough performance. (From Ref. 9.)

thawed dough is subjected to molding. Dough make-up after thawing can redevelop the weakened dough structure.

In conclusion, a compressed yeast with freeze-thaw–tolerant properties should be used for frozen dough production. A higher level of yeast is generally recommended, e.g., 4-5%, in order to compensate for the loss of activity during freezing, storage, and thawing. Much higher levels of yeast 6-8% have a detrimental effect on aroma and flavor of the baked product [18].

## D.   Shortening

A higher shortening level is recommended for frozen dough production [19]. Generally, shortening protects dough structure from damage due to ice crystallization. Inoue et al. [20] found that incorporation of shortening as an oil-in-water O/W

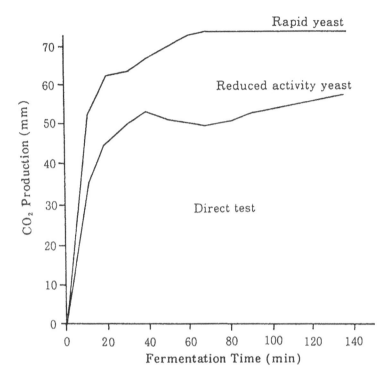

**Figure 6** Influence of fermentation rate on $CO_2$ production (rheofermentometer analysis). (From Ref. 9.)

emulsion system remarkably improved frozen dough stability (Table 4). Although the mechanism of this improvement is not understood clearly, it is speculated that the O/W emulsion system can protect the dough structure by a more uniform and complete distribution of shortening in the dough.

## E. Oxidants

Oxidant level should be increased in frozen dough formulations, as oxidants increase dough strength. It is well known that fermentation before dough make-up increases dough strength and improves the gas-retention properties of the dough in the oven. Accordingly, minimized fermentation before dough make-up should be compensated for by an increased level of oxidants. Also, added dough strength enables it to tolerate the weakening effects of freezing, frozen storage, and thawing. Potassium bromate and ascorbic acid have been widely used as additives in the baking industry. A combination of the two has been found to be

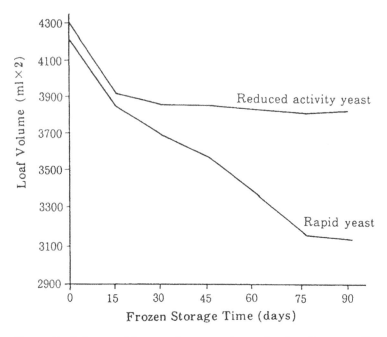

**Figure 7**   Influence of fermentation rate on frozen dough performance. (From Ref. 9.)

effective in maintaining frozen dough stability [12,15,21]. In countries where the use of potassium bromate for baking is permitted, the addition of 30–45 ppm of potassium bromate (flour basis) and 100 ppm of ascorbic acid is recommended for frozen dough. In countries where the use of potassium bromate is prohibited, the use of 0.5–2.0% enzyme-active soy or fava bean flour in frozen dough preparation is recommended for dough-strengthening purposes [9]. The beneficial effect of the bean flours is thought to be due to their lipoxygenase activity [22]. When using bean flours as the source of lipoxygenase, the beneficial dough strengthening effect must be balanced against the detrimental effects on flavor.

## F.   Surfactants

In general, the types of surfactants that act as dough strengtheners can also improve frozen dough stability. Sodium or calcium stearoyl lactylate (SSL or CSL) and diacetyl tartaric acid ester of monoglyceride (DATEM) have been shown to be effective in maintaining loaf volume and crumb softness of breads made from frozen doughs [15,23,24]. Hosomi et al. [25] have shown that a hydrophilic sugar ester improves the baking and rheological properties of frozen doughs. In countries

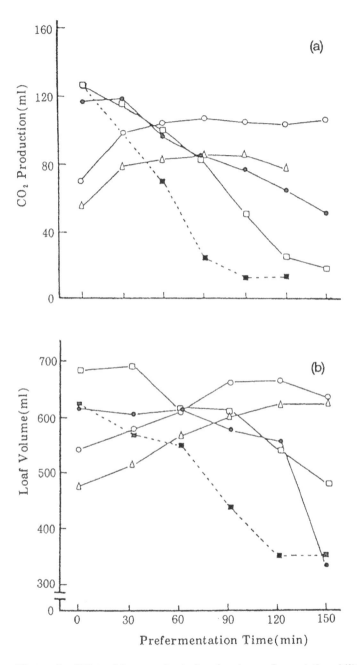

**Figure 8** Effect of fermentation before freezing on fermentative ability of frozen dough stored at $-20°C$ for 7 days. (a) $CO_2$ production for frozen dough made with 20 g flour. (b) Loaf volume for frozen dough made with 100 g flour. ■, commercial yeast; ●, commercial freeze-tolerant yeast; □, FRI 802 yeast; ○, FRI 501 yeast; △, FRI 413 yeast. (From Ref. 11.)

**Table 4** Effects of O/W Emulsion Shortening System[a] on Baking Properties of Frozen Dough

| Frozen storage period[b] | Control[c] | | 40% O/W[d] | | 60% O/W[e] | |
|---|---|---|---|---|---|---|
| | FP[f] (min) | LV[g] (cc) | FP (min) | LV (cc) | FP (min) | LV (cc) |
| Nonfrozen | 55 ± 0 | 855 ± 13 | 55 ± 0 | 853 ± 10 | 55 ± 0 | 875 ± 13 |
| 1 day | 80 ± 1 | 838 ± 10 | 82 ± 1 | 837 ± 20 | 82 ± 2 | 860 ± 15 |
| 28 days | 97 ± 5 | 802 ± 8 | 94 ± 2 | 847 ± 8 | 97 ± 3 | 862 ± 10 |
| 70 days | 120 ± 2 | 767 ± 16 | 106 ± 2 | 830 ± 5 | 105 ± 2 | 852 ± 16 |

[a]10% (flour basis) of hydrogenated canola oil was emulsified with a part of water and 0.2% of polyglycerin monostearate into an oil-in-water (O/W)-type emulsion before mixing.
[b]Stored at –20°C.
[c]Contains 10% (flour basis) of hydrogenated canola oil.
[d]10% of hydrogenated canola oil was added to the dough as O/W emulsion (40% water + 60% oil).
[e]10% of hydrogenated canola oil was added to the dough as O/W emulsion (60% water + 40% oil).
[f]Final proof time, thawed frozen dough pieces were proofed to the same height with the height of the final proofed nonfrozen dough pieces, mean ± standard deviation of three replicates.
*Source*: Ref. 20.

where the use of surfactants is permitted, the addition should be at a level of about 0.5%. Table 5 shows the effects of CSL and DATEM on frozen dough baking and rheological properties [20]. These surfactants improved the gas-retention properties of the control dough as indicated by increased loaf volume. The mechanism for the dough-strengthening effect by surfactants at the molecular level is not understood clearly. One theory, which appears rational, is that the surfactants enhance interactions between gluten proteins [26,27]. Monoglycerides can be used for frozen doughs as antistaling and crumb-softening agents in the same way that they are used in doughs for conventional baking.

## G.  Flavoring Substances

Breads from frozen doughs possess less flavor and aroma than breads from conventional doughs because of the minimized fermentation before dough make-up and freezing. Accordingly, the addition of fermented products to improve bread flavor is recommended for the production of high-quality bread from frozen doughs. It should be noted that fermentation products that contain alcohol or activated yeast, like liquid ferments, decrease freeze-thaw stability of the yeast during freezing [3,4]. Therefore, organic acid-based fermented products are recommended. The use of the new freeze-tolerant yeast, which can be frozen in the presence of a high alcohol concentration without significantly decreasing its activity [3], makes the use of alcohol-fermented products possible.

**Table 5** Effects of Surfactants on Baking and Rheological Properties of Frozen Dough

| Frozen storage period[a] | Control[b] | | CSL[c] | | DATEM[d] | |
|---|---|---|---|---|---|---|
| | LV[e] (cc) | E[f] (mm) | LV (cc) | E (mm) | LV (cc) | E (mm) |
| Nonfrozen | 855 ± 13 | 123 ± 3 | 915 ± 13 | 117 ± 2 | 917 ± 18 | 117 ± 3 |
| 1 day | 838 ± 10 | 124 ± 5 | 890 ± 13 | 118 ± 2 | 893 ± 16 | 118 ± 2 |
| 28 days | 802 ± 8 | 125 ± 5 | 858 ± 14 | 118 ± 5 | 880 ± 9 | 119 ± 5 |
| 70 days | 767 ± 16 | 130 ± 5 | 832 ± 16 | 122 ± 3 | 843 ± 15 | 119 ± 2 |

[a]Stored at –20°C.
[b]Contains 10% (flour basis) of hydrogenated canola oil.
[c]Contains 9.5% of hydrogenated canola oil and 0.5% of calcium stearoyl-2-lactylate.
[d]Contains 9.5% of hydrogenated canola oil and 0.5% of diacetyl tartaric acid esters of monoglyceride.
[e]Loaf volume, mean ± standard deviation of three replicates.
[f]Extensibility measured by modified extensigraphy after final proofing, mean ± standard deviation of three replicates.
*Source*: Ref. 20.

## V. FROZEN DOUGH PROCESSING

### A. Mixing

In order to obtain a stable frozen dough, it is necessary to adequately develop by mixing a rather stiff dough due to decreased water absorption and low temperature. High-speed mixers are suitable for this purpose. Mixed dough temperature must be low enough to minimize fermentation before freezing. Generally, a final dough temperature of 20°C is recommended compared with 27–30°C for conventional doughs. Consequently, the use of chilled water or ice in the dough formulation and an effective cooling of the mixer jacket during mixing are necessary. When the cooling system of a mixer jacket is not sufficient or an additional lowering of dough temperature is required, cryogenic cooling of the dough can be used. This technique uses liquid nitrogen or carbon dioxide. The cryogen is sprayed onto flour in the mixer before mixing is started. One gram of liquid nitrogen and liquid carbon dioxide can remove about 98 and 153 kcal, respectively, through vaporization and ventilation to the atmosphere or sublimation and ventilation to the atmosphere [22]. The rather high mixing energy required to fully develop the dough for freezing tends to increase the dough temperature. Therefore, delayed addition of salt after full dough development or the use of reducing agents like cysteine or glutathione may facilitate dough development and thereby maintain a lower dough temperature by shortening the mixing time. Delayed addition of yeast, which reduces yeast activation before freezing, has been recommended for frozen dough preparation [9, 21].

After mixing, dough is divided into loaf pieces and rounded. The dough pieces should be made up with minimal rest time in order to minimize yeast activation.

Usually, the rounded dough pieces are rested for 15-20 minutes before make-up. It is well known that doughs mixed at lower temperature require a shorter rest time. The overall effect is to increased frozen dough stability. However, drastic reduction of fermentation before make-up yields immature doughs. Immature doughs lack gas-retention properties and cannot expand satisfactorily in the oven.

Neyreneuf and van der Plaat [9] studied the effects of dough temperature on frozen French bread performance. Their results (Fig. 9) showed that a dough temperature of 18°C did not produce a dough with adequate gas-retention capacity. Oxidation from added oxidants and physical development from carbon dioxide released by the yeast on gluten structure that occurs during fermentation yields doughs with optimal gas-retention properties. Accordingly, when an excessive reduction of fermentation before freezing is employed in order to obtain a highly stable frozen dough, other means must be used to compensate for the abbreviated maturation of the dough. This is usually achieved by the addition of oxidants.

Table 6 shows the effect of extremely reduced fermentation due to low temperature (18°C) of the dough on the loaf volume of white bread (Y. Inoue, unpublished). The decrease in loaf volume was quite drastic, and it was not possible to improve loaf volume by increasing the mixing energy input (result not shown). However, addition of dough strengtheners like DATEM and sugar esters improved

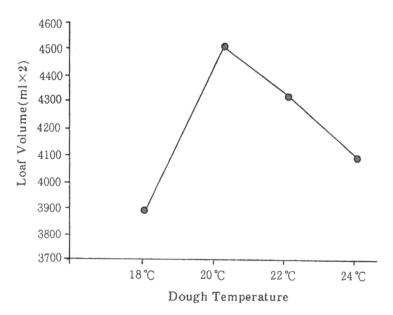

**Figure 9**   Influence of dough temperature on frozen French bread dough performance (90 days storage). (From Ref. 9.)

**Table 6**   Effect of Mixed Dough Temperature and Dough
Sheeting and Folding on Loaf Volume of Bread Produced by
a Short-Time Dough Procedure[a]

|  | Mixed dough temperature | | |
| --- | --- | --- | --- |
|  | 30°C | 18°C | 18°C (+sheeting)[b] |
| Loaf volume (cc)[c] | 2446 | 2260 | 2530 |

[a]20 min of floor time before dividing.
[b]Sheeted and folded three times immediately after mixing and before
dividing.
[c]Baked from 450-g dough piece.
*Source*: Y. Inoue, unpublished results.

the loaf volume slightly. A marked improvement was obtained by sheeting and
folding the cool dough three times immediately after mixing and before dividing
(Table 6). The sheeting and folding did not affect dough temperature significantly
and appeared to compensate for the reduced dough maturity. The flattening action
of the sheeting combined with rotation of the axis of the dough promotes a
two-dimensional structure in the dough rather than the unidirectional structure
produced by ordinary dough mixers [8]. The folding of the dough probably further
promotes development of the three-dimensional structure of the dough. However,
this sheeting and folding process may not be practical because of its complexity.
However, the effect of the process demonstrates that dough development is needed
for improvement of frozen dough stability.

## B.  Dough Make-Up

The time required for dividing and making up a batch of dough for freezing should
be minimized in order to minimize fermentation before freezing. Therefore, smaller
batches of dough are recommended for frozen dough preparation compared with
conventional baking practice. Dough pieces should be made up as tightly as
possible to counter the dough weakening that occurs during freezing, storage, and
thawing. Room temperature for make-up should be low enough to minimize
fermentation. The dough pieces should be frozen immediately after making up.

## C.  Freezing

Made-up dough pieces should be frozen slowly to maintain gas-producing properties (see Sec. II) but rapidly enough to maintain the gas-retention properties of the
dough (see Sec. III). Generally, the velocity to cool a 20°C dough core to −10°C in
a freezer should be between 0.3 and 1.2 °C/min [6,7]. Since freezing time regulates
frozen dough production, the highest possible cooling velocity is generally used.

Five types of freezers can be used for frozen dough production: (1) quiescent freezer, (2) blast freezer, (3) contact freezer, (4) impingement freezer, and (5) cryogenic freezer. Generally, freezing rate increases in the order of 1 to 5. Blast and contact freezers are used most widely.

Weight and shape of made-up dough pieces also affect freezing rates. For example, a smaller dough piece is frozen more rapidly than a larger dough piece, if the same freezing condition is applied. Accordingly, temperature, conveyer speed, or airflow of a freezer should be adjusted for each type of frozen dough product to ensure the appropriate freezing rate. Dough pieces are frozen until the core temperature is below freezing (usually −10 to −15°C). The freezing points of a leaner dough like a French bread is about −5°C, while that of rich formula doughs like dinner rolls is about −10°C.

The frozen dough pieces are bagged in polyethylene bags and packaged in cardboard boxes. The polyethylene bags to be used for frozen dough should possess the following characteristics [19]: (1) high moisture loss protection, (2) high oxygen barrier characteristics, (3) physical strength against brittleness and breakage at low temperature, (4) stiffness to work on automatic machinery, and (5) good heat sealability.

The packaged frozen dough pieces are stored in a storage freezer at −20°C. The temperature of the frozen dough piece decreases gradually and equilibrates throughout the dough to −20°C.

## D.  Storage and Transport of Frozen Doughs

Properly prepared frozen doughs can be stored at −20°C for up to several months without significant deterioration of quality. However, frozen doughs deteriorate significantly if temperature fluctuation occurs during storage. Temperature fluctuation causes vapor pressure variations among ice crystals in the frozen doughs, which promotes growth of large ice crystals in the doughs by migration of moisture through sublimation. Growth of ice crystals is harmful for both the dough structure and the yeast cells [12,15]. Consequently, a storage freezer at a frozen dough plant should have refrigeration capacity high enough to minimize temperature fluctuations due to the introduction of partially frozen doughs or warm air during the removal of frozen doughs.

Transportation of frozen dough products from plants to retail bakeries also constitutes a part of frozen storage. Trucks used must have refrigeration capability sufficient to minimize temperature fluctuation of the frozen doughs. Frozen doughs must be transferred without serious temperature fluctuations from a storage freezer to the trucks and from the trucks to the storage freezer at the bakery. Compared with the storage freezer at plants, smaller storage freezers with smaller refrigeration capacities are generally used at retail bakeries. To minimize temperature fluctuations in the storage and handling of frozen doughs, the following precautions should be taken at a retail bakery: (1) delivered frozen dough products should be placed into the storage freezer quickly, (2) frozen doughs should be baked as quickly as possible, (3) time that the storage freezer door is open should be minimized, (4) an adequate quantity of frozen dough should be stored, (5) left-over frozen dough pieces should be returned immediately to the storage freezer, and (6) left-over frozen dough pieces should not be stored uncovered. The quality of frozen doughs can easily be destroyed by mishandling during the transportation stage or at a retail bakery. Therefore, careful supervision of transportation and frozen storage at retail bakeries are necessary to obtain satisfactory results from frozen doughs.

## E.  Thawing

Lorenz [19] suggested that frozen dough should be thawed rapidly in order to maintain the yeast activity in the dough. This suggestion is based on the hypothesis that slow thawing may be more harmful than rapid thawing because it allows more time for the ice crystals formed in the yeast cell to enlarge to a lethal size. However, slow thawing, for example, thawing in a retarder at $0°C$ overnight, does not affect yeast activity in frozen dough [13, 21]. This hypothesis applies to rapidly frozen doughs in which internal ice crystallization in yeast cells has occurred. Accordingly, frozen dough can be thawed in a retarder at $-3$ to $5°C$, at room temperature (around $25°C$), or in a final proofer at $30–40°C$ [21]. Prior thawing in a retarder shortens final proof time because of the higher internal temperature of the dough when placed in the proofer [21]. The baking schedule can be started from final proofing after thawing overnight. It should be noted that considerable fermentation takes place in the dough as soon as the temperature reaches the melting point during thawing. This early fermentation may cause undesirable structural change in the dough [14]. Therefore, longer thawing times should be in a retarder at low temperature in order to minimize fermentation. Thawing of large frozen dough pieces (300-500 g) takes a long time at room temperature. The use of the final proofer for thawing can reduce the thawing time. However, this type of thawing tends to produce uneven temperature distribution in the thawing dough pieces. The outer layer of the dough pieces is thawed and warmed up much faster than the core of the dough. Fermentation in the thawed outer layers will begin while the core is still frozen. This leads to uneven crumb grain, i.e., a finer center portion and a coarser outer portion. Therefore, the use of the retarder for slower thawing is recommended, especially for loaf-size frozen dough pieces.

Appearance of unsightly blisters on bread crust is a potential problem associated with frozen dough. Moisture in the air condenses on the frozen dough surface during thawing due to a large temperature difference between the dough and the atmosphere. Condensation of moisture on the surface of frozen doughs produces areas where the dough is more extensible. On expansion during baking, these areas tend to blister. Consequently, it is recommended that frozen doughs be covered with a plastic sheet during thawing. Most of the moisture will then condense on the surface of the sheet instead of on the surface of the dough. The cover will also prevent drying out of the doughs during the latter stage of thawing. The dough surface should be as even as possible in order to minimize the blister problem. To achieve this, doughs should be fully developed and made up quickly prior to freezing, before significant fermentation occurs.

## F. Final Proofing

The temperature of the thawed frozen doughs going into the final proofer is usually lower than that of conventional doughs. Accordingly, a lower final proofing temperature and humidity should be used in order to reduce the temperature difference between thawed doughs and the final proofing temperature. When frozen doughs are thawed in a retarder, the temperature of the dough is still cold enough to cause extreme condensation on the surface and uneven fermentation in the final proofer. Therefore, it is recommended that thawed doughs be held at room temperature for 30–60 minutes, covered with a plastic sheet, before being placed in the proofer. Final proof time is generally adjusted to produce doughs of desired height.

## G. Baking and Postbaking Handling

Baking and postbaking handling of the final product from frozen doughs proceed according to common practice used in conventional baking. The usual precautions of proper cooling before slicing and wrapping must be exercised.

## VI. FROZEN BREAD (BAKED GOODS)

## A. Introduction

As mentioned earlier, the quality of baked goods, especially bread, which generally has a higher moisture content than other baked products, deteriorates rapidly during storage after baking. This change is commonly referred to as staling. The losses resulting from bread staling have a significant economic impact. Freezing represents at present the most effective practical means for either inhibiting or greatly retarding the staling process [29]. During the late 1950s and early 1960s, Pence and coworkers [30–37] established the guidelines for bread freezing that have been used by the baking industry to facilitate production and delivery schedules without decreasing product quality.

Bread does not require yeast activity or gas-retention properties, which are important quality factors in unbaked dough. Accordingly, bread can be frozen, stored in the frozen state, and thawed more easily than dough. However, successful freezing of bread also requires careful and informed attention to all aspects of the operation. Improper freezing, frozen storage, and thawing of bread may enhance staling. This section discusses the effects of the key processing steps on bread quality after freezing and thawing.

## B. Effect of Age of Bread at Time of Freezing

Bread staling can be separated into crumb staling and crust staling. Crumb staling is more complex than crust staling, and the changes, like crumb hardness and crumbliness, that occur are of primary concern to the consumer [29]. Consequently this section focuses on crumb staling.

It is well known that all breads stale if they contain sufficient moisture. Bread staling is a very complex phenomenon, and the mechanism involved is not understood clearly. However, there is little doubt that starch retrogradation is the most important single factor in crumb firmness [38–40]. Crystallization of partially gelatinized starch due to molecular aggregation is considered to be the major factor in staling [39, 40]. Crystallization occurs most rapidly during the first few hours after baking (Table 7). Therefore, it is recommended that bread be frozen as soon as possible after baking in order to maintain its freshness [32].

## C. Effects of Freezing

Katz [38] found the rate of bread staling to be temperature dependent as follows: crumb firming was most rapid at −2.8 to −1.7°C, and the rate decreased as the temperature was raised or lowered. This result was later confirmed by Pence and Standrige [32] (see Fig. 10), who suggested that bread should be frozen as rapidly as possible. Pence et al. [31] confirmed this suggestion by studying the effect of freezing rate on crumb firmness. Their results (Table 8) clearly show that the best retention of crumb softness and resiliency requires freezing to be accomplished as rapidly as possible after baking.

Crumb-firming rates decrease significantly when bread core temperatures decrease below bread-freezing points (−5 to −10°C). Accordingly, bread should be

**Table 7** Effect of Storage Time on the Soluble Starch, Amylose, and Amylopection from Bread Crumb

| | Storage times (h) | | | | | | | | |
|---|---|---|---|---|---|---|---|---|---|
| | 0.16 | 1 | 2 | 5 | 12 | 24 | 48 | 72 | 96 |
| Soluble starch, % | 2.06 | 1.76 | 1.36 | 1.28 | 1.23 | 1.11 | 1.09 | 0.86 | 0.81 |
| Amylose, % | 0.93 | 0.64 | 0.49 | 0.41 | 0.40 | 0.31 | 0.27 | 0.22 | 0.19 |
| Amylopectin, % | 1.13 | 1.12 | 0.87 | 0.86 | 0.83 | 0.80 | 0.82 | 0.64 | 0.62 |

*Source*: Ref. 40.

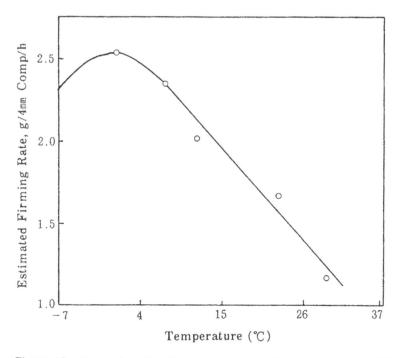

**Figure 10** The relationship of temperature to the estimated average rate of firming of a commercial white pan bread. The temperatures used were 30, 23, 12, 8, 1, and -7°C. (From Ref. 33.)

**Table 8** Effect of Freezing Time on Crumb Firmness[a]

| Freezing time (h) | Av. increase in firmness, (g/4 mm compression)[b] |
|---|---|
| 0.5 | 16 |
| 1.5 | 32 |
| 3 | 57 |
| 7 | 71 |

[a]Defrosting time was approximately 5 h.
[b]Minimum significant difference at 5% point = 13 g/4 mm compression.
*Source*: Ref. 31

frozen as rapidly as possible until its core temperature drops below the freezing point. Cryogenic freezing with liquid nitrogen can achieve very rapid freezing, but is quite expensive. In practice, a freezing time of about 90 minutes, attainable at –29°C and moderate air velocity, will adequately meet the requirements [37]. Subsequent cooling of the bread to a more desirable storage temperature (–18°C or lower) can be achieved safely at slower, less costly rates [42].

Unwrapped bread is frozen considerably faster (Fig. 11) than is wrapped bread (Fig. 12). Pence et al. [30] demonstrated that the freezing rate of unwrapped bread responds to differences in air velocity and to orientation of the loaf in the air stream, as well as to freezer temperature. Moisture loss during freezing of unwrapped bread can be neglected [30]. Consequently, it is recommended that bread be frozen unwrapped in order to minimize quality loss during freezing. However, it is practically difficult to freeze unwrapped sliced bread [37].

Although the rate of freezing of wrapped bread is slower, freezing of bread after slicing and wrapping generally facilitates the operation of frozen bread production. Pence et al. [30] examined factors affecting the freezing rate of wrapped bread and found freezer temperature to be the factor with the greatest influence on freezing rate (Fig. 12). On the other hand, air velocity had little effect on freezing rate (Fig.

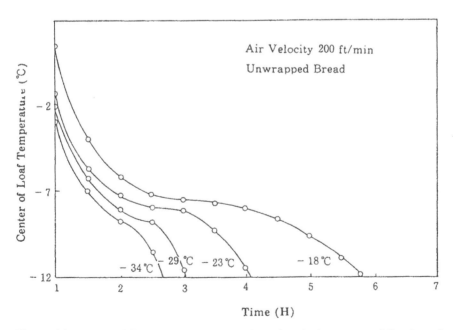

**Figure 11** Effect of freezer temperature at a low air velocity on rate of freezing of unwrapped bread. (From Ref. 31.)

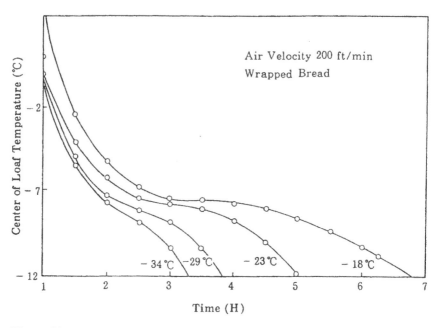

**Figure 12**   Effect of freezer temperature at a low velocity on rate of freezing of wrapped bread. (From Ref. 31.)

13). Therefore, a 5–6°C lower freezing temperature for wrapped bread is generally recommended. Pence [33] observed that breads wrapped in waxed paper, cellophane, and aluminum foil all froze at similar rates, whereas loosely wrapped loaves froze significantly more slowly than either normal or snugly wrapped loaves. Bread intended to be frozen should be wrapped as tightly as possible in order to minimize the insulating effect of the air space between the loaf and the wrapper. Freezing of bread packed in delivery cartons is not recommended because the freezing rate of the bread is substantially decreased (Fig. 14) due to the insulating effect of dead air space [30]. Consequently, bread should be packed after freezing. Freezing times, temperatures, and air flows for commercial freezing of bread and rolls recommended by Bamford [43] are listed in Table 9.

## D.   Effects of Frozen Storage

Under proper storage conditions, bread can be stored in the frozen state for many weeks without significant quality loss [44–46]. Although moisture losses from unwrapped breads during freezing are negligible, as mentioned above, serious moisture losses can occur during prolonged frozen storage [30]. Thus, bread to be stored in a freezer over an extended period should be wrapped. Common types of wrapping materials are quite adequate to prevent moisture loss [37].

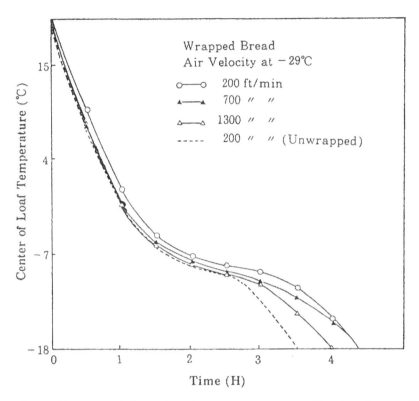

**Figure 13** Effect of air velocity at low temperature on rate of freezing of wrapped bread. (From Ref. 31.)

Pence et al. [31] demonstrated that frozen storage temperature seriously affected crumb firming and flavor of frozen bread. Rapid crumb firming and loss of flavor occurred during storage at –9°C (Tables 10 and 11). However, these quality losses were negligible after 4 weeks of storage of frozen bread at –18°C. Similar results were obtained with soft dinner and cinnamon rolls [47] and Danish pastry [48]. Such results clearly indicate that frozen bread should be stored at –18°C or below for best retention of quality. Additionally, Pence et al. [31] observed that freezing did not prevent absorption of foreign odors and flavors by bread. However, these occur more slowly at frozen storage temperatures. Consequently, they recommended that frozen storage of bread probably should not exceed one month, even if the lower storage temperature is used.

It is well known that unstable storage temperatures or refreezing substantially decrease the quality of frozen foods. In the case of frozen bread, crumb firmness increases substantially under improper storage conditions [31].

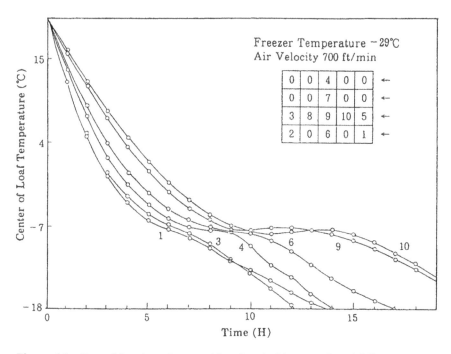

**Figure 14**   Rate of freezing of wrapped bread packed in a pasteboard delivery carton at low temperature and intermediate air velocity. Numbers indicate position of test loaves in carton and order in which they froze. Arrows indicate directions of air flow. (From Ref. 31.)

Pence et al. [31, 36] reported the formation of a narrow whitened or opaque ring (reminiscent in appearance of freezer-burn in meats) in the crumb just beneath the crust during storage of frozen bread. Length of storage before initial occurrence and subsequent rate of inward growth of the ring were dependent on the storage temperature. The rings began to appear after 2 weeks of frozen storage at –9°C, while the appearance of the rings required 10 weeks or more of frozen storage at –18°C. Pence et al. [36] indicated that the rings resulted from a movement of moisture by sublimation and diffusion from the interior of the crumb to a region of lower moisture content just beneath the crust. The rings are a useful index of improper storage as follows: wide clear zones outside the rings indicate the bread was appreciably stale before freezing, and wide white rings indicate a large loss of moisture from a loaf during storage because of improper packaging, according to Pence et al. [36].

## E.  Effects of Thawing

Theoretically, thawing rate should be as rapid as possible in order to minimize crumb firming during thawing. Pence et al. [34] demonstrated that both relative

**Table 9** Freezing Conditions for Bread and Rolls

| Products | Packaging | Method of freezing | Freezing time (h) | Temperature (°C) | Air flow (ft/min) |
|---|---|---|---|---|---|
| 1 lb bread | Bare | Blast or contact | 2.75 | −29 | 600 |
| | Single wrapped | Blast or contact | 3.25 | −29 | 600 |
| | Double wrapped | Blast or contact | 3.50 | −29 | 600 |
| | Double wrapped (bag and wrapped) | Blast or contact | 3.50 | −29 | 600 |
| 2 lb bread | Bare | Blast or contact | 3 | −29 | 600 |
| | Single wrapped | Blast or contact | 3.50 | −29 | 600 |
| | Double wrapped | Blast or contact | 4 | −29 | 600 |
| Dinner rolls (rich formula) | Plastic bag foil pan | Blast or contact | 1 | −29 | 600 |
| Cinnamon rolls (rich formula) | Plastic bag foil pan | Blast or contact | 1.50 | −29 | 600 |

*Source:* Ref. 43

**Table 10**   Effect of Temperature of Frozen Storage on
Firmness of Bread (g/4 mm Compression)[a]

| Days in storage | Temperature of storage (°C) | | | Nonfrozen control |
|---|---|---|---|---|
| | −18 | −12 | −9 | |
| 0 | 90[3] | — | — | — |
| 7 | 96 | 110 | 134 | 100 |
| 14 | 100 | 111 | 149 | 103 |
| 28 | 96 | 116 | 153 | 92 |
| 46 | 103 | 121 | 161 | — |

[a]Minimum significant difference at 5% point = 12 g/4 mm compression.
[b]Nonfrozen bread stored approximately 30 h at room temperature.
[c]The fresh nonfrozen bread used had an average firmness of 71 g/4 mm compression; after freezing, overnight equilibration to −18°C, and immediate defrosting, it had a value of 90 g/4 mm compression.
*Source*: Ref. 30

humidity and air temperature significantly affected the rate of thawing. A more or less regular decrease in thawing time resulted from each increase in air temperature at a given relative humidity, but effects from increases in relative humidity at a given temperature were less consistent and generally smaller. However, Pence et al. [31] found that the effect of rate of thawing was less than rate of freezing on crumb firming. However, an appreciable increase in firmness was obtained when thawing time was 5 hours or greater (Table 12). Pence and Standrige [32] found that freezing (90 min at −29°C) and thawing (5 h at 23°C) caused firming approx-

**Table 11**   Effect of Temperature of Frozen Storage on
Average Rank in Freshness of Flavor in Bread as Judged
by Taste Panel[a]

| Days in storage | Temperature of storage (°C) | | | Nonfrozen control[b] |
|---|---|---|---|---|
| | −18° | −12° | −9° | |
| 7 | 2.0 | 2.1 | 3.5 | 2.5 |
| 14 | 2.0 | 2.2 | 3.4 | 2.5 |
| 28 | 2.0 | 2.5 | 3.7 | 2.0 |

[a]Minimum significant differences at 5% point = 0.7, 0.9, and 0.9 for the 7-, 14-, and 28-day comparisons, respectively. Comparisons in rank between different storage periods are valid only as they relate to the nonfrozen control bread.
[b]Nonfrozen bread stored approximately 30 hours at room temperature.
*Source*: Ref. 31.

imately equal to the 24 hours of storage of nonfrozen bread at 23°C. However, 48 hours after thawing, the frozen bread was equal in firmness to unfrozen bread 48 hours after baking. Such results suggest that frozen bread can be thawed at room temperature in practice without significantly decreasing overall quality.

Gain or loss of moisture by the product during thawing can be substantial and serious [37]. Although high-relative-humidity air can increase thawing rate, it tends to cause condensation of moisture on the crust or the wrappers. The condensed moisture may cause undesirable moisture change in the breads or damage the wrappers. Accordingly, for best results, the relative humidity in air for thawing should be in the range of 40–60% [34, 49, 50]. Close control of humidity is also very important in thawing of unwrapped bread. Pence et al. [34] observed unwrapped bread to lose substantial moisture during thawing with dry air. On the other hand, thawing at higher relative humidities can cause an uptake of sufficient moisture to change the crisp crust into a rubbery soggy state. The most desirable conditions at low air velocity (about 150 ft/min) would appear to be close to 50% relative humidity.

Frozen bread can be thawed very rapidly by the use of microwave heating. This method has not been used extensively because of the high initial cost of the equipment. It was anticipated that microwave heating might refreshen the bread by reversing firmness increases unavoidably imparted during the freezing process [34]. However, Pence et al. [34] observed rapid thawing of bread to an internal temperature of 50–60°C by microwave heating produced incomplete refreshening of the bread. Further research on the use of microwave heating for thawing of frozen bread is needed to take full advantage of this new technology.

## VII. SUMMARY

Freshness is the most important attribute of baked goods. Consequently, inconvenient and uneconomical bakery operations like night or early morning work have

**Table 12** Effect of Defrosting Time on Crumb Firmness[a]

| Defrosting time (h) | Average increase in firmness (g/4 mm compression)[b] |
|---|---|
| 2.3 | 13 |
| 5.5 | 32 |
| 10.8 | 31 |

[a]Freezing time was about 80 minutes.
[b]Minimum significant difference at 5% point = 10 g/4 mm compression.
*Source*: Ref. 31.

been necessary for bakeries to provide fresh-baked goods to consumers. The use of frozen dough and frozen baked goods can facilitate bakery operations. Technological developments in frozen dough and frozen baked goods have improved the quality of baked goods. However, strictly speaking, baked goods, especially breads, still require further improvement in quality. For example, breads from frozen doughs have less flavor than breads made by conventional baking, because considerable reduction in fermentation before molding in frozen dough preparation is necessary to make the dough stable during freezing, frozen storage, and thawing. Substantial staling during the freezing and thawing process occurs in thawed frozen breads. Further research on frozen dough and baked goods is needed to minimize the quality gaps between baked goods made from frozen dough or frozen baked goods and those made by conventional baking practices.

## REFERENCES

1. P. P. Merritt, The effect of preparation on the stability and performance of frozen, unbaked yeast-levened doughs, *Bakers Dig.* 52(5):18 (1960).
2. L. Kline and T. F. Sugihara, Factors affecting the stability of frozen bread doughs. I. Prepared by straight dough method, *Bakers Dig.* 42(5):44 (1968).
3. Y. Tanaka, Freezing injury of baker's yeast in frozen dough, *Nippon Shokuhin Kogyo Gakkaishi* 28:100 (1981).
4. K. H. Hsu, R. C. Hoseney, and P. A. Seib, Frozen dough. I. Factors affecting stability of yeasted doughs, *Cereal Chem.* 56:419 (1979).
5. K. H. Hsu, R. C. Hoseny, and P. A. Seib, Frozen dough. II. Effects of freezing and storing conditions on the stability of yeasted doughs, *Cereal Chem.* 56:424 (1979).
6. T. A. Lehmann and P. Dreese, Stability of frozen bread dough—effects of freezing temperatures, *AIB Tech. Bull.* III.(7): (1981).
7. O. Neyreneuf and B. Delpuech, Freezing experiments on yeasted dough slabs. Effects of cryogenic temperatures on the baking performance, *Cereal Chem.* 70:109 (1993).
8. P. Mazur, Physical and temporal factors involved in the death of yeast at sub-zero temperatures. *Biophys. J.* 1:247 (1963).
9. O. Neyreneuf and J. B. Van Der Plaat, Preparation of frozen French bread dough with improved stability, *Cereal Chem.* 68:60 (1991).
10. Y. Oda, K. Uno, and S. Otha, Selection of yeasts for bread making by the frozen dough method, *Appl. Environ. Microbiol.* 52:941 (1986).
11. A. Hino, H. Takano, and Y. Tanaka, New freeze-tolerant yeast for frozen dough preparation, *Cereal Chem.* 64:269 (1987)
12. A. Inoue and W. Bushuk, Studies on frozen doughs. I. Effects of frozen storage and freeze-thaw cycles on baking and rheological properties, *Cereal Chem.* 68:627 (1991).
13. Y. Inoue and W. Bushuk, Studies on frozen doughs. II. Flour quality requirements for bread production from frozen dough, *Cereal Chem.* 69:423 (1992).
14. Y. Inoue, H. D. Sapirstein, S. Takayanagi, and W. Bushuk, Studies on frozen doughs. III. Factors involved in dough weakening during frozen storage and thaw-freeze cycles, *Cereal Chem.*

15. E. Varriano-Marston, K. H. Hue, and J. Mhadi, Rheological and structural changes in frozen dough, *Bakers Dig.* 54(1):32 (1980).

16. P. T. Berglund, D. R. Shelton, and T. P. Freeman, Frozen bread dough untrastructure as affected by duration of frozen storage and freeze-thaw cycles, *Cereal Chem.* 68:105 (1991).

17. M. J. Wolt and B. L. DAppolonia, Factors involved in the stability of frozen dough. II. The effects of yeast type, flour type, and dough additives on frozen dough stability, *Cereal Chem.* 61:213 (1984).

18. K. Lorenz and W. C. Bechtel, Frozen dough variety breads: Effect of bromate level on white bread, *Bakers Dig.* 39(4):53 (1965).

19. K. Lorenz, Frozen dough. Present trend and future outlook, *Bakers Dig.* 4(2):14 (1974).

20. Y. Inoue, H. D. Sapirstein, and W. Bushuk, Studies on frozen doughs. IV. Effect of shortening systems on baking and rheological properties, *Cereal Chem.*

21. D. K. Dubois and D. Blockcolsky, Frozen bread dough, Effect of additives, *AIB Tech. Bull.* VIII(4): (1986).

22. C. E. Stauffer, Frozen dough production, Advances in Baking Technology (B. S. Kamel and C. E. Stauffer, ed.), Blackie Academic and Professional, Glasgow, United Kingdom, 1992, p. 88.

23. E. W. Davis, Shelf-life studies on frozen doughs, *Baker's Dig.* 55(3):12 (1981).

24. P. E. Marston, Frozen dough for bread making, *Baker's Dig.* 52(5):18 (1978).

25. K. Hosomi, M. Uozumi, K. Nishio, and H. Matsumoto, Studies on frozen dough baking. The effects of sugar esters with various HLB-values, *Nippon Shokuhin Kogyo Gakkaishi* 39:806 (1992).

26. R. C. Hoseny, K. F. Finney, and Y. Pomeranz, Functional (breadmaking) and biochemical properties of wheat flour components. IV. Gliadin-lipid-glutenin interaction in wheat gluten, *Cereal Chem.* 47:135 (1970).

27. V. A. De Stefainis, J. G. Ponte, Jr., F. H. Chung, and N. A. Ruzza, A binding of crumb softeners and dough strengtheners during breadmaking, *Cereal Chem.* 54:13 (1977).

28. F. MacRitchie, Physicochemical processes in mixing, *Chemistry and Physics of Baking* (J. M. V. Blanshard, P. J. Frazier, and T. Galliard, ed.), Royal Society of Chemistry, London, 1986, p. 132.

29. E. J. Pyler, Keeping properties of bread, *Baking Science and Technology*, 3rd. ed., Vol. II, Sosland Publishing Company, Merriam, KS, 1988, p. 815.

30. J. W. Pence, T. M. Lubisich, D. K. Mecham, and G. S. Smith, Effects of temperature and air velocity on rate of freezing of commercial bread, *Food Technol.* 9:342 (1955).

31. J. W. Pence, N. N. Standridge, T. M. Lubisich, D. K. Mecham, and H. S. Olcott, Studies on the preservation of bread by freezing, *Food Technol.* 9:495 (1955).

32. J. W. Pence and N. N. Standridge, Effects of storage temperature and freezing on the firming of a commercial bread, *Cereal Chem.* 32:519 (1955).

33. J. W. Pence, The freezing, storage, and defrosting of commercial bread, *Proceedings of the 31st Annual Meeting of the American Society of Bakery Engineers*, Chicago, IL, 1955, p. 106.

34. J. W. Pence, N. N. Standridge, and M. J. Copley, Effect of temperature and relative humidity on the rate of defrosting of commercial bread, *Food Technol.* 10:492 (1956).

35. J. W. Pence, N. N. Standridge, D. K. Mecham, T. M. Lubisich, and H. S. Olcott,

Moisture distribution in fresh, frozen, and frozen-defrosted bread, *Food Technol.* 10:76 (1956).

36.  J. W. Pence, N. N. Standridge, D. R. Black, and F. T. Jones, White rings in frozen bread, *Cereal Chem.* 35:15 (1958).

37.  J. W. Pence, Bread and rolls, *The Freezing Preservation of Foods*, Vol. IV (D. K. Tressler, W. B. Van Arsdel, and M. J. Copley, ed.), The AVI Publishing Company, Westport, CT, 1968, p. 386.

38.  J. R. Katz, Gelatinization and retrogradation of starch in relation to the problem of bread staling, *A Comprehensive Survey of Starch Chemistry*, Vol. 1 (R. P. Walton, ed.), Chemical Catalog Co., New York, 1928, p. 28.

39.  T. J. Schoch, Starch in bakery products, *Baker's Dig.* 39(2):27 (1965).

40.  S. K. Kim and B. L. DAppolonia, The role of wheat flour constituents in bread staling, *Baker's Dig.* 51(1):38 (1977).

41.  M. M. Morad and B. L. D'Appolonia, Effect of surfactants and baking procedure on total water-solubles and soluble starch in bread crumb, *Cereal Chem.* 57:141 (1980).

42.  J. L. Cauble and R. S. Murdough, Freezing baked goods automatically, *Baker's Dig.* 30(2):56 (1956).

43.  R. Bamford, Freezing and thawing of bakery products, *Baker's Dig.* 49(3):40 (1975).

44.  W. H. Cathcart and S. V. Luber, Freezing as a means of retarding bread staling, *Ind. Eng. Chem.* 31:362 (1939).

45.  W. H. Cathcart, Further studies on the retardation of the staling of bread by freezing, *Cereal Chem.* 18:771 (1941).

46.  P. D. Arnold, Two years of handling frozen breads and rolls, *Proceedings of the 32nd Annual Meeting of the American Society of Bakery Engineers*, Chicago, IL, 1956, p. 165.

47.  K. Kulp and W. G. Bechtel, The effect of freezing, defrosting, and storage conditions on the freshness of dinner rolls and cinnamon rolls, *Cereal Chem.* 37:170 (1960).

48.  K. Kulp and W. G. Bechtel, The effect of freezing and frozen storage on the freshness and firmness of Danish pastry, *Food Technol.* 15:273 (1961).

49.  E. J. Barta, Principles of the defrosting of frozen bakery products, *Baker's Dig.* 32(4):50.

50.  B. Belderok and W. H. G. Weibols, Studies on the defrosting of frozen bread, *Food Technol.* 18:1813 (1964).

# Index